Mathematics and Visualization

More information about this series at http://www.springer.com/series/4562

Charles D. Hansen · Min Chen
Christopher R. Johnson
Arie E. Kaufman · Hans Hagen
Editors

Scientific Visualization

Uncertainty, Multifield, Biomedical, and Scalable Visualization

Springer

Editors
Charles D. Hansen
Christopher R. Johnson
School of Computing, Scientific Computing and Imaging Institute
University of Utah
Salt Lake City, UT
USA

Min Chen
e-Research Centre
University of Oxford
Oxford
UK

Arie E. Kaufman
Department of Computer Science
Stony Brook University
New York, NY
USA

Hans Hagen
Fachbereich Informatik
Technische Universität Kaiserslautern
Kaiserslautern
Germany

ISSN 1612-3786 ISSN 2197-666X (electronic)
ISBN 978-1-4471-7104-1 ISBN 978-1-4471-6497-5 (eBook)
DOI 10.1007/978-1-4471-6497-5

Mathematics Subject Classification: 65D18, 68U05, 97R60

Springer London Heidelberg New York Dordrecht

Softcover reprint of the hardcover 1st edition 2014

Printed on acid-free paper

Springer is part of Springer Science+Business Media (www.springer.com)

Preface

Scientific Visualization is the transformation of abstract data, derived from observation or simulation, into readily comprehensible images, and has proven to play an indispensable part of the scientific discovery process in many fields of contemporary science. Since its inception two decades ago, the techniques of Scientific Visualization have aided scientists, engineers, medical practitioners, and others in the study of a wide variety of data sets including, for example, high-performance computing simulations, measured data from scanners (CAT, MR, confocal microscopy), Internet traffic, and financial records. One of the important themes being nurtured under the aegis of Scientific Visualization is the utilization of the broad bandwidth of the human sensory system in steering and interpreting complex processes and simulations involving voluminous data sets across diverse scientific disciplines. Since vision dominates our sensory input, strong efforts have been made to bring the mathematical abstraction and modeling to our eyes through the mediation of computer graphics.

In June 2011, we organized a Dagstuhl seminar, with 54 participants, that focused on the four parts of this book. The seminar comprised talks from leaders in the field and breakout sessions on the four specific topics: Uncertainty Visualization, Multifield Visualization, Biomedical Visualization, and Scalable Visualization. This book is a culmination of the four topics with contributed chapters from the participants for each of the four parts of the book.

We would like to thank all of the authors for their thoughtful and insightful contributed chapters. We would also like to thank Catherine Waite and Lynn Brandon from Springer UK for their assistance and patience in generating this book.

Oxford — Min Chen
Kaiserslautern — Hans Hagen
Salt Lake City — Charles D. Hansen
Christopher R. Johnson
Stony Brook — Arie E. Kaufman

Contents

Part IV Scalable Visualization

Part I
Uncertainty Visualization

Chapter 1
Overview and State-of-the-Art of Uncertainty Visualization

Georges-Pierre Bonneau, Hans-Christian Hege, Chris R. Johnson, Manuel M. Oliveira, Kristin Potter, Penny Rheingans and Thomas Schultz

What is not surrounded by uncertainty cannot be the truth.
—*Richard Feynman*

True genius resides in the capacity for evaluation of uncertain, hazardous, and conflicting information.
—*Winston Churchill*

Abstract The goal of visualization is to effectively and accurately communicate data. Visualization research has often overlooked the errors and uncertainty which accompany the scientific process and describe key characteristics used to fully understand the data. The lack of these representations can be attributed, in part, to the

G.-P. Bonneau (✉)
The University of Grenoble, Grenoble, France
e-mail: Georges-Pierre.Bonneau@ujf-grenoble.fr

H.-C. Hege
Zuse Institute Berlin, Berlin, Germany
e-mail: hege@zib.de

C.R. Johnson · K. Potter
Scientific Computing and Imaging Institute, University of Utah, Salt Lake City, UT, USA
e-mail: crj@sci.utah.edu

K. Potter
e-mail: kpotter@sci.utah.edu

M.M. Oliveira
Instituto de Informática, UFRGS, Porto Alegre, RS, Brazil
e-mail: oliveira@inf.ufrgs.br

P. Rheingans
University of Maryland Baltimore County, Baltimore, MD, USA
e-mail: rheingan@cs.umbc.edu

T. Schultz
University of Bonn, Bonn, Germany
e-mail: schultz@cs.uni-bonn.de

T. Schultz
MPI for Intelligent Systems, Tübingen, Germany

C.D. Hansen et al. (eds.), *Scientific Visualization*, Mathematics and Visualization,
DOI 10.1007/978-1-4471-6497-5_1

inherent difficulty in defining, characterizing, and controlling this uncertainty, and in part, to the difficulty in including additional visual metaphors in a well designed, potent display. However, the exclusion of this information cripples the use of visualization as a decision making tool due to the fact that the display is no longer a true representation of the data. This systematic omission of uncertainty commands fundamental research within the visualization community to address, integrate, and expect uncertainty information. In this chapter, we outline sources and models of uncertainty, give an overview of the state-of-the-art, provide general guidelines, outline small exemplary applications, and finally, discuss open problems in uncertainty visualization.

1.1 Introduction

Visualization is one window through which scientists investigate, evaluate and explore available data. As technological advances lead to better data acquisition methods, higher bandwidth, fewer memory limits, and greater computational power, scientific data sets are concurrently growing in size and complexity. Because of the reduction of hardware limitations, scientists are able to run simulations at higher resolution, for longer amounts of time, using more sophisticated numerical models. These advancements have forced scientists to become increasingly reliant on data processing, feature and characteristic extraction, and visualization as tools for managing and understanding large, highly complex data sets. In addition, there is becoming a greater accessibility to the error, variance, and uncertainty not only in output results but also incurred throughout the scientific pipeline.

With increased size and complexity of data becoming more common, visualization and data analysis techniques are required that not only address issues of large scale data, but also allow scientists to understand better the processes that produce the data, and the nuances of the resulting data sets. Information about uncertainty, including confidence, variability, as well as model bias and trends are now available in these data sets, and methods are needed to address the increased requirements of the visualization of these data. Too often, these aspects remain overlooked in traditional visualization approaches; difficulties in applying pre-existing methods, escalating visual complexity, and the lack of obvious visualization techniques leave uncertainty visualization an unsolved problem.

Effective visualizations present information in a manner that encourages data understanding through the appropriate choice of visual metaphor. Data are used to answer questions, test hypotheses, or explore relationships and the visual presentation of data must facilitate these goals. Visualization is a powerful tool allowing great amounts of data to be presented in a small amount of space, however, different visualization techniques are better than others for particular types of data, or for answering specific questions. Using the most befitting visualization method based on the data type and motivated by the intended goals of the data results in a powerful tool for scientists and data analysts.

The effective visualization of uncertainty, however, is not always possible through the simple application of traditional visualization techniques. Often, the visualization of the data itself has a high visual complexity, and the addition of uncertainty, even as a scalar value, complicates the display. Issues of visual clutter, data concealment, conflicts in how the data and the uncertainty are represented, and unintentional biases are just some of the problems incurred when visualizing data accompanied by uncertainty. Also, the complexity of these data sets may not lend themselves to the straightforward application of existing visualization methods, and thus, the added burden of uncertainty can be overwhelming.

Uncertainty data are becoming more prevalent and can be found in fields such as medical imaging, geoscience, and mechanical engineering. The simulation of complex systems, compilation of sensor data, and classification of tissue type are but a few sources of uncertainty data and their expression, size, and complexity can drastically vary. Uncertainty can arise in all stages of the analysis pipeline, including data acquisition, transformation, sampling, quantization, interpolation, and visualization. It can be a single scalar value presented alongside the original data, or can be an integral aspect of the data, derived from the description of the data itself. In any case, uncertainty is an imperative component of scientific data sets and should not be disregarded in visualizations.

1.1.1 Sources of Uncertainty

Uncertainty can mean very different things in different situations, with each driven by different key characteristics and goals. The uncertainty in a data set may result from the process through which the data was gathered or generated, or it may represent variability in the phenomenon represented by the data. We divide data uncertainty sources into three broad classes: uncertainty observed in sampled data, uncertainty measures generated by models or simulations, and uncertainty introduced by the data processing or visualization processes. Variability in the underlying phenomenon could manifest itself in sampled data or be incorporated into models or simulations. A particular data set might be subject to one form of uncertainty or multiple. Different types of uncertainty offer different challenges to effective and truthful visualization. While most of the visualization literature about uncertainty concentrates on issues of visual representation rather than source, a few papers have made a thoughtful analysis of the source of uncertainty, as well [9, 30, 74, 75, 98]. Other useful discussions of the sources of uncertainty can be found in the geo-spatial visualization and GIS literatures [11, 19, 20, 62]. The discussion below draws from all these sources.

1.1.1.1 Uncertainty in Sampled Data

Uncertainty in data that is gathered through a sampling process might give the appearance of too little information, too much information, or information that just cannot

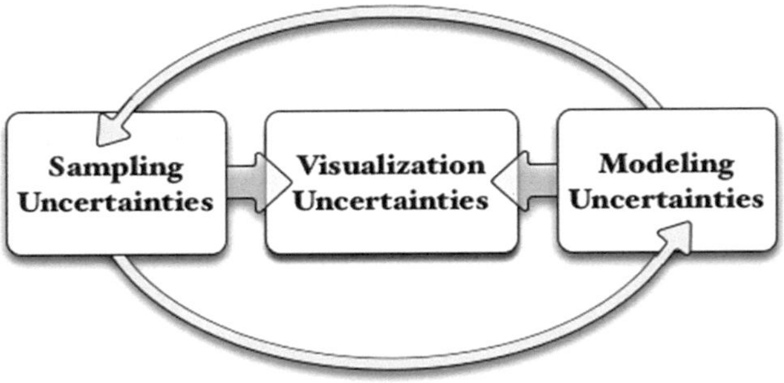

Fig. 1.1 Sources of uncertainty. Both sampling and modeling uncertainties affect each other and add to visualization uncertainties

be trusted. Data sets where missing or incomplete instances provide too little information present challenges to many visualization methods. Filtering out data with missing elements can ignore valuable information and produce awkward holes. Filling in missing values or instances by interpolation, imputation, or other estimation techniques from known values can introduce error. In such cases, data quality metrics might indicate the confidence in estimated quantities. For instance, estimating a single missing data value from a dense set of similar instances would be expected to produce a smaller error than an estimation from a sparser or more disparate set. Data sets where multiple, contradictory measurements seem to provide too much data also offer challenges for visualization. Such situations can be caused by noisy data, noisy instruments, human error in the data gathering process, or sampling at a scale different than that natural to the phenomenon. One special case of error in data measurements is that of spatial data where the error might be in the position of a sampled location, rather than in its measured values, resulting in uncertainty about where values should be displayed. Similarly, data with contradictory values might be characterized by data quality metrics based on sample value range, variance, or another measure of variability. Finally, metadata about a data source may cast doubt on its certainty. For instance, data that is old, from an untrusted source, or gathered through a nonstandard process might be regarded with some skepticism (Fig. 1.1).

1.1.1.2 Models Containing Uncertainty

Sophisticated computational models may contain elements designed to estimate the uncertainty or variability in the model predictions. The sources of this type of uncertainty include residual variability from simplifying abstractions, variability in the mechanism or magnitude of causality and relationships, potential error in model inputs, incorrect model parameters, and imprecision in tacit knowledge incorporated in the model. The range of predictions made by model ensembles, where different component models may make different assumptions or use different parameters, illustrate the potential variability in even the best models.

The output from such a model may include information about estimated error in the form of a single error measure, ranges for expected values, or predicted distributions for values or errors. These measures are applicable to numeric quantities. Alternatively, a model that makes nominal or categorical predictions may also indicate the degree of confidence in its predictions by producing multi-value predictions, where each possible value or classification is associated with a likelihood.

1.1.1.3 Uncertainty from the Visualization Process

Finally, we should understand how the visualization process impacts the propagation, magnification, perception, and impact of uncertainty. In order to do this, we must understand computational sources and magnifiers of error and uncertainty in input values, perceptual and cognitive influences on the understanding of uncertainty visualization, effects of differences in audience abilities and cultures, requirements imposed by different application tasks and goals, and competing positive and negative consequences of showing uncertainty.

1.2 Perceptual Uncertainty

Logically, it seems sensible to display information about uncertainty in a manner consistent with our cognitive models of which perceptual elements contain variability or uncertainty. A number of approaches to uncertainty visualization seem to build on this principle, representing uncertainty with such visual elements as blur, flicker, reduced saturation, sketched outlines, or transparency.

There have been relatively few careful evaluations of the effectiveness of uncertainty visualization and its impact on the decision-making process that have appeared in the visualization literature. In some cases, researchers have used quantitative evaluations or user studies to evaluate the ability of subjects to understand uncertain information [33, 109]. Zuk and Carpendale [111] present a framework for the heuristic evaluation of uncertainty visualizations from the perceptual and cognitive principles described by Bertin [6], Tufte [101], and Ware [105]. They use this framework to analyze eight uncertainty visualizations of different types and from different domains. They propose this sort of heuristic evaluation as a rough substitute when more specific evaluations are not practical.

Additional insight into the perceptual and cognitive elements of effective uncertainty representations can be found in the GIS literature. Harrower surveys a collection of evaluations of methods for representing uncertainty in map-based visualizations [38]. He observes that the most common characteristics used to judge a technique are its effects on confidence, speed, and accuracy of judgements. Two principles which may be derived from that set of evaluations are the superiority of displays that integrate value and certainty information over those that show each in a separate display and the preference for static displays over those that toggle between value and certainty. Deitrick describes experiments that show how inclusion of information about uncertainty changes the judgements made by subjects [21].

The field of medical decision-making has also considered the role of uncertainty in the decision-making process. Politi et al. [78] studied the effect of communication of uncertainty on patients engaged in shared decision-making. They reported an aversion to ambiguity in this situation, leading some patients to avoid making decisions in the presence of uncertainty while others engaged in additional information-seeking behaviors. They observed interactions between level of education and decision-making under uncertainty. In particular, less educated patients were more likely to conclude that the inclusion of visual depictions of uncertainty made data less trustworthy. Patients also tended to interpret uncertain situations in a way that reinforced their initial values and preferences. Finally, Politi et al. suggest that communication of uncertainty may lead to greater ultimate satisfaction in the decision process and a lower likelihood of regret about a decision.

There is evidence that decision-making in the presence of uncertainty takes place in different regions of the brain than decision-making in more certain conditions. Specifically, Paulus et al. [76] observed different patterns of brain activity under fMRI during different decision-making conditions. They suggest that the more complex task of decision-making under uncertainty requires more complex strategies and is more influenced by experiences in the past. The physiological evidence supports this theory by showing increased involvement of brain areas important to strategy formation and adjustment, in particular the prefrontal and parietal cortex, when uncertainty is present.

1.3 Formal Description

The consideration and quantification of uncertainties is of great importance in many practical applications and is part of the data analysis chain to support decision making. For this reason, we need to understand the data including its shortcomings, value, and relevance, which largely depends on the presence or absence of uncertainty. Our goals are to understand quantified uncertainty and deal with it, as well as independently perform uncertainty quantification ourselves.

1.3.1 What is Uncertainty?

Uncertainty is the lack of information. It can be due to randomness, such as results by chance, for example the roll of the dice or knowing the exact daily quantity of rain in Seattle. This type of uncertainty is called aleatoric and is objective in that results differ each time an experiment is run. These types of phenomenon are truly random in that the results depend on chance, and thus use probabilistic modeling to describe. Uncertainty can also be due to a lack of knowledge, that is, "knowledge that can in principle could be known," but in practice is not. This type of uncertainty is called epistemic and is subjective, such as not knowing the birth date of the last Chinese Emperor. These uncertainties are due to errors that practically cannot be controlled and can be described by non-probabilistic modeling.

1.3.2 Mathematical Modeling of Uncertainty

A variety of types of uncertainties occur in practice, including mixtures of different types of uncertainty. Quantification of uncertainties, including mixtures, requires a unifying mathematical framework, which is very difficult to establish and not yet fully accomplished.

1.3.2.1 Fundamental Setting

From a fundamental standpoint, we are interested in the situation with possible outcomes or occurrences of "events" A, B, C, where A, B, and C are subsets of the set of all elementary events in the universe. The task at hand is to then *measure* the evidence that A ever happened, the degree of truth of that statement "event A happened", and the probability that event A will happen. The question is then, how do we measure and what is measurement?

In mathematics, measurement means to assign real numbers to sets. For example, the classical task in metric geometry is to assign numbers to geometric objects for length, area, or volume. The requirement in the measurement task is that the assigned numbers should be invariant under displacement of the respective objects.

In ancient times, the act of measuring was equivalent to comparing with a standard unit. However, it soon became apparent that measurement was more complicated than initially thought in that it involves finite processes and sets. The first tool to deal with this problem was the Riemann integral which enabled the computation of length, areas, and volumes for complex shapes (as well as other measures). However, the Riemann integral has a number of deficiencies, including its applicability only to functions with a finite number of discontinuities, fundamental operations of differentiation and integration are, in general, not reversible, and limit processes, in general, can not be interchanged. In 1898, Émile Borel developed classical measure theory which includes σ-algebra to define a class of sets that is closed under set union of countably many sets and set complement, and defined as additive measure μ that associates a number $\in \mathbb{R}_0^+$ with each bounded subset in the σ-algebra. Around 1899–1902, Henry Lebesgue defined integrals based on a measure that subsumes the Borel measure, based on a special case. He connected measures of sets and measures of functions.

1.3.2.2 Quantification

Probability measure was then developed in 1933 by Andrey Nikolaevich Kolmogorov, which used classical measure theory and added the measure of 1 assigned to the universal set. This is thought of as classical probability theory.

The classical probability theory has since become the dominant approach to examine uncertainty and randomness. Extensive mathematical studies followed and resulted in highly sophisticated theories. Its foundation rests on the definition of *probability space*, which was Kolmogorov's big achievement. A probability space is a triplet (Ω, F, P). Here Ω is a countable event space containing all possible outcomes of a random event. F is the so-called σ-algebra of Ω and it represents all combinations of the outcomes from Ω. Its construction satisfies:

- It is not empty: $\emptyset \in F$ and $\Omega \in F$.
- If a set $A \in F$, then its complement $A^c \in F$.
- If sets $A_1, A_2, \ldots, \in F$, then $\bigcup_{i=1}^{\infty} A_i \in F$, and $\bigcap_{i=1}^{\infty} A_i \in F$.

P is the well known *probability measure* and it is used to assign a real number, i.e., the probability, on the occurrence of any outcomes of the events (from Ω) and their potential combinations (from F). It satisfies the following important and well known principles.

1. $0 \leq P(A) \leq 1$, for any $A \in F$.
2. $P(\Omega) = 1$. That is, the probabilities of all outcomes add up to one.
3. For $A_1, A_2, \cdots \in F$ and $A_i \cap A_j = \emptyset$, for any $i \neq j$,

$$P\left(\bigcup_{i=1}^{\infty} A_i\right) = \sum_{i=1}^{\infty} P(A_i).$$

About 50 years later, the additivity requirement became a subject of controversy in that it was too restrictive to capture the full scope of measurement. For example, it works well under idealized, error-free measurements, but is not adequate when measurement errors are unavoidable. In 1954, Gustave Choquet developed a (potentially infinite) family of non-additive measures (capacities), and for each given capacity, there exists a dual "alternating capacity". An integral based on these measures is non-additive, can be computed using Riemann or Lebesgue integration and is applied specifically to membership functions and capacities.

In 1967, Arthur P. Dempster introduced imprecise probabilities based on the motivation that the precision required in classical probability is not realistic in many applications. Imprecise probabilities deal with convex sets of probability measures rather than single measures. For each given convex set of probability measures he also introduced 2 types of a non-additive measures: lower and upper probabilities, and super- and supra-additive. This allow probabilities to be represented imprecisely by intervals of real numbers.

In 1976, Glenn Shafer analyzed special types of lower and upper probabilities and call then belief and plausibility measures. The theory based on these measures became known as Dempster-Shafer theory (DST) or evidence theory. DST is capable of dealing with interval-based probabilities, such that belief or probability measures are equal to the ranges of admissible probabilities. As it turns out, belief measures are equivalent to Choquet capacities of order inf and plausibility measures are equivalent to alternating capacities of order inf.

The comparison of membership functions of fuzzy sets and probabilities was investigated in 1978 by Michio Sugeno and found to be not directly possible. This led to the generalization of additive measures analogous to generalization such that crisp sets generalize to fuzzy sets, and additive measures generalize to (non-additive) fuzzy measures or monotone measures. The Sugeno integral was then introduced with respect to a monotone measure. That same year, Lofti Zadeh defined a *possibility function* associated with each fuzzy set that is numerically a membership function, and a *possibility measure* that is a supremum of the possibility function in each set of concern, for both crisp and fuzzy sets. This is one of several interpretations of the "theory of graded possibilities". Its connection to DST is that constant plausibility measures are equivalent to possibility measures and constant belief measures are necessity measures.

In summary, the three most utilized uncertainty theories are the Classical Probability Theory, the Dempster-Shafer Theory, and Possibility Theory and can be divided into two classes. The first class uses additive measures in which the addition equal to the union expresses no interaction between events and can be thought of as classical probability combined with measure theory. The second class uses non-additive measures, in which addition greater than the union expresses positive interaction between events, such as synergy, cooperation, coalition, enhancement or amplification, while addition less than the union expresses negative interaction between events such as incompatibility, rivalry, inhibition, downgrading, or condensation. This class combines one of many uncertainty theories with generalized measure theory.

1.4 Evaluation

Visualization research is too often neglected by industry and other potential expert users. One of the reasons is the lack of a proper evaluation of the results. This lack of evaluation was especially obvious in historical visualization fields such as volume rendering or fluid flow visualization. In the more recent domain of uncertainty visualization, researchers have made a significant effort into the assessment of the proposed techniques. The types of evaluation may be classified into three groups:

- Theoretical evaluation: the method is analyzed to see if it follows established graphical design principles,
- Low-level visual evaluation: a psychometric visual user study is performed to evaluate low-level visual effects of the method,

- Task oriented user study: a cognitive, task-based user study is conducted to assess the efficiency or the usability of the method.

1.4.1 Theoretical Evaluation

General guidelines and rules regarding visual depiction of data have been established, that have proven their efficiency. Bertin in [5], later translated in [6], has introduced the concept of visual variables. These include among others the location, size, orientation, shape, focus and realism. Furthermore he defined four visual properties, natural ordering, the ability to quantify, the ability to focus user attention (selectivity) and the ability to associate similar elements (associativity). He studied which of these properties are verified by the visual variables. Tufte in [102], through his concepts of graphical excellence and integrity, has proposed a number of guidelines to enhance the precision and the usability of graphical depiction. Chambers et al. in [14] have studied the relative influence of specific patterns on the visual perception, for example straight lines versus curves, dark versus light objects or small versus large patterns. This study leads the authors to define general rules for plot construction.

These graphical design principles may be used to conduct a theoretical evaluation of new uncertainty visualization techniques. As already mentioned in Sect. 1.2, Zuk and Carpendale in [111] have done such an evaluation for eight uncertainty visualization techniques. The same type of theoretical evaluation was followed by Riveiro in [87] to evaluate three uncertainty visualization techniques in the context of information fusion and decision making.

1.4.2 Low-Level Visual Evaluation

Barthelmé and Mamassian in [2] studied the influence on noise uncertainty in a decision-making task. Based on psychometric and simple task experiments, he proved that users can reliably measure the visual uncertainty and use this information in their decision-making. Coninx et al. in [18] conducted psychometric experiments to measure the impact of contrast sensitivity on the visibility of uncertain noisy patterns. He used this information in order to control the visibility of uncertainty data in a visualization technique based on the perturbation of colormaps by Perlin noise.

1.4.3 Task-Oriented User Study

Task oriented cognitive user studies are by far the most common way of assessing the efficiency and usability of uncertainty visualization techniques. In this type of evaluation a panel of users is typically asked to perform a task that requires not only low-level visual processing but also high-level cognitive treatment of the visual information. Standard tasks as an example may consist in counting the number of local minima in a dataset, find the location of the maximum or minimum value, find the direction of rotation of a vortex. The task completion time, task completion accuracy, user's rating of efficiency and usability may be recorded. A statistical analysis of the recorded data is done. Typical analyses include analysis of variance (ANOVA), used to check in particular if the difference in the mean value of two distributions is significant. Examples of uncertainty visualization papers with a task-based evaluation include [20, 21, 69, 91].

1.5 Review of Current State of the Art

The goal of visualization is to effectively present large amounts of information in a comprehensible manner, however, most visualizations lack indications of *uncertainty* [42, 43, 63, 83].

1.5.1 Traditional Representations

Tukey [103] proposed graphical techniques to summarize and convey interesting characteristics of a data set not only to facilitate an understanding of the given data but also to further investigation and hypothesis testing. These tested graphical methods, such as the boxplot, histogram, and scatter plot, provide identifiable representations of a data distribution, and their simplicity allows for quick recognition of important features and comparison of data sets. In addition, they can be substituted for the actual display of data, specifically when data sets are too large to plot efficiently.

1.5.1.1 1D

One of the most ubiquitous approaches to displaying uncertainty information is the boxplot [28, 34, 94, 103], which is the standard technique for presenting the *five-number summary*, consisting of the minimum and maximum range values, the upper and lower quartiles, and the median, as illustrated in Fig. 1.2a. This collection of values quickly summarizes the distribution of a data set, including range and expected value, and provides a straightforward way to compare data sets. In addition, the reduced representation afforded by the five-number summary provides a concise tool for data analysis, since only these characteristic values need to be analyzed. Figure 1.2b and c show visual modifications of the boxplot. Surveys on the introduction and evolution of the boxplot can be found in [16, 81].

The box plot is often adapted to include information about the underlying distribution, as demonstrated in Fig. 1.2d–g. The most common modification adds density

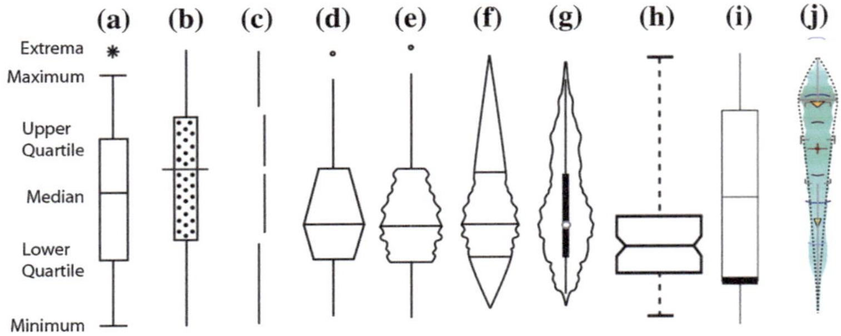

Fig. 1.2 Variations of the boxplot. **a** The construction of the boxplot [103]. **b** Range plot [94]. **c** Innerquartile plot [102]. **d** Histplot [4]. **e** Vaseplot [4]. **f** Box-percentile plot [26]. **g** Violin plot [39]. **h** Variable width notched boxplot [67]. **i** Skewplot [16]. **j** Summary plot [82]

information, typically through changes to the sides of the plot. The hist plot [4] extends the width of the cross bars at the quartiles and median to express density at these three locations. The vase plot [4] instead varies the "box" continuously to reflect the density at each point in the innerquartile range. Similarly, the box-percentile plot [26] and violin plot [39] show density information for the entire range of the data set. Density can also be shown by adding dot plots [106], which graph data samples using a circular symbol. The sectioned density plot [17] completely reconstructs the box plot by creating rectangles whose colors and size indicate cumulative density, and placement express the location of the quartiles. Sample size and confidence levels can be expressed through changing or notching the width of the plot [67] (Fig. 1.2h) or by using dot-box plots, which overlay dot plots onto box plots [107]. Other descriptors, such as skew and modality, can be added by modifying the width of the median line [67], thickening the quartile lines [16], (Fig. 1.2i) adding beam and fulcrum displays [23] alongside, or overlaying additional glyphs [82] (Fig. 1.2j).

1.5.1.2 2D

Standard implementations of the boxplot focus on univariate data distributions. The five-number summary is a useful descriptor of not only univariate, but also bivariate data distributions. The main challenge in extending the boxplot for use with higher dimensional data is how to translate the five-number summary values, which are vector values in the bivariate case, into visual metaphors with meaningful spatial positions, while maintaining the simplicity of the original boxplot. A rangefinder boxplot [3], as seen as the solid back lines in Fig. 1.3a, is a simple extension of the boxplot into 2D which determines boxplots for the two dimensions independently and draws lines to show the interquartile ranges and extrema of those plots. This idea was further improved upon, as shown as the thick gray lines in Fig. 1.3a, to emphasize the quartiles rather than the range, by moving the perpendicular lines from the extrema values to the upper and lower quartile positions and extending whisker lines to the

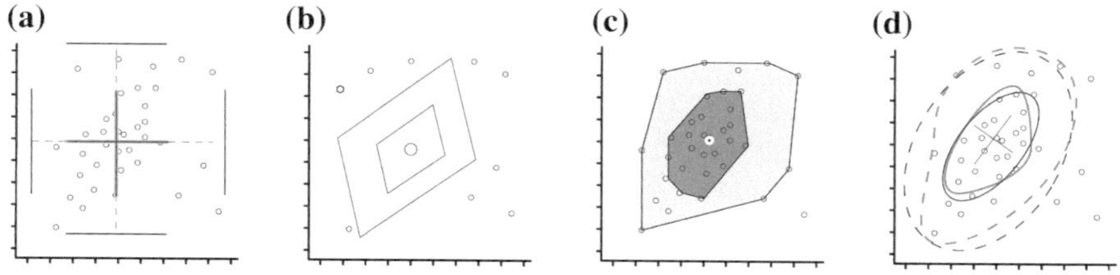

Fig. 1.3 Bivariate extensions of the boxplot. **a** The rangefinder boxplot [3]. **b** The 2D boxplot [99]. **c** The bagplot [88]. **d** The quel-(*gray*) and rel-(*black*) plots [32]

extrema value of the variables [53]. Other techniques for extending the boxplot into 2D all use the notion of a hinge that encompasses 50 % of the data and a fence that separates the central data from potential outliers. The distinctions between each of these methods are the way the contour of the hinge and fence are represented, and the methods used to calculate the contours. The 2D boxplot [99], as seen in Fig. 1.3b, computes a robust line through the data by dividing the data into three partitions, finding the median value of the two outer partitions, and using these points as the line. Depending on the relationship between the slope of the line and each variable, the quartile and fence lines are drawn either parallel to the robust line, or parallel to the variables coordinate axis. The lines not comprising the outer-fence and the inner-hinge boxes are removed. The bagplot [88] uses the concept of halfspace depth to construct a bivariate version of the boxplot, as seen in Fig. 1.3c. The relplot and the quelplot [32] use concentric ellipses to delineate between the hinge and fence regions. Both the relplot and quelplot can be seen in Fig. 1.3d.

1.5.1.3 PDFs

There is a body of research investigating methods for displaying probability distribution functions with spatial positions. Each of these methods takes an exploratory approach to the presentation of the data by filtering down the amount of data, and then providing a user interface for the scientist to explore the data sets. Ehlschlaeger et al. [25] present a method to smoothly animate between realizations of surface elevation. Bordoloi et al. [7] use clustering techniques to reduce the amount of data, while providing ways to find features of the data sets such as outliers. Streamlines and volume rendering have been used by Luo et al. [61] to show distributions mapped over two or three dimensions.

Kao et al. [48] uses a slicing approach to show spatially varying distribution data. This approach is interesting in that a colormapped plane shows the mean of the PDFs, and cutting planes along two edges allow for the interactive exploration of the distributions. Displaced surfaces as well as isosurfaces are used to enhance the understanding of the density of the PDFs.

Case studies of specific data have been performed by Kao et al. [46, 47]. Their data sets come from NASAs Earth Observing System (EOS) Satellite images and Light Detection And Ranging (LIDAR) data. The methods used to show this data include encoding the mean as a 2D color map, and using standard deviation as a displacement value. Histograms are also employed to understand better the density of the PDFs. To explore the mode of specific distributions, a small set of PDFs are plotted onto a color mapped spatial surface.

1.5.2 Uncertainty Visualization

Many visualization techniques that incorporate uncertainty information treat uncertainty like an unknown or fuzzy quantity; [75] is a survey of such techniques. These methods employ the meaning of the word uncertainty to create the interpretation of uncertainty or unknown to indicate areas in a visualization with less confidence, greater error, or high variation. Ironically, while blurring or fuzzing a visualization accurately indicates the lowered confidence in that data, it does not lead to more informed decision making. On the contrary, it obfuscates the information that leads to the measure of uncertainty. Because it obscures rather than elucidates the quantitative measures leading to the uncertain classification, such a solution to the problem of adding qualitative information to visualization misses important information.

1.5.2.1 Comparison Techniques

Often, uncertainty describes a comparison that can most clearly be understood visually, such as the difference between surfaces generated using different techniques, or a range of values that a surface might fall in. A simple approach to the visualization of this type of information is a side-by-side comparison of data sets. An example of this type of visualization is presented in Jiao et al. [41] where streamlines computed from various fiber tracking algorithms are interactively displayed along with the global and local difference measures. Another example is the time window, presented in [112], in which temporal uncertainty around archeological sites is displayed, using various visual clues, in an interactive, exploratory system.

However, this approach may not clearly manifest subtle differences when the data are nearly the same, and it becomes harder to perform this comparison as the visualization becomes more complicated. Another simple approach is to overlay the data to be compared [45]. With this technique, the addition of transparency or wire frame can produce a concise, direct comparison of the data sets. A similar approach uses difference images to display areas of variation [108]. These approaches are less effective, however, when the uncertainty can be categorized as more of a range of values rather than just two distinct ones. In such cases, a surface sweep, known as a fat surface [75], can be used to indicate all possible values. Another approach is the integration of isosurface and volume rendering. Here, an opaque isosurface can be used to indicate the most likely value, and a transparent volume rendering surrounding the isosurface can indicate the range of possible values [43]. Uncertainty information for large collections of aggregated data can be presented using hierarchical parallel coordinates [29]. Lee et al. [52] visualize differences in location and sub-tree structure between two hierarchies through color and transparency. Finally, bounded uncertainty, while not effectively visualized in 3D, can be expressed through the ambiguation of boundaries and edges of pie charts, error bars, and other 2D abstract graphs [70] or as modifications to line charts [96].

1.5.2.2 Attribute Modification

Another standard method to visualize uncertainty involves mapping it to free variables in the rendering equation or modifying the visual attributes of the data. Such methods include modifying the bidirectional reflectance function (BRDF) to change surface reflectance, mapping uncertainty to color or opacity [65, 91, 97], or pseudo-coloring using a look-up table [75]. This technique has been used as a means for conveying uncertainty in the areas of volume rendering [22, 51, 89], point cloud surface data [77], isosurfacing [45, 79, 80, 86] and flow fields [8], and is often combined with other uncertainty visualization methods. An example technique colormaps flowline curvature onto volume rendered surfaces, highlighting areas in which small changes in isovalue lead to large changes in isosurface orientation and thus indicating areas where the isosurface is a poor representation of material boundary [49]. Another example uses height as a free parameter to display uncertainty in 2D vector fields [72]. Texture can be used similarly to convey uncertainty and is also often modified by opacity, hue, or texture irregularities [18, 40, 74]. Sound has also been used as another channel for expressing uncertainty [58].

1.5.2.3 Glyphs

Glyphs are symbols used in visualization to signify data through parameters such as location, size, shape, orientation, and color. Because of the multivariate nature of glyphs, they can be used in visualization to map uncertainty to a free parameter. One such approach uses glyphs to present the distribution of multivariate aggregated data over a range of values [15]. These glyphs show the average, standard deviation, and distribution of three attributes of the data set. Conical glyphs have also been used to portray fiber tracks from DTI, leveraging the radius of the cone to encode uncertainty in the orientation of bundles [44]. An approach that modifies attributes of glyphs already present in the visualization is presented as a procedural generation algorithm [13]. In this work, the data is sampled on a regular grid and the size, color, and placement of glyphs are taken directly from the data samples. The uncertainty is then used to distort the glyphs so that glyphs with low uncertainty are very sharp, with the sharpness level decreasing as the uncertainty level increases. This distortion provides a clear indication of uncertainty and error while not placing heavy emphasis on areas of high uncertainty. In a similar fashion, contours already present in the visualization can be used [84, 85] or modified [71, 92] to express uncertainty.

Because not all data is visualized effectively using glyphs, the addition of glyphs to convey only uncertainty information is often a preferable approach. A specific example is the UISURF system [45], which visually compares isosurfaces and the algorithms used to generate them. In this system, glyphs are used to express positional and volumetric differences between isosurfaces by encoding the magnitude of the differences in the size of the glyphs. Similarly, line, arrow, and ellipsoidal glyphs can be used to depict uncertainty in radiosity solutions, interpolation schemes, vector fields, flow solvers, astrophysical data and animations through variation of placement, magnitude, radii, and orientation [54, 55, 57, 75, 91, 93, 109, 110, 113].

1.5.2.4 Image Discontinuity

Uncertainty visualization often relies on the human visual systems ability to quickly pick up an images discontinuities and to interpret these discontinuities as areas with distinct data characteristics. Techniques that utilize discontinuities rely on surface roughness, blurring, oscillations [13, 33, 56, 108], depth shaded holes, noise, and texture [22], as well as on the translation, scaling, rotation, warping, and distortion of geometry already used to visualize the data [75], to visualize uncertainty. Animation can highlight the regions of distortion or blur or highlight differences in visualization parameters [30, 60, 66]. Such techniques have been applied to multivaritate data displayed through scatter plots or parallel coordinates [27, 36].

1.6 Examples

1.6.1 Medical Visualization

A fundamental task in medical visualization is segmentation, the partitioning of a given image into regions that correspond to different materials, to different anatomical structures, or to tumors and other pathologies. Medical image acquisition typically introduces noise and artifacts, and we may wish to segment structures for which the data itself provides little contrast. This is a source of data uncertainty. In many cases, segmentation also involves complex computational models and numerous parameters, which introduces model uncertainty.

Traditional volume rendering classifies materials based on scalar intensity or feature vectors that account for first and second derivatives [50]. Lundström et al. [60] introduce probabilistic transfer functions that assign material probabilities to model cases in which the feature ranges of different materials overlap. This results in a distribution of materials at each location in space, which is visualized by an animation in which each material is shown for a duration that is proportional to its probability.

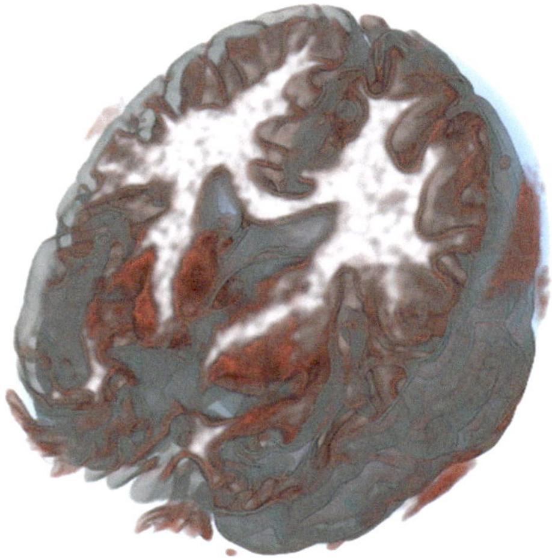

Fig. 1.4 A visualization of the brain using transfer functions that express the risk associated with classification

More complex segmentation tasks cannot be achieved based on local image properties alone. They require models that account for more global assumptions or more complex prior knowledge. Such models are also more computationally demanding and are typically run as a pre-process of the visualization. Some of them output class probabilities, from which Kniss et al. [51] derive measures that can be used to define transfer functions that enable exploring the risk associated with binary classifications, or to visualize spatial decision boundaries. Figure 1.4 shows the use of such transfer functions in a visualization of a segmented brain.

The framework of Saad et al. [90] combines volume rendering with tables that list groups of voxels for which the same materials have been found to be most, second most, and third most likely. They demonstrate several examples in which these tuples can be used to detect anomalous subregions within areas that share the most likely material. Follow-up work [89] has concentrated on identifying anomalies or misclassification by considering regions in which the image-based likelihood disagrees with shape and appearance priors.

Finally, work by Torsney-Weir et al. [100] addresses the model uncertainty in segmentation methods by providing a systematic framework to explore the impact of model parameters. This should facilitate finding settings that produce the desired segmentation, and for which the results do not change significantly when slightly changing the exact parameter values.

Fiber tracking, the reconstruction of nerve fiber bundles from diffusion MRI, is another subfield of medical visualization in which uncertainty plays an important role. It is treated in detail in Chap. 8 of this book.

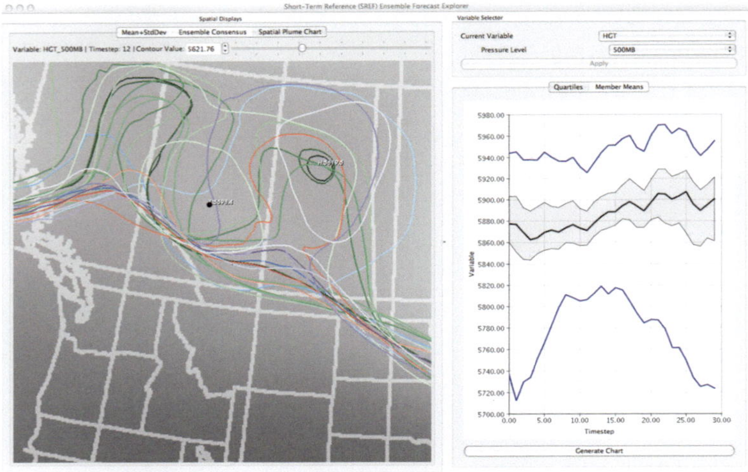

Fig. 1.5 The EnsembleVis tool [84] for exploring short-range weather forecast data

1.6.2 Weather and Climate

Uncertainties are prolific in weather and climate applications and arise not only from insufficient models, but also from our inability to accurately measure current weather conditions and obtain precise knowledge on parameter settings. The typical approach for mitigating uncertainties in weather and climate applications is to perform multi-run simulations, often using a collection of models, parameter perturbations, and initial conditions to generate outcome results for multiple variables and time steps. While the variables contained in the output of both weather and climate simulations are similar, the main differences between the two domains are the spatial region of interest and the duration of time covered. Weather applications are typically only interested in a small subsection of the planet, such as North America, and run to cover time steps within the near future. In contrast, climate modeling has a spatial interest of the whole planet and is run over hundreds of years.

The uncertainty resulting from these multi-run simulations are typically captured in what is know as "Ensemble data sets". These ensembles combine the multiple runs such that notions of probability of outcome can be explored. An ensemble consists of multiple simulation realizations and are often generated by pre-defined parameter perturbations. The visualization and analysis of these data sets aims to understand variations between models and effects of parameters and initial conditions, culminating in an understanding of the phenomenon leading to weather and climate events.

An example of a visualization and analysis tool can be seen in Fig. 1.5, which shows a screen shot of the EnsembleVis framework for the exploration of short-range weather forecasting data [84]. This tool uses a multiwindow approach to provide a collection of views for the end user, an approach used by other tools [92]. This approach allows the user to see overviews of a single time step, the progression of the data over time, drill downs to explore interesting spatial locations, including direct data display, and finally query-based exploration for more complex analyses.

1.6.3 Security and Intelligence

Security and intelligence uncertainty factors are a natural fit for security visualization, where making well-informed decisions is the primary goal. Enforcing security has become a top priority among a wide range of real-life applications, for instance large corporate or government/military networks. However, the task of decision making is notoriously difficult due to the malicious, hidden nature of attacks, sparse sampling of real-time environment, and time-critical requirements. Therefore, in security analysis uncertainty often exists among decisions at all levels, ranging from global scale such as "is there any malicious activity?" to finer scale such as "which entities are malicious?" or "in what order did these events actually occur?". The results of these decisions are used to make recommendations which can have significant operational impact, as nodes identified as malicious will be quarantined or removed from the network. Previously, both automated attack mitigation and interactive visualization approaches have been developed for security visualization. These existing techniques serve as a good platform for the integration of uncertainty visualizations and interactions. For example, several visual abstractions have been explored for detecting the sybil attack, which is a coordinated attack that can subvert many types of networks [24]. Sybil attacks are challenging to detect due to their variable attack forms and patterns. Because of this, traditional signature-based or behavior-based methods are ineffective, and security analysts must often find these nodes through manual analysis of their network. Visual abstractions from both adjacency matrix of the network connections [59] and spectral space [37] are explored, which can elucidate the signature patterns of an attack and apply automatic pattern matching algorithms or interactive analysis methods to search for similar patterns. As the short paper (Chap. 7) in this chapter describes, the factors of uncertainty can be introduced to existing detection mechanisms to improve the continuing analytic process. Since uncertainty is prevalent in security applications, the impact of uncertainty should be integrated into the entire procedure of data analysis and interactive exploration. Many current security visualization approaches can and should be augmented with interactions and visualizations for specifying and managing analytic uncertainty. By integrating analytic uncertainty in security visualization, analysts are able to make better-informed decisions regarding critical network infrastructure issues.

1.7 Open Problems

1.7.1 Perceptual and Cognitive Implications

Since visualization often relies heavily on the use of colors to convey information, it can be quite challenging for individuals with color vision deficiency. For them, even interpreting visualizations that would pose no problems for individuals with normal color vision can be a difficult task. In this case, however, the resulting ambiguity, and therefore, uncertainty, is inherent to the observer, falling outside the broad sources of uncertainty discussed in Sect. 1.1.1 (i.e., uncertainty observed in sampled data, uncertainty measures generated by models or simulations, and uncertainty introduced by the data processing or visualization processes). Thus, individuals with color vision deficiency have to constantly deal with uncertainty visualizations and make decisions based on ambiguous information. For those individuals, the display of additional data that tries to express the amount of uncertainty from various sources may even generate further ambiguities. The issues involving uncertainty visualization and color vision deficiency are discussed in Chap. 2.

1.7.2 Comparative Visualizations

The visualization of uncertainty may involve a comparison of different results, such as a weather forecast generated with different parameters. To detect similarities or differences in the results a comparative visualization technique [73] can be employed. In 3D a visualization via fusion [10, 12] is not feasible beyond a small number (2 or 3) of data sets, due to clutter and inter-dependence of the different data sets. An alternative to fusion is a side-by-side view of the data sets. This may be problematic in 3D since it is hard to find corresponding reference points in more than two volumes. As an example to control a 3D comparison Balabanian et al. [1] propose to integrate volume visualization into a hierarchical graph structure. These integrated views provide an interactive side-by-side display of different volumes while the parameter space can be explored through the graph structure. In 2D a blending of different results has basically the same issues as a fusion in 3D [31, 35]. There are techniques which allow a comparative visualization of different data sets in a single image. Urness et al. [104] introduced color weaving for flow visualization to compare different flow fields in a single 2D view. In contrast to blending, each pixel of the resulting image represents an unmodified value from one of the data sets. The generated pattern provides a good overview to detect similar or varying regions in the data sets. To compare certain regions in more detail, e.g., borders, it is better to consider larger comparison areas than individual pixels. In this context it is crucial that data sets which should be compared are visualized next to each other to get a direct comparison for a certain area. For only two data sets a checkerboard pattern can be used to achieve screen door transparency [95]. The white squares show one data set and the black squares show the other data set. The attribute block by Miller [68] allows a simultaneous comparison of four data sets. A repeating 2×2 pattern provides a shared border between all four data sets. An extension to this approach is the comparative visualization technique of Malik et al. [64]. Instead of a rectangular pattern a hexagonal pattern is used to more finely subdivide the image space. This allows the comparison of a larger number of data sets to one central data set since the hexagonal pattern can be subdivided according to the number of data sets to compare. Uncertainty of a measurement, simulation, or process provides an additional data stream which generates further visualization challenges. Uncertainty may be shown at discrete positions through glyphs or icons. For a dense representation of uncertainty, comparative visualization seems to be a promising emerging area. Topics of research will be: integrated views; sparsification of many data sets which shall be shown simultaneously; comparative navigation; visualization of competing, contradictive, or conflicting features.

Acknowledgments The authors gratefully acknowledge research support from the National Science Foundation, Department of Energy, the National Institutes of Health, and the King Abdullah University for Science and Technology.

References

1. Balabanian, J., Viola, I., Gröller, E.: Interactive illustrative visualization of hierarchical volume data. In: Proceedings of Graphics Interface 2010, Ottawa, Ontario, Canada, pp. 137–144 (2010)
2. Barthelmé, S., Mamassian, P.: Evaluation of objective uncertainty in the visual system. PLoS Comput. Biol. **5**(9), e1000504 (2009)
3. Becketti, S., Gould, W.: Rangefinder box plots. Am. Stat. **41**(2), 149 (1987)
4. Benjamini, Y.: Opening the box of a boxplot. Am. Stat. **42**(4), 257–262 (1988)
5. Bertin, J.: Sémiologie graphique: Les diagrammes—Les réseaux—Les cartes. Editions de l'Ecole des Hautes Etudes en Sciences (1967)
6. Bertin, J.: Semiology of Graphics. The University of Wisconsin Press (1983) (Translated by William Berg)
7. Bordoloi, U., Kao, D., Shen, H.W.: Visualization techniques for spatial probability density function data. Data Sci. J. **3**, 153–162 (2005)
8. Botchen, R.P., Weiskopf, D., Ertl, T.: Texture-based visualization of uncertainty in flow fields. In: IEEE Visualization 2005, pp. 647–654 (2005)
9. Boukhelifa, N., Duke, D.J.: Uncertainty visualization: why might it fail? In: CHI Extended Abstracts'09, pp. 4051–4056 (2009)
10. Bürger, R., Hauser, H.: Visualization of multi-variate scientific data. Comput. Graph. Forum **28**(6), 1670–1690 (2009)
11. Buttenfield, B., Ganter, J.: Visualization and gis: what should we see? What might we miss? In: 4th International Symposium on Spatial Data Handling, vol. 1, pp. 307–316 (1990)
12. Cai, W., Sakas, G.: Data intermixing and multi-volume rendering. Comput. Graph. Forum **18**(3), 359–368 (1999)
13. Cedilnik, A., Rheingans, P.: Procedural annotation of uncertain information. In: IEEE Proceedings of Visualization 2000, pp. 77–84 (2000)
14. Chambers, J.M., Cleveland, W.S., Kleiner, B., Tukey, P.A.: Graphical Methods for Data Analysis. Wadsworth (1983)

15. Chlan, E.B., Rheingans, P.: Multivariate glyphs for multi-object clusters. In: Proceedings of InfoVis '05, pp. 141–148 (2005)
16. Choonpradub, C., McNeil, D.: Can the box plot be improved? Songklanakarin J. Sci. Technol. **27**(3), 649–657 (2005)
17. Cohen, D.J., Cohen, J.: The sectioned density plot. Am. Stat. **60**(2), 167–174 (2006)
18. Coninx, A., Bonneau, G.P., Droulez, J., Thibault, G.: Visualization of uncertain scalar data fields using color scales and perceptually adapted noise. In: Applied Perception in Graphics and Visualization (2011)
19. Couclelis, H.: The certainty of uncertainty: GIS and the limits of geographic knowledge. Trans. GIS **7**(2), 165–175 (2003)
20. Deitrick, S., Edsall, R.: The influence of uncertainty visualization on decision making: An empirical evaluation. In: Progress in Spatial Data Handling, pp. 719–738. Springer, Berlin (2006)
21. Dietrick, S.: Uncertainty visualization and decision making: Does visualizing uncertain information change decisions? In: Proceedings of the XXII International Cartographic Conference (2007)
22. Djurcilov, S., Kim, K., Lermusiaux, P., Pang, A.: Visualizing scalar volumetric data with uncertainty. Comput. Graph. **26**, 239–248 (2002)
23. Doane, D.P., Tracy, R.L.: Using beam and fulcrum displays to explore data. Am. Stat. **54**(4), 289–290 (2000)
24. Douceur, J.R.: The sybil attack. In: The First International Workshop on Peer-to-Peer Systems, pp. 251–260 (2002)
25. Ehlschlaeger, C.R., Shortridge, A.M., Goodchild, M.F.: Visualizing spatial data uncertainty using animation. Comput. GeoSci. **23**(4), 387–395 (1997)
26. Esty, W.W., Banfield, J.D.: The box-percentile pot. J. Stat. Softw. **8**(17), 1–14 (2003)
27. Feng, D., Kwock, L., Lee II, Y.: R.M.T.: matching visual saliency to confidence in plots of uncertain data. IEEE Trans. Visual. Comput. Graph. **16**(6), 980–989 (2010)
28. Frigge, M., Hoaglin, D.C., Iglewicz, B.: Some implementations of the box plot. Am. Stat. **43**(1), 50–54 (1989)
29. Fua, Y.H., Ward, M., Rundensteiner, E.: Hierarchical parallel coordinates for exploration of large datasets. In: Proceedings of Vis '99, pp. 43–50 (1999)
30. Gershon, N.D.: Visualization of fuzzy data using generalized animation. In: Proceedings of the IEEE Conference on Visualization, pp. 268–273 (1992)
31. Gleicher, M., Albers, D., Walker, R., Jusufi, I., Hansen, C., Roberts, J.: Visual comparison for information visualization. Inf. Visual. **10**(4), 289–309 (2011)
32. Goldberg, K.M., Iglewicz, B.: Bivariate extensions of the boxplot. Technometrics **34**(3), 307–320 (1992)
33. Grigoryan, G., Rheingans, P.: Point-based probabilistic surfaces to show surface uncertainty. In: IEEE Trans. Visual. Comput. Graph. **10**(5), 546–573 (2004)
34. Haemer, K.W.: Range-bar charts. Am. Stat. **2**(2), 23 (1948)
35. Hagh-Shenas, H., Kim, S., Interrante, V., Healey, C.: Weaving versus blending: a quantitative assessment of the information carrying capacities of two alternative methods for conveying multivariate data with color. IEEE Trans. Visual. Comput. Graph. **13**(6), 1270–1277 (2007)
36. Haroz, S., Ma, K.L., Heitmann, K.: Multiple uncertainties in time-variant cosmological particle data. In: IEEE Pacific Visualization Symposium, pp. 207–214 (2008)
37. Harrison, L., Hu, X., Ying, X., Lu, A., Wang, W., Wu, X.: Interactive detection of network anomalies via coordinated multiple views. In: Proceedings of the 7th International Symposium on Visualization for Cyber Security, VizSec '10 (2010)
38. Harrower, M.: Representing uncertainty: Does it help people make better decisions? In: UCGISWorkshop: Geospatial Visualization and Knowledge Discovery Workshop (2002)
39. Hintze, J.L., Nelson, R.D.: Violin plots: a box plot-density trace synergism. Am. Stat. **52**(2), 181–184 (1998)
40. Interrante, V.: Harnessing natural textures for multivariate visualization. IEEE Comput. Graph. Appl. **20**(6), 6–11 (2000)

41. Jiao, F., Phillips, J.M., Stinstra, J., Krüger, J., Varma, R., Hsu, E., Korenberg, J., Johnson, C.R.: Metrics for uncertainty analysis and visualization of diffusion tensor images. Lect. Notes Comput. Sci. **6326**(2010), 179–190 (2010)
42. Johnson, C.R.: Top scientific visualization research problems. IEEE Comput. Graph. Appl. **24**(4), 13–17 (2004)
43. Johnson, C.R., Sanderson, A.R.: A next step: visualizing errors and uncertainty. IEEE Comput. Graph. Appl. **23**(5), 6–10 (2003)
44. Jones, D.K.: Determining and visualizing uncertainty in estimates of fiber orientation from diffusion tensor mri. Magn. Reson. Med. **49**, 7–12 (2003)
45. Jospeh, A.J., Lodha, S.K., Renteria, J.C., Pang., A.: Uisurf: Visualizing uncertainty in isosurfaces. In: Proceedings of the Computer Graphics and Imaging, pp. 184–191 (1999)
46. Kao, D., Dungan, J.L., Pang, A.: Visualizing 2d probability distributions from eos satellite image-derived data sets: A case study. In: Proceedings of the Conference on Visualization '01, VIS '01, pp. 457–460 (2001)
47. Kao, D., Kramer, M., Love, A., Dungan, J., Pang, A.: Visualizing distributions from multi-return lidar data to understand forest structure. Cartograph. J. **42**(1), 35–47 (2005)
48. Kao, D., Luo, A., Dungan, J.L., Pang, A.: Visualizing spatially varying distribution data. In: Information Visualization '02, pp. 219–225 (2002)
49. Kindlmann, G., Whitaker, R., Tasdizen, T., Moller, T.: Curvature-based transfer functions for direct volume rendering: Methods and applications. In: Proceedings of the 14th IEEE Visualization 2003 (VIS'03), pp. 67–74 (2004)
50. Kniss, J., Kindlmann, G., Hansen, C.: Multidimensional transfer functions for interactive volume rendering. IEEE Trans. Visual. Comput. Graph. **8**(3), 270–285 (2002)
51. Kniss, J.M., Uitert, R.V., Stephens, A., Li, G.S., Tasdizen, T., Hansen, C.: Statistically quantitative volume visualization. In: Proceedings of IEEE Visualization 2005, pp. 287–294 (2005)
52. Lee, B., Robertson, G.G., Czerwinski, M., Parr, C.S.: Candidtree: visualizing structural uncertainty in similar hierarchies. Inf. Visual. **6**, 233–246 (2007)
53. Lenth, R.V.: Comment on rangefinder box plots. Am. Stat. **42**(1), 87–88 (1988)
54. Li, H., Fu, C.W., Li, Y., Hanson, A.J.: Visualizing large-scale uncertainty in astrophysical data. IEEE Trans. Visual. Comput. Graph. **13**(6), 1640–1647 (2007)
55. Lodha, S., Sheehan, B., Pang, A., Wittenbrink, C.: Visualizing geometric uncertainty of surface interpolants. In: Proceedings of the Conference on Graphics Interface '96, pp. 238–245 (1996)
56. Lodha, S.K., Faaland, N.M., Charaniya, A.P.: Visualization of uncertain particle movement. In: Proceedings of the Computer Graphics and Imaging Conference, pp. 226–232 (2002)
57. Lodha, S.K., Pang, A., Sheehan, R.E., Wittenbrink, C.M.: Uflow: Visualizing uncertainty in fluid flow. In: Proceedings Visualization '96, pp. 249–254 (1996)
58. Lodha, S.K., Wilson, C.M., Sheehan, R.E.: Listen: sounding uncertainty visualization. In: Proceedings Visualization '96, pp. 189–195 (1996)
59. Lu, A., Wang, W., Dnyate, A., Hu, X.: Sybil attack detection through global topology pattern visualization. Inf. Visual. **10**(1), 32–46 (2011)
60. Lundström, C., Ljung, P., Persson, A., Ynnerman, A.: Uncertainty visualization in medical volume rendering using probabilistic animation. IEEE Trans. Visual. Comput. Graph. **13**(6), 1648–1655 (2007)
61. Luo, A., Kao, D., Pang, A.: Visualizing spatial distribution data sets. In: Proceedings of the Symposium on Data Visualisation 2003, VISSYM '03, pp. 29–38 (2003)
62. MacEachren, A., Robinson, A., Hopper, S., Gardner, S., Murray, R., Gahegan, M., Hetzler, E.: Visualizing geospatial information uncertainty: what we know and what we need to know. Cartograph. Geograph. Inf. Sci. **32**(3), 139–160 (2005)
63. MacEachren, A.M., Robinson, A., Hopper, S., Gardner, S., Murray, R., Gahegan, M., Hetzler, E.: Visualizing geospatial information uncertainty: what we know and what we need to know. Cartograph. Geograph. Inf. Sci. **32**(3), 139–160 (2005)
64. Malik, M.M., Heinzl, C., Gröller, M.E.: Comparative visualization for parameter studies of dataset series. IEEE Trans. Visual. Comput. Graph. **16**(5), 829–840 (2010)

65. Masuch, M., Freudenberg, B., Ludowici, B., Kreiker, S., Strothotte, T.: Virtual reconstruction of medieval architecture. In: Proceedings of EUROGRAPHICS 1999, Short Papers, pp. 87–90 (1999)
66. Masuch, M., Strothotte, T.: Visualising ancient architecture using animated line drawings. In: Proceedings of the IEEE Conference on Information Visualization, pp. 261–266 (1998)
67. McGill, R., Tukey, J.W., Larsen, W.A.: Variations of box plots. Am. Stat. **32**(1), 12–16 (1978)
68. Miller, J.: Attribute blocks: visualizing multiple continuously defined attributes. IEEE Comput. Graph. Appl. **27**(3), 57–69 (2007)
69. Newman, T.S., Lee, W.: On visualizing uncertainty in volumetric data: techniques and their evaluation. J. Vis. Lang. Comput. **15**, 463–491 (2004)
70. Olston, C., Mackinlay, J.D.: Visualizing data with bounded uncertainty. In: Proceedings of the IEEE Symposium on Information Visualization (InfoVis'02), pp. 37–40 (2002)
71. Osorio, R.A., Brodlie, K.: Contouring with uncertainty. In: 6th Theory and Practice of Computer Graphics Conference, pp. 59–66 (2008)
72. Otto, M., Germer, T., Hege, H.C., Theisel, H.: Uncertain 2d vector field topology. Comput. Graph. Forum **29**(2), 347–356 (2010)
73. Pagendarm, H., Post, F.: Comparative visualization—approaches and examples. In: 5th Eurographics Workshop on Visualization in Scientific Computing, Rostock, Germany (1994)
74. Pang, A., Furman, J.: Data quality issues in visualization. In: SPIE Visual Data Exploration and Analysis, vol. 2278, pp. 12–23 (1994)
75. Pang, A., Wittenbrink, C., Lodha, S.: Approaches to uncertainty visualization. Vis. Comput. **13**(8), 370–390 (1997)
76. Paulus, M., Hozack, N., Zauscher, B., McDowell, J., Frank, L., Brown, G., Braff, D.: Prefrontal, parietal, and temporal cortex networks underlie decision=making in the presence of uncertainty. NeuroImage **13**, 91–100 (2001)
77. Pauly, M., Mitra, N.J., Guibas, L.: Uncertainty and variability in point cloud surface data. In: Symposium on Point-Based Graphics, pp. 77–84 (2004)
78. Politi, M., Han, P., Col, N.: Communicating the uncertainty of harms and benefits of medical interventions. Med. Decis. Mak. **27**(5) 681–695 (2007)
79. Pöthkow, K., Hege, H.C.: Positional uncertainty of isocontours: condition analysis and probabilistic measures. IEEE Trans. Visual Comput. Graph. **PP**(99), 1–15 (2010)
80. Pöthkow, K., Weber, B., Hege, H.C.: Probabilistic marching cubes. Comput. Graph. Forum **30**(3), 931–940 (2011)
81. Potter, K.: Methods for presenting statistical information: The box plot. In: Hagen, H., Kerren, A., Dannenmann, P. (eds.) Visualization of Large and Unstructured Data Sets, GI-Edition, Lecture Notes in Informatics (LNI) S-4, pp. 97–106 (2006)
82. Potter, K., Kniss, J., Riesenfeld, R., Johnson, C.R.: Visualizing summary statistics and uncertainty. In: Computer Graphics Forum, Proceedings of Eurovis 2010, vol. 29(3), pp. 823–831 (2010)
83. Potter, K., Rosen, P., Johnson, C.R.: From quantification to visualization: A taxonomy of uncertainty visualization approaches. IFIP Advances in Information and Communication Technology Series p. (To Appear) (2012). (Invited Paper)
84. Potter, K., Wilson, A., Bremer, P.T., Williams, D., Doutriaux, C., Pascucci, V., Johhson, C.R.: Ensemble-vis: A framework for the statistical visualization of ensemble data. In: IEEE Workshop on Knowledge Discovery from Climate Data: Prediction, Extremes., pp. 233–240 (2009)
85. Praßni, J.S., Ropinski, T., Hinrichs, K.: Uncertainty-aware guided volume segmentation. IEEE Trans. Visual Comput. Graph. **16**(6), 1358–1365 (2010)
86. Rhodes, P.J., Laramee, R.S., Bergeron, R.D., Sparr, T.M.: Uncertainty visualization methods in isosurface rendering. In: EUROGRAPHICS 2003 Short Papers, pp. 83–88 (2003)
87. Riveiro, M.: Evaluation of uncertainty visualization techniques for information fusion. In: 10th International Conference on Information Fusion, pp. 1–8 (2007)
88. Rousseeuw, P.J., Ruts, I., Tukey, J.W.: The bagplot: a bivariate boxplot. Am. Stat. **53**(4), 382–387 (1999)

89. Saad, A., Hamarneh, G., Möller, T.: Exploration and visualization of segmentation uncertainty using shape and appearance prior information. IEEE Trans. Visual. Comput. Graph. **16**(6), 1366–1375 (2010)
90. Saad, A., Möller, T., Hamarneh, G.: Probexplorer: uncertainty-guided exploration and editing of probabilistic medical image segmentation. Comput. Graph. Forum **29**(3), 1113–1122 (2010)
91. Sanyal, J., Zhang, S., Bhattacharya, G., Amburn, P., Moorhead, R.J.: A user study to compare four uncertainty visualization methods for 1d and 2d datasets. IEEE Trans. Visual. Comput. Graph. **15**(6), 1209–1218 (2009)
92. Sanyal, J., Zhang, S., Dyer, J., Mercer, A., Amburn, P., Moorhead, R.J.: Noodles: a tool for visualization of numerical weather model ensemble uncertainty. IEEE Trans. Visual. Comput. Graph. **16**(6), 1421–1430 (2010)
93. Schmidt, G.S., Chen, S.L., Bryden, A.N., Livingston, M.A., Rosenblum, L.J., Osborn, B.R.: Multidimensional visual representations for underwater environmental uncertainty. IEEE Comput. Graph. Appl. **24**(5), 56–65 (2004)
94. Spear, M.E.: Charting Statistics. McGraw-Hill, New York (1952)
95. Stokking, R., Zubal, I., Viergever, M.: Display of fused images: methods, interpretation, and diagnostic improvements. Semin. Nucl. Med. **33**(3), 219–227 (2003)
96. Streit, A., Pham, B., Brown, R.: A spreadsheet approach to facilitate visualization of uncertainty in information. IEEE Trans. Visual. Comput. Graph. **14**(1), 61–72 (2008)
97. Strothotte, T., Puhle, M., Masuch, M., Freudenberg, B., Kreiker, S., Ludowici, B.: Visualizing uncertainty in virtual reconstructions. In: Proceedings of Electronic Imaging and the Visual Arts, EVA Europe '99, p. 16 (1999)
98. Thomson, J., Hetzler, B., MacEachren, A., Gahegan, M., Pavel, M.: A typology for visualizing uncertainty. In: Proceedings of SPIE. vol. SPIE-5669, pp. 146–157 (2005)
99. Tongkumchum, P.: Two-dimensional box plot. Songklanakarin J. Sci. Technol. **27**(4), 859–866 (2005)
100. Torsney-Weir, T., Saad, A., Möller, T., Hege, H.C., Weber, B., Verbavatz, J.M.: Tuner: principled parameter finding for image segmentation algorithms using visual response surface exploration. IEEE Trans. Vis. Comput. Graph. (TVCG) **17**(12), 1892–1901 (2011)
101. Tufte, E.: The Visual Display of Quantitative Information, 2nd edn. Graphics Press, Cheshire (2001)
102. Tufte, E.R.: The Visual Display of Quantitative Information. Graphics Press, Cheshire (1983)
103. Tukey, J.W.: Exploratory Data Analysis. Addison-Wesley, Reading (1977)
104. Urness, T., Interrante, V., Marusic, I., Longmire, E., Ganapathisubramani, B.: Effectively visualizing multi-valued flow data using color and texture. Proc. IEEE Visual. Conf. **03**, 115–121 (2003)
105. Ware, C.: Information Visualization: Perception for Design, 2nd edn. Morgan Kaufmann Publishers, Los Altos (2004)
106. Wilkinson, L.: Dot plots. Am. Stat. **53**(3), 276–281 (1999)
107. Wilkinson, L.: The Grammar of Graphics. Springer, New York, Inc. (1999)
108. Wittenbrink, C., Pang, A., Lodha, S.: Verity visualization: Visual mappings. Technical Report, University of California, Santa Cruz (1995)
109. Wittenbrink, C.M., Pang, A.T., Lodha, S.K.: Glyphs for visualizing uncertainty in vector fields. IEEE Trans. Visual. Comput. Graph. **2**(3), 266–279 (1996)
110. Zehner, B., Watanabe, N., Kolditz, O.: Visualization of gridded scalar data with uncertainty in geosciences. Comput. Geosci. **36**(10), 1268–1275 (2010)
111. Zuk, T., Carpendale, S.: Theoretical analysis of uncertainty visualization. In: SPIE vol. 6060: Visualization and Data Analysis, vol. 2006, pp. 66–79 (2006)
112. Zuk, T., Carpendale, S., Glanzman, W.D.: Visualizing temporal uncertainty in 3d virtual reconstructions. In: Proceedings of the 6th International Symposium on Virtual Reality, Archaeology and Cultural Heritage (VAST 2005), pp. 99–106 (2005)
113. Zuk, T., Downton, J., Gray, D., Carpendale, S., Liang, J.: Exploration of uncertainty in bidirectional vector fields. In: Society of Photo-Optical Instrumentation Engineers (SPIE) Conference Series, vol. 6809 (2008). Published online

Chapter 2
Uncertainty Visualization and Color Vision Deficiency

Manuel M. Oliveira

Abstract Color vision deficiency (CVD) affects a large number of individuals around the world, compromising their ability to effectively interpret color-coded information. This directly impacts the way these individuals perceive visualizations, often introducing ambiguities and uncertainties. This article provides an overview of the causes of color vision deficiency and discusses the main tools and techniques available for helping designers to create more effective visualizations for individuals with CVD. It also discusses the limitations of the existing techniques and presents some open questions for guiding research efforts in improving visualization experiences for larger audiences.

2.1 Introduction

Current estimates indicate that approximately 200 million individuals worldwide have some form of color vision deficiency (CVD) [10, 11]. Such condition compromises their ability to effectively perform color-related tasks, which impacts their private lives and professional activities [8]. Since visualizations tend to make intensive use of colors to convey information, many visualizations are not perceived by individuals with CVD as they are intended to be (e.g., Figs. 2.1 right, and 2.2b). This leads to uncertainties, forcing those individuals to make important decisions based on ambiguous information, which may have catastrophic implications. Thus, the perceptual limitations imposed by color vision deficiency is a relevant subject to the visualization community, but one that has not yet received all the attention it deserves. To produce more effective visualizations, we need to devise techniques that avoid excluding this significant fraction of the population. This article briefly discusses the the causes of color vision deficiency and the main techniques available to help the affected individuals to recover, as much as possible, the loss of color contrast. After pointing out the inherent limitations of these techniques, the article presents some open questions that should guide research efforts in this area.

M.M. Oliveira (✉)
Instituto de Informática—UFRGS, Porto Alegre, RS, Brazil
e-mail: oliveira@inf.ufrgs.br

C.D. Hansen et al. (eds.), *Scientific Visualization*, Mathematics and Visualization,
DOI 10.1007/978-1-4471-6497-5_2

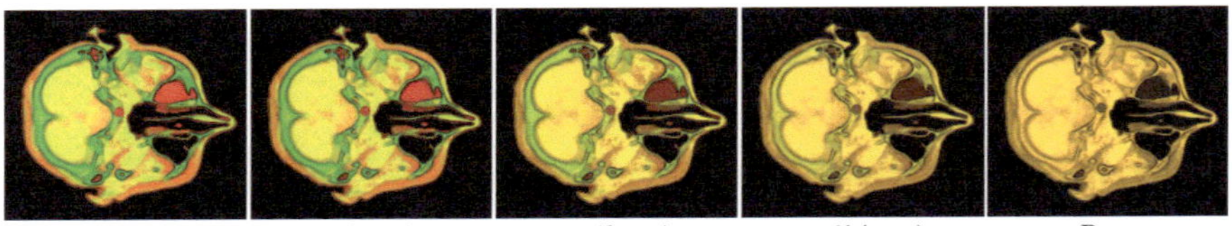

Fig. 2.1 Simulation of the color perception of individuals with CVD. A reference image (*left*) is followed by the simulation of the perceptions of anomalous trichromats (protanomalous) with various degrees of severity (spectral shifts of 2, 8, and 14 nm). The perception of a protanope is shown on the *right*. All images were simulated using the model described in [4]

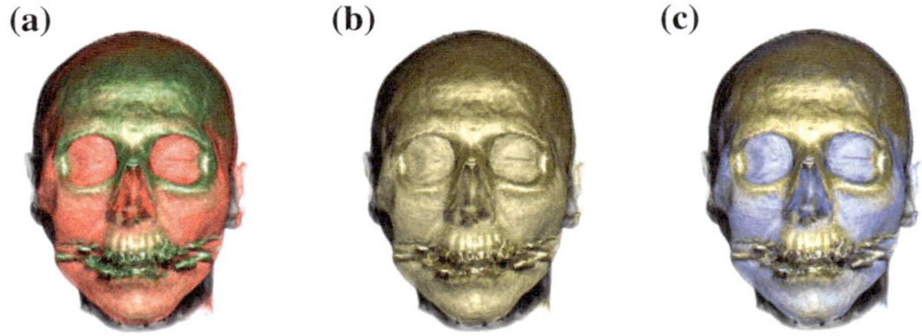

Fig. 2.2 Example image recoloring: **a** reference image. **b** Simulated perception of a deuteranope, using [4]. **c** Recolored version of the reference image for deuteranopes, using [5]. Note the significant enhancement of color contrast with respect to **b**. Since the color gamut of deuteranopes is a subset of the RGB color space, this image is perceived the same way both by deuteranopes and by normal trichromats. The case of recoloring for other dichromats is similar

2.1.1 Color Vision Deficiency

Human normal color vision requires three kinds of retinal photoreceptors. These are called *L*, *M*, and S cone cells, and have higher sensitivity to the long, medium, and short wavelengths of the visible spectrum, respectively. The specific type of photopigment contained in each kind of cone cell determines its spectral response. Some natural variations in the composition of these photopigments can shift their spectral sensitivities to different bands of the visible spectrum [11]. In this case, the affected individuals are called *anomalous trichromats*, and can be further classified as *protanomalous*, *deuteranomalous*, or *tritanomalous*, if the affected photopigment is associated with the L, M, or S cones, respectively. The bigger the shift, the more the individual's color perception will vary with respect to the perception of an individual with normal color vision (*normal trichromat*). In case one type of photopigment is missing, the individual is called a *dichromat*. Likewise, (s)he can be classified as *protanope*, *deuteranope*, or *tritanope*, according to the type of missing photopigment (L, M, or S, respectively). Much rarer conditions include the cases of individuals with a single kind of photopigment (*cone monochromats*) or no functional cone cells at all (*rod monochromats*).

As a consequence of the existence of three types of photoreceptors, normal color vision spans a 3-D color space. The color gamut of a dichromat, on the other hand, is only two-dimensional and can be represented by a surface patch in the same 3-D color space. Such a reduced gamut is the cause of the ambiguity experienced by dichromats: many different colors are perceived as the same, when projected onto such patches. For anomalous trichromats, the color gamut falls in between these two extremes, moving towards the gamut of a dichromat as the degree of severity of the anomaly increases. For spectral shifts of approximately 20 nm, the perception of an anomalous trichromat becomes similar to the perception of a dichromat [6, 11].

Currently, there is no clinical or surgical treatment for color vision deficiency. Given the relevance of the problem, a few techniques have been recently proposed to simulate the perception of individuals with CVD [2, 4, 7], and to enhance image contrast through recoloring [3, 5, 9]. Next, I briefly discuss these techniques, showing how they can assist the design of more inclusive visualization experiences, but also discussing their inherent limitations, which calls for more research.

2.2 Tools for More Inclusive Visualizations

The first step to produce more effective visualizations for individuals with CVD is to understand their perceptual limitations. Meyer and Greenberg [7], and Brettel et al. [2] presented simulation techniques for the color perception of dichromats. Machado et al. [4] introduced a physiologically-based model that supports the simulation of dichromatic as well as anomalous trichromatic vision (with arbitrary degrees of severity) in a unified way. This simulation model works in real time and can be quickly incorporated into existing systems. Thus, a visualization designer can get instantaneous feedback on how it would be perceived by individuals with CVD (Fig. 2.1). Such knowledge allows the designer to refine the visualization, making it more effective for wider audiences. While simulation models help to increase the awareness of the perceptual limitations due to CVD, they do not directly help the affected individuals to recover the loss of color contrast.

To address the problem of enhancing color contrast, a few automatic image-recoloring techniques for dichromats have been proposed in recent years [3, 5, 9]. Essentially, all these approaches define ways of mapping the colors in the original image to a new set of colors in the dichromat's gamut. This is done while trying to preserve the perceptual color differences among all pairs of colors in the original image. Rasche et al. [9] proposed an approach that uses a constrained multivariate optimization procedure applied to a reduced set of quantized colors. The resulting algorithm does not scale well with size of the input image and the number of quantized colors, and is not applicable to interactive applications. Kuhn et al. [3] present a solution based on a mass-spring optimization that achieves interactive rates, and tries to preserve the naturalness of the original images (i.e., preserve the colors that can already be perceived by dichromats). More recently, Machado and Oliveira [5] introduced a projection-based recoloring approach that works in real time, enforces

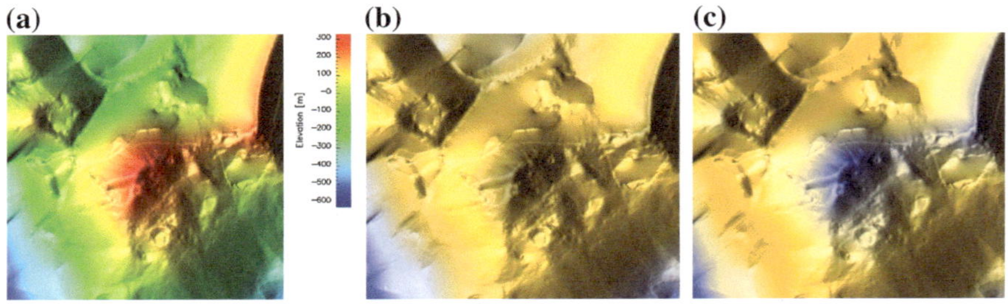

Fig. 2.3 Limitation of recoloring techniques. **a** Original image with colors over most of the RGB color space. **b** Simulated perception of a deuteranope. **c** Due to the dichromat's limited color gamut, by trying to solve some color ambiguities, recoloring techniques may introduce new ones

temporal coherence, and can be easily integrated with existing visualization applications. This makes it suitable for use in interactive visualizations. Figure 2.2 shows an example of image recoloring produced by this technique.

All recoloring techniques for dichromats share an inherent limitation: *they define mappings from a 3-D color space to a 2-D color gamut*. Thus, such techniques tend to become ineffective as the original image content spans the entire or most of the 3-D color space. In those situations, trying to solve some ambiguity by rearranging colors on the dichromat's 2-D color space might introduce new ambiguities (Fig. 2.3). Moreover, current recoloring techniques are restricted to the set of colors found in each input image or video. Thus, mappings between pairs of colors in one image or video may not be preserved in different ones (e.g., some cell structures may be recolored in blue in one image, while appearing yellow in another).

2.2.1 Open Research Questions

In order to address the limitations discussed in the previous paragraph, we need to consider the following open research questions:

- Q1 How can one enhance visualizations by encoding additional information in order to compensate for the reduced color gamut of dichromats? Or, in other words, how can one lift the two-dimensional color gamut restriction?
- Q2 Can the experiences learned from addressing the previous question also be exploited to enhance visualizations for normal trichromats?
- Q3 How can one obtain content-independent solutions that can be consistently used over different images and videos?
- Q4 Is it possible to satisfactorily extend these solutions to also represent natural scenes, where colors have some associated meanings to the viewers?

We have started to explore some of these questions. In one initial effort, we have investigated augmenting colors with simple patterns to encode information for dichromats [1]. Our results suggest that such combination can improve the performance of individuals with CVD in some visualization tasks, besides increasing their confidence in making color-based choices. We have also noticed that the use of patterns can help normal trichromats to fine tune color-related decisions. The use of patterns is, however, just one option in wide space of possibilities, and many creative solutions are waiting to be discovered.

2.3 Conclusion

Color-vision-deficient individuals routinely experience uncertainty visualizations, both in their private lives and professional activities. This article discussed the causes of such perceptual limitations, and briefly described the tools and techniques currently available that try to address this issue. Most of the illustrations and discussions focused on the case of dichromats, since, in general, they face stronger restrictions than anomalous trichromats. After analyzing the limitations of the existing techniques, the article presented a list of open questions that need to be considered in our quest for more inclusive visualizations. By understanding how to effectively deal with the restrictions faced by individuals with CVD, we should also be able to produce richer visualizations experiences for normal trichromats.

Acknowledgments This work was sponsored by CNPq-Brazil (Processo No. 308936/2010-8). Figure 2.1 (left) was kindly provided by Francisco Pinto.

References

1. Behzad, S., Majumder, A., Oliveira, M.M., Schneider, R.G., Raskar, R.: Using patterns to encode color information for dichromats. IEEE TVCG **19**(1), 118–129 (2013)
2. Brettel, H., Viénot, F., Mollon, J.D.: Computerized simulation of color appearance for dichromats. J. Opt. Soc. Am. **14**(10), 2647–2655 (1997)
3. Kuhn, G.R., Oliveira, M.M., Fernandes, L.A.F.: An efficient naturalness-preserving image-recoloring method for dichromats. IEEE TVCG **14**(6), 1747–1757(2008)
4. Machado, G.M., Oliveira, M.M., Fernandes, L.A.F.: A physiologically-based model for simulation of color vision deficiency. IEEE TVCG **15**(6), 1291–1298 (2009)
5. Machado, G.M., Oliveira, M.M.: Real-time temporal-coherent color contrast enhancement for dichromats. Comput. Graph. Forum **29**(3), 933–942. In: Proceedings of EuroVis 2010
6. McIntyre, D.: Colour Blindness: Causes and Effects. Dalton Publ. (2002)
7. Meyer, G.W., Greenberg, D.P.: Color-defective vision and computer graphics displays. IEEE Comput. Graph. Appl. **8**(5), 28-40 (1988)
8. Olsen, M.M., Harris, K.R.: Color Vision Deficiency and Color Blindness: An Introduction to the Problem. Fern Ridge Press (1988)
9. Rasche, K., Geist, R., Westall, J.: Re-coloring images for gamuts of lower dimension. Comput. Graph. Forum **24**(3), 423-432 (2005)
10. Rigden, C.: The eye of the beholder—Designing for colour-blind users. Br. Telecommun. Eng. **17**, 2–6 (1999)
11. Sharpe, L., Stockman, A., Jägle, H., Nathans, J.: Opsin genes, cone photopigments, color vision, and color blindness. In: Gegenfurtner, K.R., Sharpe, L.T. (eds.) Color Vision: From Genes to Perception. Cambridge University Press, Cambridge (1999)

Chapter 3
Analysis of Uncertain Scalar Data with Hixels

Joshua A. Levine, David Thompson, Janine C. Bennett, Peer-Timo Bremer, Attila Gyulassy, Valerio Pascucci and Philippe P. Pébay

Abstract One of the greatest challenges for today's visualization and analysis communities is the massive amounts of data generated from state of the art simulations. Traditionally, the increase in spatial resolution has driven most of the data explosion, but more recently ensembles of simulations with multiple results per data point and stochastic simulations storing individual probability distributions are increasingly common. This chapter describes a relatively new data representation for scalar data, called hixels, that stores a histogram of values for each sample point of a domain. The histograms may be created by spatial down-sampling, binning ensemble values, or polling values from a given distribution. In this manner, hixels form a compact yet information rich approximation of large scale data. In essence, hixels trade off data size and complexity for scalar-value "uncertainty".

We summarize several techniques for identifying features in hixel data using a combination of topological and statistical methods. In particular, we show how to approximate topological structures from hixel data, extract structures from multi-modal

---

J.A. Levine (✉)
School of Computing, Clemson Univeristy, Clemson, SC, USA
e-mail: levinej@clemson.edu

D. Thompson
Kitware, Inc., Carrboro, NC, USA
e-mail: david.thompson@kitware.com

J.C. Bennett · P.P. Pébay
Sandia National Laboratories, Livermore, CA, USA
e-mail: jcbenne@sandia.gov

P.P. Pébay
e-mail: pppebay@sandia.gov

P.-T. Bremer
Lawrence Livermore National Laboratory, Livermore, CA, USA
e-mail: bremer5@llnl.gov

A. Gyulassy · V. Pascucci
Scientific Computing and Imaging Institute, University of Utah, Salt Lake City, UT, USA
e-mail: jediati@sci.utah.edu

V. Pascucci
e-mail: pascucci@sci.utah.edu

C.D. Hansen et al. (eds.), *Scientific Visualization*, Mathematics and Visualization,
DOI 10.1007/978-1-4471-6497-5_3

distributions, and render uncertain isosurfaces. In all three cases we demonstrate how using hixels provides the capability to recover prominent features that would otherwise be either infeasible to compute or ambiguous to infer. We use a collection of computer tomography data and large scale combustion simulations to illustrate our techniques.

3.1 Foundations

The concepts presented in this chapter rely on mathematical foundations from both the topology and statistics communities. A summary of the statistical methods used in this chapter can be found in [3]. The topological tools presented in this paper are based on Morse theory, a mathematical tool to study how the "shape" of a function is related to the shape of its domain [8, 9]. Morse theory is a well understood concept in the context of smooth scalar fields, and has been effectively extended to piecewise-linear [2] and discrete [4] domains. Algorithms for computing the Morse-Smale (MS) complex have been presented in the piecewise linear context [1, 2] as well as the discrete context [5–7]. In this chapter we summarize the method presented in [10] to extend the use of the MS complex to hixels.

We are interested in characterizing an uncertain scalar field defined at many points in a metric space, $\mathbb{M}$. A *hixel* is a point $x_i \in \mathbb{M}$ with which we associate a histogram of scalar values, $h(x_i)$. In our setting, the $h(x_i)$ could either represent a collection of values in a block of data, collections of values at a location over a set of runs in an ensemble, or uncertainty about the potential values a location may represent. Figure 3.1 shows several empirical distributions with maxima identified.

3.1.1 Bucketing Hixels

When a hixel is defined empirically as a number of samples n_{f_i} on a finite support $\{f_j \mid j \in \{1, 2, \ldots, N_f\}\}$, we call each entry of the support a *bin*. The probability distribution (specifically here a probability mass function) is thus given by:

$$h\colon\ f_i \longmapsto \frac{n_{f_i}}{\sum_{k=1}^{N_f} n_{f_k}}$$

for each possible value f_j. Whether this distribution is defined empirically or analytically, for instance as a weighted sum of Gaussians, we are interested in identifying regions of high probability associated with peaks in the probability density. For that we will perform topological segmentation of the histogram to identify peaks as well as a range of function values associated with each peak. This range of function values is called a *bucket*. A bucket aggregates one or more bins and is assigned a

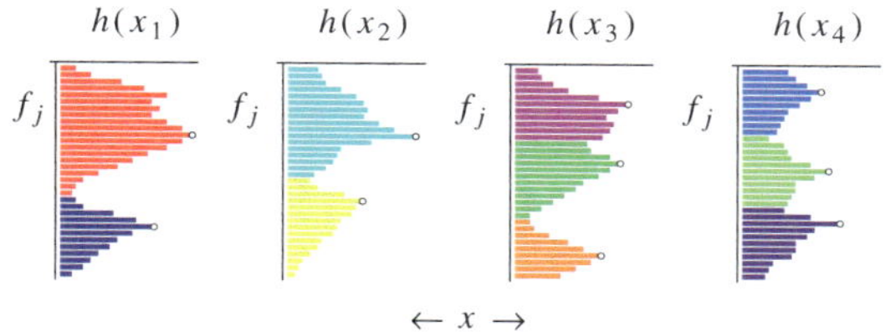

Fig. 3.1 Four probability distributions represented as histograms $h(x_i)$ with 32 bins f_j (rotated 90°). Maxima (identified with *black circles*) indicate function values with high probability. *Colors* indicate bucketing, the aggregation of bins of the histograms into modes based on the stable manifolds of persistent maxima. ©IEEE reprinted, with permission, from Thompson et al. [10]

probability given by the cumulative distribution function over that range of function values. Figure 3.1 illustrates how the distributions have been bucketed by merging maxima of h with their lowest persistence [2].

Because our scalar function f is represented by probability distribution h and we are interested in identifying regions of high probability, we use a variant on the notion of persistence. Typically, persistence ranks maxima by the difference in function value between their value and their paired minima. Instead, we assign a value equal to the area of the histogram between the pair (we call this ranking *areal persistence*). By ordering intervals between maxima and minima according to the area underneath them, peaks in probability density may be eliminated according to the probability associated with them. The decision of which of the two possible minima (assuming the maximum is interior) should be merged with the peak is made using regular persistence: the smaller difference in function value indicates the region to which corresponds the peak to be eliminated. Buckets can be merged in this fashion until the probability of the smallest bucket is above some threshold. When the number of samples is small, this threshold must be close to 1 since our confidence will be low. Assuming that f has a finite variance (so that the central limit theorem holds), the threshold may be lowered as the number of samples increases. Eventually, each hixel will have one or more buckets corresponding to probable function values associated with a peak in the distribution function; each bucket thus corresponds to an estimated *mode* of the distribution.

Figure 3.2 shows the bucket counts for the jet dataset as areal persistence thresholds are varied. At low thresholds, hixels that encompass areas of turbulent behavior have high bucket counts. As persistence simplification is applied, but increasing the threshold of areal persistence, buckets are merging indicating the most probable modes of the dataset.

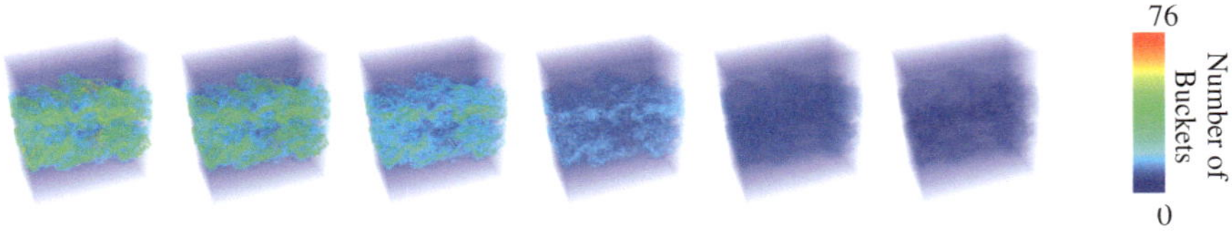

Fig. 3.2 Varying areal persistence and its effect on bucketing for the Jet dataset. Using hixels with size 16^3 and 256 bins/histogram, we vary areal persistence for all powers of two between 16 and 512, inclusive, from *left to right*. *Color* indicates how many buckets at that hixel's position in (x, y, z) space. At low levels of persistence, as many as 76 buckets can be selected in the hixel, but as persistence increases, most hixels have only 1 or 2 buckets. ©IEEE reprinted, with permission, from Thompson et al. [10]

3.2 Analysis of Hixel Data

In this section we summarize algorithms (1) to extract approximations of common topological structures from hixel data; (2) to segment multi-modal data by splitting individual histograms into their modes and correlate neighboring modes; and (3) to define and render uncertain isosurfaces.

3.2.1 Sampled Topology

Hixels encode potential scalar values along with their distributions at sample locations, and thus can aid visualization of the uncertainty in topological segmentations of down-sampled data. We use a three step process where we (1) sample the hixels to generate individual instances of the coarser representation, (2) compute the Morse complex on the instance, and (3) aggregate multiple instances of the segmentation to visualize its variability. We generate an instance V_i of the down-sampled data by picking values at each sample location from the co-located hixel. The value is picked at random, governed by the distribution encoded by the hixel. By picking values independently from neighboring values, we can simulate any possible down-sampling of the data, assuming each hixel's distribution is independent.

For each sampled field V_i we compute the Morse complex of the instance using a discrete Morse theory based algorithm [5], and identify basins around minima for varying persistence simplification thresholds. We next create a binary field C_i that encodes the geometric information of the arcs of the complex. Each sample location in V_i contributes a value of 1 to C_i if the sample is on the boundary of two or more basins, otherwise it contributes 0 if the sample is in the interior of a basin.

To visualize the variability of the topological segmentation of sub-sampled data, we repeatedly sample the hixels producing V_i's, and compute their basin boundary representations C_i. After n iterations, an aggregate function is computed over the boundary representations, recording the fractional identification of a sample location

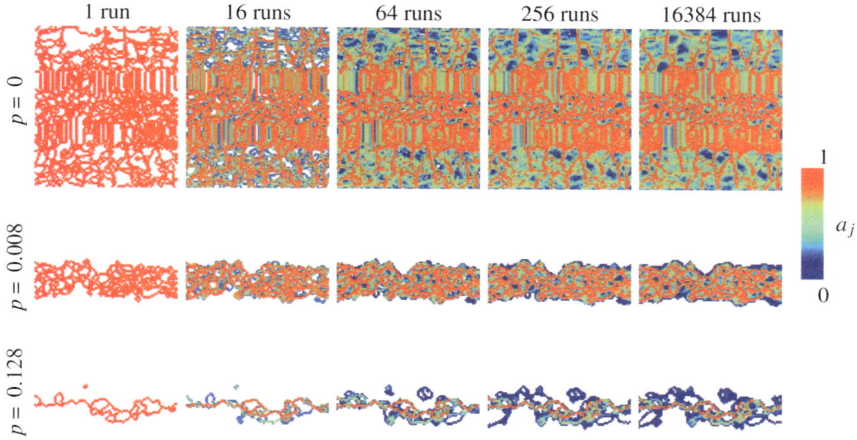

Fig. 3.3 We sample the hixel data for an 8×8 blocking of combustion data, and compute the aggregate segmentation for a number of iterations, also varying the level of persistence simplification. Adjacent white pixels are identified in the interior of the same basin in every single run. The images converge as the number of iterations increases *left to right*. ©IEEE reprinted, with permission, from Thompson et al. [10]

as a basin boundary. Formally, at each sampled location we compute the aggregate function $a_j = \frac{1}{n}\sum_{C_i} c_j$. Note that a_j can take values between one and zero, where one indicates it was identified as the boundary of basins in every instance, and zero meaning it was identified as interior in every instance. In this manner, we visualize rasterizations of the geometry of the Morse complex.

One point of interest is the amount of sampling required to capture a reasonable aggregate field. Figure 3.3 shows each aggregate slice for the 8×8 block size, as number of iterations and topological persistence are varied. The convergence of these sequences indicates that the distribution represented by the hixels produces stable modes of segmentation.

3.2.2 Topological Analysis of Statistically Associated Buckets

We next describe a novel statistical technique for recovering prominent topological features from ensemble data stored in hixel format. This computation is aided by the fact that ensemble data has a statistical dependence between runs that allows us to build a structure representing a predictive link between neighboring hixels. Our algorithm identifies subregions of space and scalar values that are consistent with positive association and we perform topological segmentation on only those regions. After bucketing all hixels, we compute a *contingency table* or tabular representation between each pair of adjacent hixels, h_i and h_j, of the counts of all observed com-

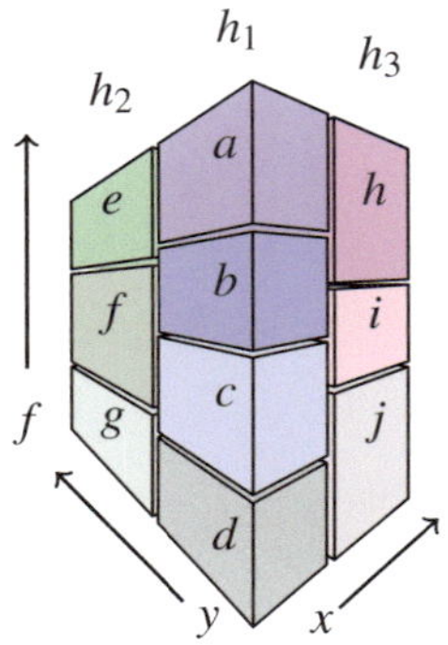

h_1–h_2	e	f	g
a	4	2	0
b	2	3	1
c	0	5	1
d	6	0	0

h_1–h_3	h	i	j
a	5	1	0
b	1	4	1
c	2	4	0
d	0	1	5

Fig. 3.4 Shown is a 2-dimensional spatial domain (x and y) and a function, f, on which the data is hixelated (*vertical axis*). Hixel h_1 has buckets a, b, c, and d; h_2 has e, f, and g; and h_3 has h, i, and j. On the *right*, two contingency tables are shown tabulating the simultaneously observed samples for $h_1 - h_2$ and $h_1 - h_3$. ©IEEE reprinted, with permission, from Thompson et al. [10]

binations of values as shown in Fig. 3.4. By considering simultaneously observed samples of h_i and h_j, it is possible to identify pairs of buckets that co-occur more frequently than if they were statistically independent by identifying those whose *pointwise mutual information* (pmi) is greater than zero. Pointwise mutual information is a statistical measure of association between realizations of discrete random variables. The pmi of a realization (x, y) of a pair of discrete random variables (X, Y) is defined as:

$$\text{pmi}(x, y) := \log \frac{p_{(X,Y)}(x, y)}{p_X(x)p_Y(y)},$$

where p_X, p_Y, and $p_{(X,Y)}$ respectively denote the probability density functions of X, Y, and the joint probability (X, Y), for all possible outcomes of X and Y. When the joint probability vanishes the pmi is set to $-\infty$. Note that if X and Y are independent, then the pointwise mutual information vanishes everywhere the joint probability does not. Naturally, as this is a pointwise quantity, a zero value of the pmi does not indicate mutual independence of the random variables.

Pairs of buckets in neighboring hixels, with a pmi greater then some $\varepsilon \geq 0$, can be treated as edges in a graph connecting buckets We call these connected components *sheets*, illustrated in Fig. 3.5. Sheets are geometrically like lower-dimensional surfaces in the product space of the spatial variables and the scalar data. Once we have selected sheets, we compute topological basins of minima and maxima on each sheet individually. We examine sheets on a mixture of 2 stochastic processes shown in Fig. 3.6a. This data highlights the fact that individual hixels can be multi-modal and can behave as both a minimum and maximum. A naive analysis that computes the mean or median followed by standard topological segmentation would fail to incorporate the multi-modal nature of the data. To addresses this issue, topological analysis is performed on sheets of the domain that have likely simultaneously observable sets of behavior.

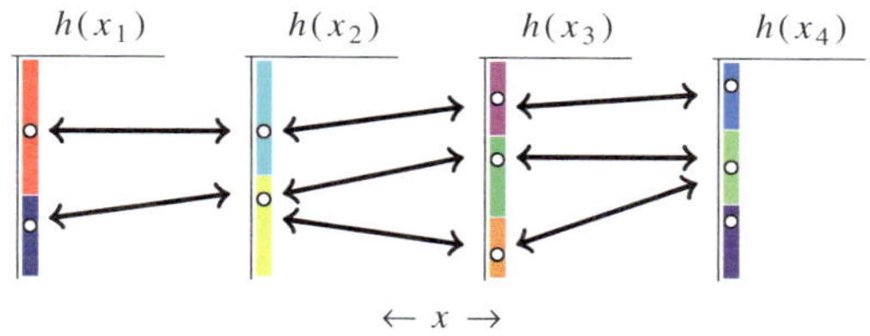

Fig. 3.5 Once the hixels have been bucketed and modes have been identified, we compute the pmis between buckets that are spatially adjacent and connect those with positive associations to form sheets. ©IEEE reprinted, with permission, from Thompson et al. [10]

Fig. 3.6 **a** Volume rendering of a hixel data set generated by sampling a mixture of Poisson and normal distributions. Basins of minima (**b**) and of maxima (**c**) are shown for sheets, identifying prominent features associated with each process in the mixture model. ©IEEE reprinted, with permission, from Thompson et al. [10]

There are 512×512 hixels in the mixture model data set, each with 128 equally-sized bins. The shortest axis in the images corresponds to histogram bins, thus a spatially higher location along that axis indicates a higher function value. The data is a mixture of two distributions at each hixel with 3,200 samples from a Poisson distribution and 9,600 samples from a Gaussian distribution. Hue and opacity are used to illustrate the density of samples. When the number of samples in a hixel bin is zero, the bin is rendered as a transparent red. When the number of samples in a bin is large, the bin is rendered as an opaque yellow.

Each hixel's Poisson and Gaussian distribution have different parameter values that vary smoothly over the image. The Poisson λ parameter is a maximum of 100 at five source points arranged in a circle and decreases to a minimum value of approximately 12 proportional to the distance to the nearest of these points. The Gaussian mean (standard deviation) is a minimum (maximum) at 4 points arranged in the same circle as the 5 Poisson source points. The mean varies from 32 to 108 while the standard deviation varies from 16 to 3.8. Topological basins of minima and maxima are shown in Fig. 3.6b, c for all sheets with pmi greater than zero. Our approach clearly extracts separate sheets belonging to the two processes, allowing topological analysis to identify the prominent features of each distribution.

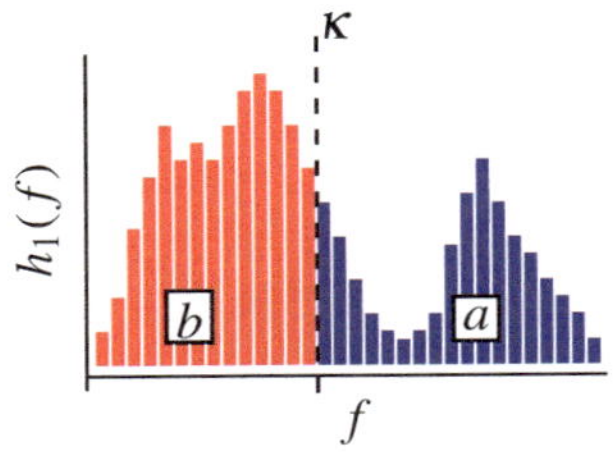

Fig. 3.7 Slicing a histogram. For an isovalue κ, we can use the histogram to count the number of voxels with function values above (a) and below (b) i. ©IEEE reprinted, with permission, from Thompson et al. [10]

3.2.3 Fuzzy Isosurfacing

When down-sampling larger datasets, hixels enable preserving the presence of an isosurface within the data. In particular, because hixels store the counts of all function values present within a block, we can compute the likelihood of the presence of an isosurface within that block. Given a hixel h_i and an isovalue κ, we *slice* the histogram at κ and compute the number of voxels above and below κ. These two counts, a and b for the count above and below, respectively, provide an indication as to how frequently the isosurface κ may exist within the block. Alternatively these values can approximate the surface area of the isosurface within the block, Fig. 3.7 visualizes this slicing process.

Using the values a and b, we can then compute a likelihood field. We let $g = \frac{a}{b} - \frac{b}{a}$. For hixels that have $a = b$, g takes on the value 0, while $g > 0$ for hixels that are strongly above κ and $g < 0$ for hixels that are strongly below. If $a = 0$, we set $g = b$, and when $b = 0$ we set $g = a$. By volume rendering the g field we can get a "fuzzy" depiction of where the isosurface exists in a hixelated field. By comparison, naive down-sampling of the scalar field could either move or destroy isovalues. By visualizing the field g we get a more honest depiction about where that isovalue was originally in the dataset, and can thus preserve that information.

Figure 3.8 shows visualizations of the stag dataset for $\kappa = 580$, down-sampled from its original size of $832 \times 832 \times 494$ to $208 \times 208 \times 123$, $104 \times 104 \times 61$, $52 \times 52 \times 30$, $26 \times 26 \times 15$, and $13 \times 13 \times 7$. Hixels of block size b^3 used $2b^2$ bins. By tracking a histogram of values, at lower resolutions we can preserve the fidelity of the isosurface and display a more expressive view of the data. Using only a single value, it is challenging to preserve the thin features of the isosurface, as the legs, antenna, and mandibles are hard to preserve. Figure 3.2 shows a side-by-side comparison of the isosurfaces produced at $\kappa = 580$ for the mean and lower-left fields as compared to the volume rendering of the g field when the hixel block size is 16^3.

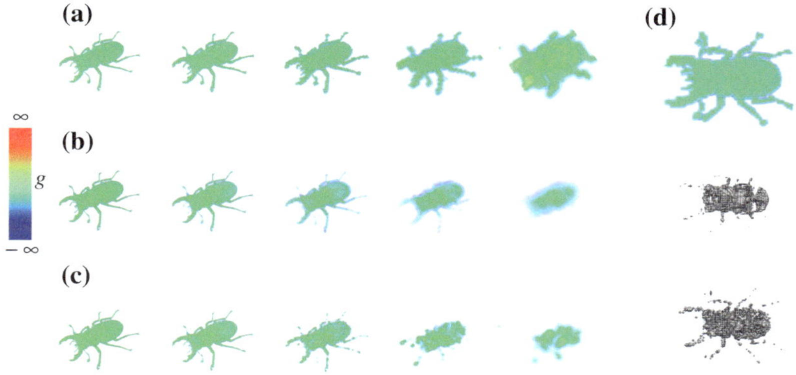

Fig. 3.8 **a–c** Volume rendering of g field for Stag, as compared to down-sampling with the mean and lower left corner of each block. The dataset is originally $832 \times 832 \times 494$, and from *left-to-right* we shown hixel sizes of 4^3 to 64^3, with all powers of 2 in between. **d** For hixel size 16^3, we compare the likelihood field g volume rendered to the isosurfaces computed for $\kappa = 580$ for the mean and lower-left down-sampling. ©IEEE reprinted, with permission, from Thompson et al. [10]

3.3 Discussion

By unifying the representations of large scalar fields from various modalities, hixels enable the analysis and visualization of data that would be otherwise be challenging to process. While hixels have utility, they present a number of challenges and open questions to explore. One important question regards information preserved by the hixels vs. resolution loss. A study is required to explore the appropriate number of bins per hixel as well as persistence thresholds for bucketing and mode seeking algorithms. The performance of hixels was not currently emphasized in our work, but the complexity of many techniques used here should allow for scaling to larger data. Additional research is required to find a balance between data storage allotted for the histograms versus feature preservation. Finally, further studies on what topological features can and cannot be easily preserved by hixelation is required.

References

1. Bremer, P.T., Edelsbrunner, H., Hamann, B., Pascucci, V.: A topological hierarchy for functions on triangulated surfaces. IEEE Trans. Vis. Comp. Graph. **10**(4), 385–396 (2004)
2. Edelsbrunner, H., Harer, J., Zomorodian, A.: Hierarchical Morse-Smale complexes for piecewise linear 2-manifolds. Discrete Comput. Geom. **30**(1), 87–107 (2003)
3. Feller, W.: An Introduction to Probability Theory and its Applications, vol. 1. Wiley, New York (1968)
4. Forman, R.: A user's guide to discrete Morse theory. In: Proceedings of the 2001 Internat. Conference on Formal Power Series and Algebraic Combinatorics. A Special Volume of Advances in Applied Mathematics, p. 48 (2001)

5. Gyulassy, A., Bremer, P.T., Hamann, B., Pascucci, V.: A practical approach to Morse-Smale complex computation: scalability and generality. IEEE Trans. Vis. Comput. Graph. **14**(6), 1619–1626 (2008)
6. King, H., Knudson, K., Mramor, N.: Generating discrete morse functions from point data. Exp. Math. **14**(4), 435–444 (2005)
7. Lewiner, T., Lopes, H., Tavares, G.: Applications of forman's discrete morse theory to topology visualization and mesh compression. IEEE Trans. Vis. Comp. Graph. **10**(5), 499–508 (2004)
8. Matsumoto, Y.: An Introduction to Morse Theory. Translations of Mathematical Monographs, vol. 208. American Mathematical Society, Providence (2002)
9. Milnor, J.: Morse Theory. Princeton University Press, Princeton (1963)
10. Thompson, D., Levine, J., Bennett, J., Bremer, P.T., Gyulassy, A., Pascucci, V., Pebay, P.: Analysis of large-scale scalar data using hixels. In: IEEE Symposium on Large-Scale Data Analysis and Visualization (LDAV) (2011)

Chapter 4
On the (Un)Suitability of Strict Feature Definitions for Uncertain Data

Tino Weinkauf

Abstract We discuss strategies to successfully work with strict feature definitions such as topology in the presence of noisy/uncertain data. To that end, we review previous work from the literature and identify three strategies: the development of fuzzy analogs to strict feature definitions, the aggregation of features, and the filtering of features. Regarding the latter, we will present a detailed discussion of filtering ridges/valleys and topological structures.

4.1 Introduction

Features are not only ubiquitous in scientific data, but they are inherent to the underlying natural phenomena. For example in fluid dynamics, vortex structures influence important properties such as the lift of an airfoil or the drag of a car. Understanding such features—when and where they occur, their strength, their dynamics—is crucial to understanding and controlling the underlying phenomena.

Definition and interpretation of features depend on the underlying application, but usually they represent important structures (vortex, stagnation point) or changes to such structures (events, bifurcations). There are at least two ways of defining features in scientific visualization:

- **Smooth Feature:** a fuzzy area of the domain where every point adheres to a given definition *to some extent*.
- **Strict Feature:** a well-defined subset of the domain which *fully* adheres to a given definition; usually a geometric object such as a point, line, or surface.

Smooth feature definitions lend themselves to interactive visual analysis approaches such as SimVis [1], where the user can create a fuzzy feature definition interactively by brushing in different views of the data and exploring the result in linked views. The result is subjective, but the approach leaves room for exploration in cases where the features cannot be described a priori (yet). Smooth feature definitions address uncertainty or noise, at minimum, by communicating it through their fuzziness.

T. Weinkauf (✉)
Max-Planck-Institut für Informatik, Saarbrücken, Germany
e-mail: weinkauf@mpi-inf.mpg.de

C.D. Hansen et al. (eds.), *Scientific Visualization*, Mathematics and Visualization,
DOI 10.1007/978-1-4471-6497-5_4

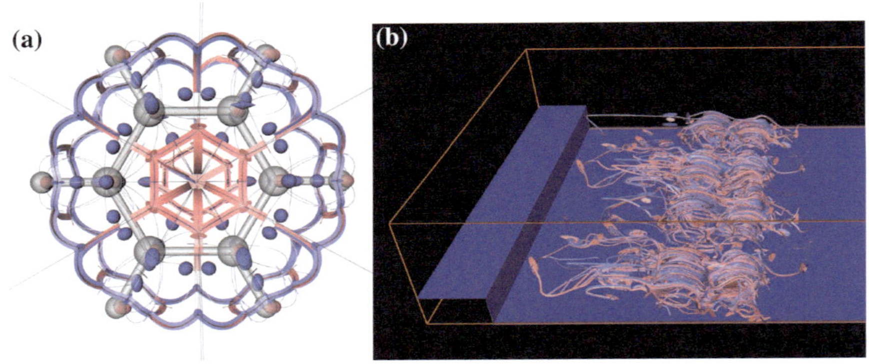

Fig. 4.1 Strict feature definitions such as topology produce a wealth of structures in the presence of noise/uncertainty

Strict feature definitions, on the other hand, lend themselves to an automated data analysis, which can be carried out on supercomputers along the simulation. This is preferable in some cases: for example, if the data set is too large to be efficiently handled by commodity hardware, or if the analysis results have some influence on the simulation itself (e.g. simulation steering). The extraction of strict features is easily automated and therefore, it is perfectly fitted for batch jobs on supercomputers. Furthermore, strict feature definitions usually depend on fewer parameters than smooth definitions or even no parameters at all. Hence, the interpretation of the results depends less on a user-defined parametrization (e.g. isovalue, transfer function, fuzzy feature definition), which in turn makes it easier to strive for a more objective analysis.

However, strict feature definitions do not incorporate fuzziness: a point is either part of a feature or not. There is no room for uncertainty. In fact, such definitions communicate certainty. Strict feature definitions usually produce a wealth of structures in the presence of noise or uncertainty. Figure 4.1 illustrates this using the topological skeletons of two vector fields: the quasi-analytically given electrostatic field around the benzene molecule, and the numerically simulated flow behind a step.

At first glance, it seems as if the concepts of strict feature definitions and uncertainty are antagonistic. In the following, we show that this is not necessarily the case. We will explore different strategies to successfully work with strict feature definitions such as topology in the presence of noisy/uncertain data.

4.2 Strict Features Versus Noisy or Uncertain Data

4.2.1 Fuzzy Analogs to Strict Feature Definitions

A way of dealing with uncertain data is to move away from strict requirements and develop fuzzy analogs to strict feature definitions. An example for this is uncertain vector field topology by Otto et al. [11–13], where the uncertainty is incorporated into the integration of stream lines by means of a Monte Carlo process. This leads to a notion of critical points as distributions rather than distinct points as in the classic case. In order to reflect the uncertainty inherent to brain diffusion MRI data, Schultz et al. [19] introduced fuzzy versions of topological features for tensor fields based on probabilistic tractography.

Strict definitions such as topology usually come as a part of a larger theoretical framework with a number of properties and guarantees. In the beginning, newly developed fuzzy analogs replace usually only certain parts of such a theoretical framework. The development effort to rebuild the whole framework in a fuzzy manner may be very high. For the field of topology, it is for example unclear whether the Morse inequalities still hold for uncertain topology. This leaves room for future research, but it also shows that fuzzy analogs are usually not a full replacement for the corresponding strict feature definitions—at least in the beginning.

Another example for a fuzzy analog is the *Probabilistic Marching Cubes* method by Pthkow et al. [15, 16], where the positional uncertainty of isocontours is evaluated. Again, it remains a task for future research how this relates for example to contour trees.

4.2.2 Aggregation of Features

As already mentioned, strict feature definitions usually produce a wealth of structures in the presence of noise or uncertainty. Aggregating features using statistical methods may help in some applications to reveal the most dominating trends in a data set. This has been done by Garth et al. [5] for tracked critical points in unsteady 3D flows. These features are curves in 4D and often prone to noise. A principal component analysis of all space-time positions of all critical points has been used to determine their principal spatial direction as well as their common center of movement. This dimensionality reduction effectively reduced the amount of information and the resulting visualizations aid in understanding the most dominating trends of the data set.

4.2.3 Filtering of Features

A common method for dealing with a large number of mainly noise-induced features is to filter them according to one or more criteria. The goal is to quantify each feature point in terms of a certain "feature strength" and to keep only the most dominant (parts of the) features.

In the following, we will detail this concept using the example of extremal structures—lines and surfaces at which the scalar function value becomes minimal or maximal with respect to the local neighborhood. These features are important in many applications. For example, vortex core lines can be found in a flow data set as lines where the Q-criterion becomes maximal [18], or pressure minimal [10]. Strongest particle separation in an unsteady flow is denoted by surfaces where the Finite Time Lyapunov Exponent (FTLE) becomes maximal [7].

There are two types of approaches for extracting extremal structures: the local analysis due to ridges/valleys and the global point of view by means of topology.

4.2.3.1 Filtering of Ridges and Valleys

The extraction of ridges and valleys requires derivatives of the examined scalar field f. The commonly used *Height Ridge* definition [2] builds on the first and second derivatives of f, i.e., the gradient $\mathbf{g}$ and the Hessian $\mathbf{H}$. As elegantly formulated by Peikert and Sadlo [14], ridge lines in a 2D scalar field are found as a subset of the zero contour of the derived field $d = \det(\mathbf{g}|\mathbf{Hg})$. Noise in the original data as well as its amplification in the derivatives usually cause a wealth of spurious extraction results for ridge and valley lines. Therefore, filtering of extraction results is mandatory. Many filtering criteria have been proposed in the literature to quantify the importance of ridges: the feature strength of a ridge, the height of a ridge, the angle between the gradient and a ridge line segment, or the length/area of a connected component [14, 17].

4.2.3.2 Filtering of Topological Structures

Topology provides a different means for extracting extremal structures. The Morse-Smale (MS) complex of a 2D scalar field f is comprised of points and lines, which provide a segmentation of the domain into monotone cells [9], i.e., regions in which f behaves monotonically increasing from a local minimum to a local maximum. Each cell is cornered by critical points (a minimum, a maximum, and saddle points). The boundaries between cells are provided by separation lines—so-called *separatrices*. They are extremal lines—the topological analog to ridges/valleys.

Two types of approaches exist to extract the MS complex. The continuous approach [8, 20] builds on the gradient $\mathbf{g}$ and Hessian $\mathbf{H}$ of f. Noise in f and noise amplification in $\mathbf{g}$ and $\mathbf{H}$ pose a numerical challenge for this approach just as much as for ridges. The discrete approach due to Forman's *discrete Morse theory* [4] works on sampled data only, but does not require any derivatives or other numerical computations, since it describes the MS complex in a purely combinatorial fashion. So while noise is less of a problem due to the exclusion of derivatives, spurious extraction results still show up because of the noise level in the original data f. Hence, filtering is necessary. A well-accepted filtering criterion for critical points is *persistence* due to Edelsbrunner et al. [3]. The separation lines of a 2D scalar field can be filtered using a closely related measure called *separatrix persistence* [21], which determines the feature strength of a separatrix or parts thereof. It was originally introduced to filter salient edges on surfaces meshes. Figure 4.2 shows an

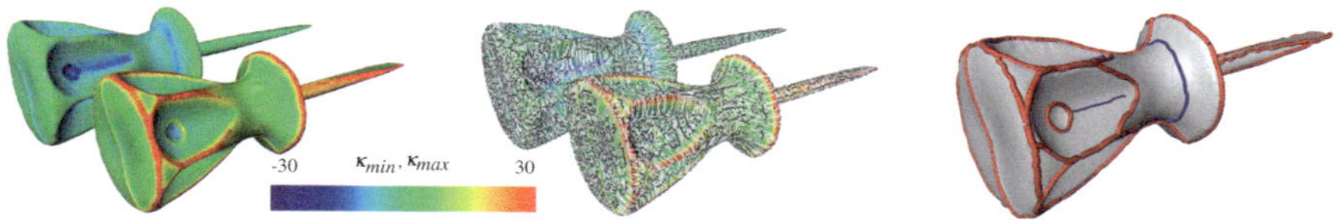

Fig. 4.2 Salient edges of a surface are detected using a topological analysis of the principal curvatures $\kappa_{min}, \kappa_{max}$. After computing the curvature fields (*left*), we extract their topological skeletons using discrete Morse theory (*middle*). It consists of critical points and separatrices—the latter being lines of minimal/maximal curvature. We quantify the significance of these lines using *separatrix persistence* and remove all parts below a certain threshold. This effectively removes all noise-induced structures and yields all perceptually salient concave and convex edges of the input surface (*right*)

example. Later, it has also been applied to filter extremal lines in general scalar fields such as the elevation maps of Mars [6].

4.3 Conclusion

While strict feature definitions introduce a notion of certainty by their mere definition, they are not necessarily antagonistic to the concept of uncertainty. We reviewed three strategies for dealing with strict feature definitions in the presence of noisy/uncertain data: the development of fuzzy analogs to strict feature definitions, the aggregation of features, and the filtering of features. The latter is usually done by means of a notion of "feature strength", which is not only useful to remove noise-induced structures, but may also be applied to build a hierarchy of features.

References

1. Doleisch, H., Gasser, M., Hauser, H.: Interactive feature specification for focus+context visualization of complex simulation data. In: Data Visualization 2003. Proceeding Visualisation Symposium 03, pp. 239–248 (2003)
2. Eberly, D.: Ridges in Image and Data Analysis. Kluwer Acadamic Publishers, Dordrecht (1996)
3. Edelsbrunner, H., Letscher, D., Zomorodian, A.: Topological persistence and simplification. Discrete Comput. Geom. **28**(4), 511–533 (2002)
4. Forman, R.: Morse theory for cell-complexes. Adv. Math. **134**(1), 90–145 (1998)
5. Garth, C., Tricoche, X., Scheuermann, G.: Tracking of vector field singularities in unstructured 3D time-dependent datasets. Proc. IEEE Vis. **2004**, 329–336 (2004)
6. Gnther, D., McGuire, P.C., Walter, S., Weinkauf, T., Hege, H.C.: Extraction of valley networks in mars elevation maps. In: European Planetary Science Congress (Poster). Rome, Italy (2010)
7. Haller, G.: Distinguished material surfaces and coherent structures in three-dimensional fluid flows. Phys. D **149**(4), 248–277 (2001)
8. Helman, J., Hesselink, L.: Visualizing vector field topology in fluid flows. IEEE Comput. Graphics Appl. **11**, 36–46 (1991)
9. Milnor, J.: Morse Theory. Princeton University Press (1963)
10. Miura, H., Kida, S.: Identification of tubular vortices in turbulence. J. Phys. Soc. Jpn. **66**(5), 1331–1334 (1997)

11. Otto, M., Germer, T., Hege, H.C., Theisel, H.: Uncertain 2D vector field topology. Computer Graphics Forum (Proceedind Eurographics 2010) 29, pp. 347–356 (2010)
12. Otto, M., Germer, T., Theisel, H.: Closed stream lines in uncertain vector fields. In: Proceeding Spring Conference on Computer Graphics (SCCG) (2011)
13. Otto, M., Germer, T., Theisel, H.: Uncertain topology of 3D vector fields. In: Proceedings of 4th IEEE Pacific Visualization Symposium (PacificVis 2011), pp. 67–74. Hong Kong, China (2011)
14. Peikert, R., Sadlo, F.: Height ridge computation and filtering for visualization. Proc. Pac. Vis. **2008**, 119–126 (2008)
15. Pthkow, K., Hege, H.C.: Positional uncertainty of isocontours: condition analysis and probabilistic measures. IEEE Trans. Vis. Comput. Graph. **17**(10), 1393–1406 (2011)
16. Pthkow, K., Weber, B., Hege, H.C.: Probabilistic marching cubes. Comput. Graph. Forum **30**(3), 931–940 (2011)
17. Sadlo, F., Peikert, R.: Efficient visualization of Lagrangian coherent structures by filtered AMR ridge extraction. In: IEEE Transactions on Visualization and Computer Graphics (Proceedings Visualization 2007) 13(6), pp. 1456–1463 (2007)
18. Sahner, J., Weinkauf, T., Hege, H.C.: Galilean invariant extraction and iconic representation of vortex core lines. In: Brodlie, K.J. K., Duke, D. (eds.) Proceeding Eurographics / IEEE VGTC Symposium on Visualization (EuroVis '05), pp. 151–160. Leeds, UK (2005)
19. Schultz, T., Theisel, H., Seidel, H.P.: Topological visualization of brain diffusion MRI data. In: IEEE Transactions on Visualization and Computer Graphics (Proceeding IEEE Visualization) 13(6), pp. 1496–1503 (2007)
20. Theisel, H., Weinkauf, T., Hege, H.C., Seidel, H.P.: Stream line and path line oriented topology for 2D time-dependent vector fields. Proc. IEEE Vis. **2004**, 321–328 (2004)
21. Weinkauf, T., Günther, D.: Separatrix persistence: extraction of salient edges on surfaces using topological methods. Computer Graphics Forum (Proceeding SGP '09) 28(5), pp. 1519–1528 (2009)

Chapter 5
The Haunted Swamps of Heuristics: Uncertainty in Problem Solving

Artem Amirkhanov, Stefan Bruckner, Christoph Heinzl and M. Eduard Gröller

Abstract In scientific visualization the key task of research is the provision of insight into a problem. Finding the solution to a problem may be seen as finding a path through some rugged terrain which contains mountains, chasms, swamps, and few flatlands. This path—an algorithm discovered by the researcher—helps users to easily move around this unknown area. If this way is a wide road paved with stones it will be used for a long time by many travelers. However, a narrow footpath leading through deep forests and deadly swamps will attract only a few adventure seekers. There are many different paths with different levels of comfort, length, and stability, which are uncertain during the research process. Finding a systematic way to deal with this uncertainty can greatly assist the search for a safe path which is in our case the development of a suitable visualization algorithm for a specific problem. In this work we will analyze the sources of uncertainty in heuristically solving visualization problems and will propose directions to handle these uncertainties.

5.1 Introduction

Solving visualization tasks is somewhat similar to finding a path through different terrain (see Fig. 5.1). This terrain consists of two areas with very different characteristics: steep mountains and dangerous swamps. The mountains represent high grounds of theory. Choosing a path through the mountains one is on stable ground, but the path may be steep, tedious, and inexperienced travelers might get lost or stuck at a dead

---

A. Amirkhanov (✉) · S. Bruckner · M.E. Gröller
Vienna University of Technology, Vienna, Austria
e-mail: artem@cg.tuwien.ac.at

S. Bruckner
e-mail: bruckner@cg.tuwien.ac.at

M.E. Gröller
e-mail: groeller@cg.tuwien.ac.at

C. Heinzl
Wels Campus, Upper Austrian University of Applied Sciences, Wels, Austria
e-mail: christoph.heinzl@fh-wels.at

C.D. Hansen et al. (eds.), *Scientific Visualization*, Mathematics and Visualization,
DOI 10.1007/978-1-4471-6497-5_5

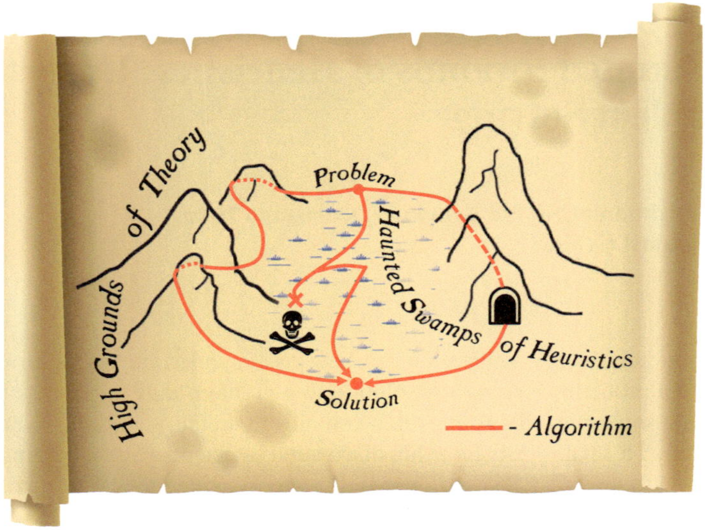

Fig. 5.1 Task solving is finding a path from a problem to a solution

end. The resulting way can be a serpentine road or may even lead through a tunnel, if enough effort is invested in its construction. Another possibility is to go through the haunted swamps of heuristics. The path through the swamps seems to be rather easy and straight-forward, as it is flat walking, but in fact it is neither smooth nor safe. There might be unexpected turns and twists. There are few and badly marked paths through these swamps. In case of even slight deviations the pioneer can easily find himself at a dead end or even get sucked into the deadly waters. Only few researchers find viable, elegant paths through the swamps. If it works out it may result in a shorter way from problem to solution as compared to going on the high grounds of theory.

Using heuristics is often considered as a bad choice in the design of a visualization algorithm. Reviewers of visualization papers tend to dislike heuristics. They comment on heuristics like: *lots of parameter tweaking*, *only heuristics*, *yet another heuristic*, *too many heuristic choices*, or *ad hoc parameter specification*. Should we try to avoid heuristics and attempt to only find the theoretically well-grounded solutions?

5.2 Heuristics

Heuristic (or heuristics; Greek: "$E\upsilon\rho\iota\sigma\kappa\omega$", meaning *to find* or *to discover*) refers to experience-based techniques for problem solving, learning, and discovering. As an adjective, heuristic pertains to the process of gaining knowledge or some desired result by intelligent guesswork rather than by following some pre-established rules, laws, or formulae. The underlying theory might not even be known. Humans often

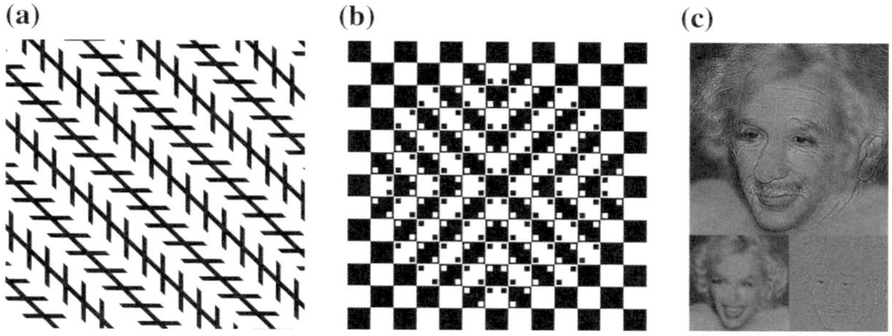

Fig. 5.2 Three well known optical illusions: Zöllner illusion (**a**), bulging checker board illusion (**b**), and blur and picture content illusion (**c**)

apply heuristic and approximate approaches if they have to solve complex problems. In many cases they do not have the complete information for a precise solution. Heuristics are about finding a good enough solution where an exhaustive search would be impractical. One of the most commonly used heuristics, which can initiate a problem solving process, is trial and error. Other common examples of heuristics are, e.g., drawing a figure for better problem understanding, working backward from an assumed solution, or examining a concrete example of an abstract problem. Previous experiences and known information result in such heuristic concepts as prejudices and stereotypes. By evolution some heuristic approaches are firmly anchored in perceptual and mental processes. Heuristic problem solving may work in many circumstances, but in some cases fails to deliver the correct solution. This can lead to cognitive biases in decision making or to imperfections in perception like optical illusions. Three of these optical illusions are shown in Fig. 5.2. Long black lines with horizontal and vertical marks in the Zöllner illusion [20] (Fig. 5.2a) are parallel to each other but do not seem to be. The apparent bulging of the checker board (Fig. 5.2b) is not real: the board is planar. The third example (Fig. 5.2c) shows how frequencies affect the perceived content of a picture. If the viewer examines the image from a short distance, Albert Einstein's face is seen, but, if looking from farther away, the face changes to that of Marilyn Monroe. These three examples are synthetic, but sometimes optical heuristics can fail in real life situations as well. For instance, the size of the moon seems to vary depending on its distance from the horizon.

Algorithm developers use heuristics for problem solving in many ways. Whenever the information available for a task is incomplete or exact solutions are too expensive, an algorithm or some algorithmic part is supplemented with heuristics. Often the heuristic portions of an algorithm are encoded in coefficients or parameters. The following section will elaborate more on the heuristic nature of parameters, issues connected to parameters and ways to explore the uncertainty of an algorithm by studying its parameters.

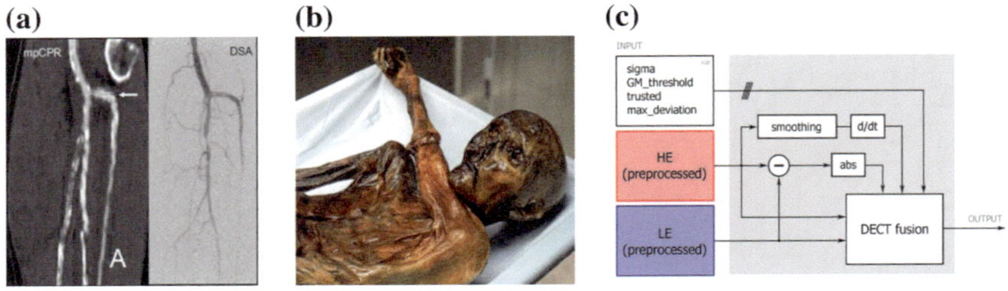

Fig. 5.3 Objects of desire in science. **a** Data (from Multipath CPR [15]). **b** Mummy [Ötzi the Iceman (© South Tyrol Museum of Archaeology—www.iceman.it)]. **c** Algorithm (courtesy of Heinzl [11])

5.3 Objects of Desire in Science

Every scientific discipline has its own object of desire, i.e., study focus. In some areas like geology or medicine these are the physical and medical phenomena and processes hidden inside the data. Data collection, classification, and analysis in order to get useful insights are in the center of the scientific activity (Fig. 5.3a). In other areas artifacts, fossils and mummies (Fig. 5.3b) are the investigated items. In yet other disciplines pieces of text or poems might be in the center of attention. In computer science algorithms (and data structures they work on) are the key entities researchers and developers are designing and investigating (Fig. 5.3c).

An algorithm is a set of instructions that operate on data which are given through constants and variables. And then there are parameters. Constants are, as the name says, constant during the execution of an algorithm. They are fixed and may be mathematical or physical quantities like the cosmological constant. Variables contain values that change during algorithm execution. So where do parameters fit into this picture? Parameters (again a Greek term) are auxiliary measures which are arbitrary but fixed. They are neither constants nor variables. If an algorithm simulates a specific model within a class of models that share the same characteristics, the parameter is fixed for this one model. Switching to another model in the class means varying the corresponding parameter. Parameters are somewhat dual in their nature. And they are just 'auxiliary'. Computer scientists and also visualization researchers are very fond of the instruction part of their algorithms, which they dedicate a lot of time to. Constraints, boundary conditions, approximations, and calibrations are issues that often are encoded in parameters. Even more: if the algorithm still does not work properly, it can only be in the parameters or more of them are needed. They are our easy back-door out. Parameters in many cases do not get the necessary attention and are all too often supposed to be specified heuristically. And it is exactly here were heuristics get a bad reputation. Sometimes the inadequacy of an algorithm is covered up by a set of unintuitive parameters which the developer himself cannot control properly. So to make a virtue out of necessity, the parameters are declared to be user-defined and this is sold as additional flexibility. In reality often this puts an

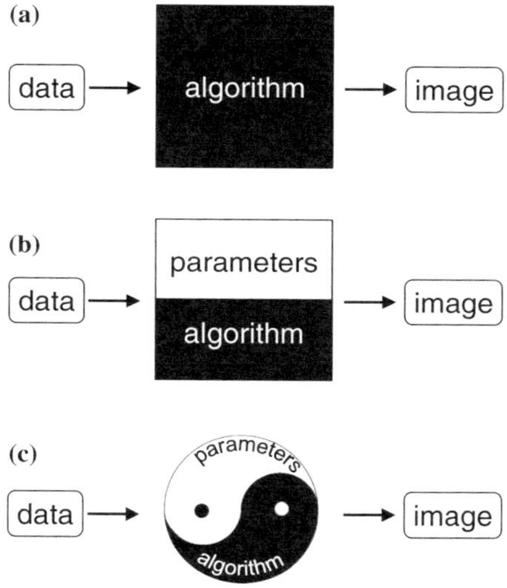

Fig. 5.4 Algorithm and its parameters. **a** Algorithm without parameters. **b** Parameters on top of the algorithm. **c** Yin and Yang union of algorithm and its parameters

undue burden on the user and impacts the usability and applicability of an algorithm. Finding a path, i.e., solving a problem, requires an algorithm *and* parameters. Visualization research is often concerned with taking data and producing an image as visual result. This mapping is realized through an algorithm. But as said above an algorithm alone (like in Fig. 5.4a) is typically not sufficient by itself. Also an algorithm and its parameters do not live side-by-side or one is on top of the other (like in Fig. 5.4b). An algorithm and its parameters are closely intertwined in a yin yang union (like in Fig. 5.4c). Changing an algorithm has an immediate impact on the pertaining parameters, some of them may even vanish or new ones might come into existence. On the other hand, changing parameters may heavily impact the functioning of an algorithm. The results may even be similar to the results from a quite different algorithm (with other parameters of its own). In various disciplines parameter-space analyses are already well established to determine the robustness and stability of processes or procedures. In the area of visualization the investigation of parameters and the spaces they live in has gained increased interest only in recent years. Knowledge-assisted visualization or the visualization of variations and ensembles go into this direction. In the next section we will discuss parameter-space analysis in more detail.

5.4 Parameter-Space Analysis

New and improved imaging modalities, like dual energy computed tomography, allow measuring the same specimen with varying parameters. Increased computing performance (multi-core CPUs, GPUs) allows not only calculating one simulation run but hundreds or even thousands of runs with changing parameter settings. This necessitates investigating and visualizing large sets of simulations or data ensembles at the

(a) (b)

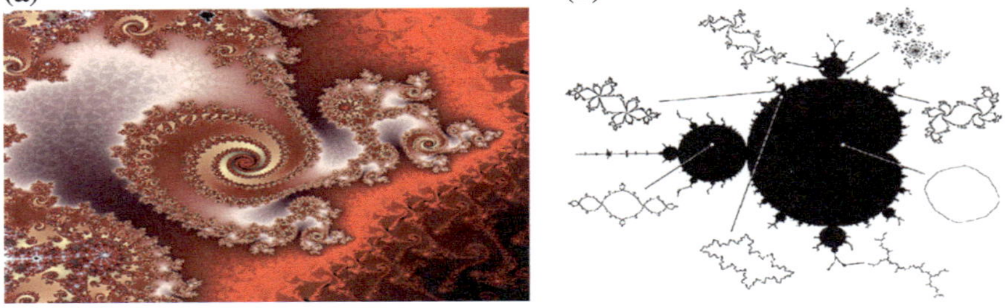

Fig. 5.5 Close-up of a Julia set (**a**) and the Mandelbrot set as parameter map of all Julia sets (**b**) (courtesy of Falconer [8] © John Wiley and Sons)

same time. Dynamical systems are an illustrative example where parameter-space analysis has already been applied for a long time. Julia sets (Fig. 5.5a) are the result of iterating a simple quadratic polynomial in the complex plane. Each polynomial is characterized by a parameter p. Different parameters lead to greatly varying results where the outcome may for example be a connected or disconnected Julia set. Doing an analysis of all possible parameters leads to a parameter space display where the beautiful and immensely intricate Mandelbrot set appears (Fig. 5.5b). The parameter p of the Julia sets turns into a variable in the parameter space where the Mandelbrot set resides. As a side note: the Mandelbrot sets comprises all those values p whose corresponding Julia sets are connected. With dynamical systems a parameter-space analysis may be local or global. A local investigation looks at small perturbations of a parameter to identify for example stability properties which could be direction dependent. A global investigation looks at larger structures in parameter space, e.g., asymptotic behavior, basins of attraction, bifurcations or topological items like separatrices. In the visualization domain parameter-space analyses become feasible as well for data ensembles and parameterized simulation runs. Parameter-space investigations from other fields might act as guiding examples, though the peculiarities of our applications have to be taken into account. While for dynamical systems parameters often change continuously, in our applications parameters may be for example discontinuous, discrete, or categorical in nature. Certain regions of parameter space may be uninteresting or even meaningless because of the physical properties of the underlying phenomenon. With the holistic view on large ensembles or simulation runs interesting questions arise:

- What is the local stability of a (visualization) parameter setting?
- How do different parameters influence each other?
- What are permissible (visualization) parameter ranges?
- How can we automatically define parameter settings that optimize certain properties?
- How sensitive is the visualization outcome on parameter perturbations?
- How to efficiently sample the high dimensional parameter spaces?
- How to do reconstruction in these parameter spaces?

In the following recent examples of parameter-space exploration in the visualization domain are shortly discussed.

Ma [12] introduced a visualization system which presents information on how parameter changes affect the result image as an image graph based on data generated during an interactive exploration process. Berger et al. [4] study continuous parameter spaces in order to guide the user to regions of interest. Their system uses continuous 2D scatter plots and parallel coordinates for a continuous analysis of a discretely sampled parameter space. Not sampled areas of the parameter space are filled with predictions. The uncertainty of the predictions is taken into account and also visualized. The stability of the results with respect to the input parameters is visualized and explored.

In the work by Amirkhanov et al. [1] parameter space exploration is carried out in order to detect the optimal specimen placement on a rotary plate for industrial 3D X-ray computed tomography. The parameter space is represented by Euler angles defining the orientation of the specimen. The parameter settings providing the optimal scanning result were determined using a visual analysis tool. The stability of the result with respect to these parameters was additionally taken into account.

Analyzing how segmentation performs when parameters change and finding the optimal set of parameters is a tedious and time-consuming task. It is usually done manually by the developers of segmentation algorithms. Torsney-Weir et al. [16] presented a system to simplify this task by providing an interactive visualization framework. The *Tuner* tool samples the parameter space of a segmentation algorithm and runs computations off-line. A statistical model is then applied for the segmentation response. Hyper slices [19] of the parameter space and 2D scatter plots are used to visualize these data. Based on the prediction model, additional samples of parameter space may be specified in the regions of interest. The tool allows finding the optimal parameter values and estimating the segmentation algorithm's robustness with respect to its parameters.

FluidExplorer by Bruckner and Möller [7] is an example of goal-driven parameter exploration. They explore the parameters of physically-based simulations for the generation of visual effects such as smoke or explosions. First, the set of simulation runs with various parameter sets is run off-line. Then sampling and spatio-temporal clustering techniques are utilized to generate an overview of the achievable results. Temporal evolution of various simulation clusters is shown. The goal is to find the set of parameters resulting in a certain visual appearance. The metric is defined via user interaction when the user explores the simulation space.

The work of Waser et al. [17] uses *World Lines* to study complex physical simulations. In such time-dependent simulations parameters can change their values at arbitrary moments in time. Decision support is provided by the ability to explore alternative scenarios. A World Line is introduced as a visual combination of user events and their effects in order to present a possible future. The proposed setup enables users to interfere and add new information quickly to find the most appropriate simulation outcome. The usefulness of the technique is shown on a flooding scenario where a smoothed particle hydrodynamics simulation is used. Waser et al. further expand their framework in [18]. The authors take uncertainty of the simulation parameters into account to provide the confidence in the simulation outcome. In the proposed solution, users can perform parameter studies through the World Lines interface to process the input uncertainties. In order to transport steering information to the underlying data-flow, a novel meta-flow (extension to a standard data-flow network) is used. The meta flow handles components of the simulation steering.

World lines are an example of how to handle uncertainty and parameter variations in a computational-steering environment. Now we move further down the visualization pipeline, take a look at the visualization-mapping stage and discuss how an integral view of parameter spaces may influence our view of ensembles of visualization algorithms.

5.5 Parameter Spaces and Visualization Algorithms

A-space [2] is a space where all visualization algorithms live. In A-space every algorithm with a specific parameter setting is represented by a unique point. Perturbing the parameters of an algorithm produces a point set (solution cloud) in A-space. The solution clouds of two quite different algorithms may overlap. This means that a visualization algorithm 1 with parameter 1 produces the same or very similar results as algorithm 2 with parameter 2.

The holistic view of visualization algorithms being embedded in a common space enables interesting investigations and may lead to novel visualization techniques. Sample questions are: What is the stability of an algorithm in A-space? Are there global structures in this space? Can there be smooth transitions between rather diverse algorithms? What would be sparse blendings between various algorithms? MIDA [6] is an interesting example where two well-established volume rendering techniques, i.e., direct volume rendering (DVR) and maximum intensity projection (MIP), are combined in a fine-grained fashion. A smooth transition between DVR, MIP and MIDA itself becomes possible and allows exploiting the strengths of DVR and MIP while avoiding their weaknesses. Another more coarse-grained combination of visualization algorithms would for example be two-level volume rendering [10].

5.6 Algorithms, Parameters, Heuristics—Quo Vadis?

Algorithms and their parameters are closely intertwined. They together constitute a path from the problem to the solution by mapping data to images. Even if parameters are 'just auxiliary measures' they definitely need our help. Heuristic parameter specification is a viable approach as long as some sort of sensitivity analysis is taken care of. This sensitivity analysis should not be only done in parameter and algorithm spaces it should also be extended to data and image spaces. Furthermore the sensitivity analysis should also be applied to interaction space, as we are often confronted with interactive visualization applications. An example in this respect is the work of Gavrilescu et al. [9]. The increased complexity of data ensembles, large simulation runs and uncertainty in the data poses interesting visualization challenges. How shall we cope with the increased data and analysis complexity? Three of several possible directions include integrated views and interaction [3], comparative visualization [13] and fuzzy visualization [14]. With fuzzy visualization, techniques of information theory will play a bigger role in coping with large parameter spaces.

Currently problem solving in visualization is typically algorithm-centric and thus imperative by definition. With increased data complexity it will probably become more declarative and thus more data and image centric, as domain experts have always been data-centric. A data-centric approach means that the user does not specify how data is mapped to images but defines which features of the data he would like to see how in the result images. This is like specifying pre- and post-conditions but not the instructions to get from the first to the second. An optimization process should then automatically figure out which algorithms and parameter settings best fulfill the user defined declarations and constraints. Semantic layers [14] is a step in this direction.

Frameless rendering [5] is about efficiently rendering animation sequences where pixels are updated on a priority basis. At no point in time all pixels of the image are up-to-date, i.e., no frame is available though the animation sequence as a whole evolves. Analogously to this concept, we foresee algorithmless visualizations in the sense that not a single algorithm is explicitly specified by the user in a specific application. For different features of the data and for different parts of the image the most appropriate algorithm among a set of possible candidates might be automatically selected. Various combinations and integrations of visualization algorithms might be possible to best achieve the user goals and declarations. Each pixel or voxel might get its own algorithm on demand.

Interval arithmetic has long been used to cope with uncertainties due to rounding, measurement and computation errors. Handling ensemble data in an analogous manner may lead to densely visualizing intervals or even distributions. While there are already some approaches to locally investigate visualization parameter spaces, not much has been done in terms of a global or topological analysis. For quantitative results visualization algorithms will have to provide more stability and robustness analyses in the future. With the increased data complexity (massive-multiple, heterogeneous data) heuristic approaches and parameter space analyses will become even more important. This raises the need to visualize uncertain, fuzzy, and even contradictory information.

Very often heuristics are useful. But even if you do not (exactly) know what you are doing (this is what heuristics is about), you should make sure that it is safe what you are doing. Safety concerns robustness, stability, and sensitivity of an algorithm and its parameters. So heuristics are great, when handled with care. This way your paths through the haunted swamps will be safe ones. We for sure agree with a statement by Voltaire: "Doubt is not a pleasant condition, but certainty is absurd."

Acknowledgments We would like to acknowledge the Bridge-Project SmartCT and the K-Project ZPT (http://www.3dct.at) of the Austrian Research Promotion Agency (FFG). We also acknowledge the ViMaL project supported by the Austrian Science Fund (FWF), grant no. P21695.

References

1. Amirkhanov, A., Heinzl, C., Reiter, M., Gröller, M.E.: Visual optimality and stability analysis of 3DCT scan positions. IEEE Trans. Visual. Comput. Graph. **16**(6), 1477–1486 (2010)
2. Balabanian, J.P., Gröller, M.E.: A. In: Hagen, H. (ed.) Scientific Visualization: Interactions, Features, Metaphors. Dagstuhl Follow-Ups, vol. 2, pp. 36–47. Schloss Dagstuhl-Leibniz-Zentrum fuer Informatik, Dagstuhl, Germany (2011)
3. Balabanian, J.P., Viola, I., Gröller, M.E.: Interactive illustrative visualization of hierarchical volume data. In: Proceedings of Graphics Interface (best student paper award) (2010)
4. Berger, W., Piringer, H., Filzmoser, P., Gröller, E.: Uncertainty-aware exploration of continuous parameter spaces using multivariate prediction. Comput. Graph. Forum **30**(3), 911–920 (2011)
5. Bishop, G., Fuchs, H., Mcmillan, L., Zagier, E.J.S.: Frameless rendering: double buffering considered harmful. In: ACM Siggraph '94 Conference, vol. 28, pp. 175–176 (1994)
6. Bruckner, S., Gröller, M.E.: Instant volume visualization using maximum intensity difference accumulation. Comput. Graph. Forum **28**(3), 775–782 (2009)
7. Bruckner, S., Möller, T.: Result-driven exploration of simulation parameter spaces for visual effects design. IEEE Trans. Visual. Comput. Graph. **16**(6), 1467–1475 (2010)
8. Falconer, K.: Fractal Geometry: Mathematical Foundations and Applications, 2 edn. Wiley, New York (2003)
9. Gavrilescu, M., Malik, M.M., Gröller, M.E.: Custom interface elements for improved parameter control in volume rendering. In: 14th International Conference on System Theory and Control, pp. 219–224 (2010)
10. Hauser, H., Mroz, L., Bischi, G.I., Gröller, M.E.: Two-level volume rendering. IEEE Trans. Visual. Comput. Graph. **7**(3), 242–252 (2001)
11. Heinzl, C.: Analysis and visualization of industrial CT data. Ph.D. thesis, Institute of Computer Graphics and Algorithms, Vienna University of Technology (2009). URL:http://www.cg.tuwien.ac.at/research/publications/2009/heinzl-2008-thesis/
12. Ma, K.L.: Image graphs—A novel approach to visual data exploration. Proc. IEEE Visual. **1999**, 81–513 (1999)
13. Malik, M.M., Heinzl, C., Gröller, M.E.: Comparative visualization for parameter studies of dataset series. IEEE Trans. Visual. Comput. Graph. **16**(5), 829–840 (2010)
14. Rautek, P., Bruckner, S., Gröller, M.E.: Semantic layers for illustrative volume rendering. IEEE Trans. Visual. Comput. Graph. **13**(6), 1336–1343 (2007)
15. Roos, J., Fleischmann, D., Koechl, A., Rakshe, T., Straka, M., Napoli, A., Kanitsar, A., Sramek, M., Gröller, M.E.: Multipath curved planar reformation of the peripheral arterial tree in CT angiography. J. Radiol. **244**(1), 281–290 (2007)
16. Torsney-Weir, T., Saad, A., Möller, T., Weber, B., Hege, H.C., Verbavatz, J.M., Bergner, S.: Tuner: principled parameter finding for image segmentation algorithms using visual response surface exploration. IEEE Trans. Visual. Comput. Graph. **17**(12), 1892–1901 (2011)
17. Waser, J., Fuchs, R., Ribičić, H., Schindler, B., Blöschl, G., Gröller, M.E.: World lines. IEEE Trans. Visual. Comput. Graph. **16**(6), 1458–1467 (2010)
18. Waser, J., Ribičić, H., Fuchs, R., Hirsch, C., Schindler, B., Blöschl, G., Gröller, M.E.: Nodes on ropes: a comprehensive data and control flow for steering ensemble simulations. IEEE Trans. Visual. Comput. Graph. **17**(12), 1872–1881 (2011)
19. van Wijk, J.J., van Liere, R.: HyperSlice: visualization of scalar functions of many variables. In: Proceedings of the 4th Conference on Visualization '93, VIS '93, pp. 119–125. IEEE Computer Society (1993)
20. Zöllner, F.: Ueber eine neue Art von Pseudoskopie und ihre Beziehungen zu den von Plateau und Oppel beschriebenen Bewegungsphänomenen. Annalen der Physik **186**(7), 500–523 (1860)

Chapter 6
Visualizing Uncertainty in Predictive Models

Penny Rheingans, Marie desJardins, Wallace Brown, Alex Morrow, Doug Stull and Kevin Winner

Abstract Predictive models are used in many fields to characterize relationships between the attributes of an instance and its classification. While these models can provide valuable support to decision-making, they can be challenging to understand and evaluate. While they provide predicted classifications, they do not generally include indications of confidence in those predictions. Typical quality measures for predictive models are the percentage of predictions which are made correctly. These measures can give some insight into how often the model is correct, but provide little help in understanding under what conditions the model performs well (or poorly). We present a framework for improving understanding of predictive models based on the methods of both machine learning and data visualization. We demonstrate this framework on models that use attributes about individuals in a census data set to predict other attributes of those individuals.

P. Rheingans (✉) · M. desJardins · W. Brown · A. Morrow · D. Stull · K. Winner
University of Maryland Baltimore County, Baltimore, MD, USA
e-mail: rheingan@cs.umbc.edu

M. desJardins
e-mail: mariedj@cs.umbc.edu

W. Brown
e-mail: brown1@umbc.edu

A. Morrow
e-mail: amo3@umbc.edu

D. Stull
e-mail: ds10@umbc.edu

K. Winner
e-mail: winnerk1@umbc.edu

C.D. Hansen et al. (eds.), *Scientific Visualization*, Mathematics and Visualization,
DOI 10.1007/978-1-4471-6497-5_6

6.1 Overview

In many scientific fields, models are used to characterize relationships and processes, as well as to predict outcomes from initial conditions and inputs. These models can support the decision-making process by allowing investigators to consider the likely effects of possible interventions and identify efficient ways to achieve desired outcomes. Machine learning research on constructing complex models (such as Bayesian networks) typically focuses on maximizing predictive accuracy, or other measures of model quality. Model confidence refers to the estimated certainty in the classifications produced by the model.

We describe a new framework for improving the understanding of complex models by drawing upon the strengths of both machine learning and data visualization. These two disciplines complement each other to combine the benefits of intelligent automatic support for design and analysis with visual representations and interactions that boost human abilities. We leverage these approaches to address the challenges of developing, understanding, and using complex models to facilitate scientific discovery and informed decision-making.

Our focus is on understanding the uncertainty that is associated with model predictions. This uncertainty arises from several sources. *Sample uncertainty* occurs when regions of the instance space are not well represented in the training data, and predictions are therefore based on sparse information. *Model instability* occurs when model predictions vary, depending on the training data that was used to construct the model. *Prediction variability* occurs when a given observation may have noisy attributes, and this input uncertainty leads to uncertainty in the model's predictions. We are developing analytical techniques to create meta-models that characterize these three forms of uncertainty. To facilitate user understanding of the nature and distribution of these multiple types of uncertainty across the model space, we are developing novel visualization methods to visualize these meta-models in a display space.

This paper describes the challenges associated with model visualization and presents our approach, which uses dimension reduction techniques to produce a glyph-based two-dimensional display of model predictions. We illustrate the use of these visualization techniques in a census data domain from the UC Irving Machine Learning Repository [9].

6.2 Models

Models posit deterministic or stochastic relationships among *domain attributes*, characterized by a set of *parameters* that are used to specify *behaviors* that generate an *output*. The *attributes* are the observed or latent variables that correspond to the entities of interest in the domain. For example, in the census domain, the *attributes* include age, level of education, and occupation. The model *behavior* specifies how these attributes interact during a single application or execution of the model. The internal model *parameters* are the numeric controls on these behaviors, and are typically inferred from observations. In the census domain, the parameters might specify the ways in which education level influences one's occupation. Finally, model *outputs* refer to the values of summary measurements of interest, such as predicted income level or the probability that an individual is in a particular occupation.

Predictive models can be constructed manually or can be learned from a collection of example instances, identifying potentially complex relationships between input attributes and output probabilities. Once these relationships are understood, a predictive model can give the probability of different outcomes, given the known values of input attributes. A predictive model might output the probability that an individual with particular attributes will be in a high-income bracket. Predictive models can be constructed using classification mechanisms (which group observations into one of a small, discrete number of classes), regression techniques (which fit mathematical relationships between attributes and continuous outcomes), or density estimation methods (which build probabilistic models that capture the distribution and relationships among objects within a domain of interest).

It is straightforward to compute and then visualize a single model output for a particular set of attribute values. In many cases, however, it is important to understand model predictions more broadly. Understanding the overall behavior of the model across the range of possible attribute values is important for understanding the model as a whole. Inspecting single predictions is a very slow and inefficient way to develop this broader understanding. Rather, a summary analysis or visualization that can convey individual predictions or probabilistic distributions of predictions across *all* sample locations would provide valuable insight into the overall model behavior.

We have identified four core discovery tasks, corresponding to four categories of questions that an analyst may wish to answer:

- What are the predicted outcomes associated with specific input attribute values, or with a region of the input space?
- What predictions and errors does the model make in input regions in which little training data is available?
- Which input values or regions result in low-confidence and/or incorrect predictions?
- Where and how should model refinement efforts (e.g., data gathering or label correction) be concentrated?

6.3 Approach

The framework that we are developing is implemented as a pipeline, constructed of a series of computational steps that "flow" from training data, through model construction, to visualization and interaction. The framework is intended to be domain-independent and applicable to a wide range of classification problems.

The construction of a *probabilistic model* begins with a set of *training instances* for a given domain. Each instance consists of a vector of attribute values (for a fixed set of attributes that are associated with that particular domain) and a class label (which may be a binary "yes/no" label, or may be one of a set of categorical values). We are currently using the Weka machine learning toolkit [6] to construct models. The model maps from an unlabeled instance (attribute vector) to a probability distribution over class values (i.e., an assignment of a real-valued probability in the range [0, 1] to each class value, such that the sum of the probabilities is one). Once the model has been built, it can be used to generate predictions for both the training data and a set of previously unseen *test instances*. The test instances also have associated class labels, so they can be used to understand prediction errors on previously unseen instances.

To begin the visualization process, a *dimension reduction* method is applied to a set of instances. This process results in a mapping from the high-dimensional attribute space to a two-dimensional display space. Ideally, the dimension reduction process will preserve important properties of the instance distribution, so that similar instances appear near each other in the display space. Finally, a set of instances (which could be the training instances, the test instances, both of these sets, or a new set of sample data generated using the model) is displayed in the display space, using glyph-based representations to show the probabilistic class predictions associated with each instance. We have developed and are currently evaluating two alternative glyph representations: pie charts and a "speckled" texturing.

6.3.1 Dimension Reduction

The first step in developing a model visualization is to project the high-dimensional instance space into a two-dimensional display space. The most effective dimension reduction methods for continuous spaces, such as the ones we are interested in, produce clusters or projections in two-dimensional space that are based on the distribution and similarity of data instances in the higher dimensions. These methods include principal components analysis, multi-dimensional scaling [3], relevance maps [1], and self-organizing maps [10, 13].

The figures in this paper show visualizations that use two dimension reduction methods: feature selection (orthogonal projection using two selected attributes as axes) and principal components analysis (a statistical method for computing an orthogonal projection using linear combinations of the original attributes). We are also implementing multidimensional scaling (a similarity-preserving iterative dimension reduction technique) and self-organizing maps (an iterative method based on neural network learning).

Figure 6.1 shows two projections of an income prediction model in the census domain. Test instances are shown with circular glyphs. In both images, individuals who are predicted to make a high income are colored white, while those predicted to make a low income are colored green. In the left image, the model is projected using feature selection, with education level on the x axis and hours worked per week on

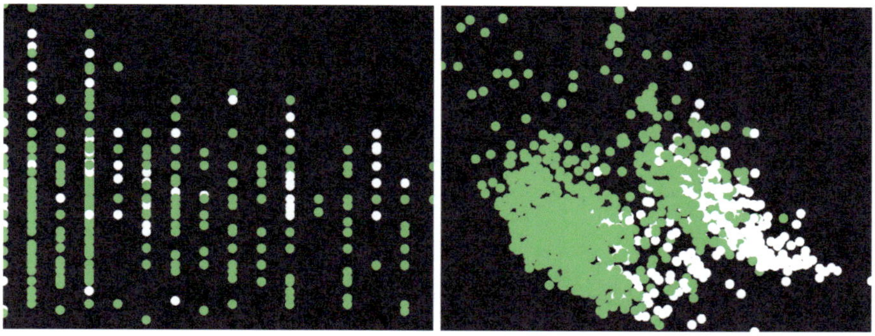

Fig. 6.1 Predictions made by the model about most likely income level for test instances in the census domain, projected using feature selection (*left*) and PCA (*right*)

the y axis. This pair of attributes does not seem to effectively group the instances by predicted income. Throughout the image, green and white glyphs are intermixed. In the right image, the model is projected using PCA. The first principal component is mapped to the x axis, while the second principal component is mapped to the y axis. High-income predictions tend to appear toward the top left of the display, while low-income predictions group to the bottom right. The center of the image contains a mixture of green and white glyphs; presumably the income of these individuals is more difficult to predict. PCA is often more effective at grouping similar instances, at the expense of easily interpretable axes.

6.3.2 Display

The next step is to create a visualization in the display space that effectively conveys information about the model and its properties. Some inspiration can be drawn from techniques designed to incorporate data certainty or multivalued data [2, 4, 7, 11, 12]. In particular, approaches that show data ranges [14], provide annotations with data certainty information [2], or display multiple possible data values associated with locations [5, 8] are relevant.

Instances are displayed using a glyph that conveys the distribution of class predictions made by that model. *Pie chart* glyphs consist of slices summing to a filled circle around a center point. Each slice represents a possible class to which the instance could belong. The arc length of each slice is found by using the probability value associated with that slice's class (i.e., the probability that the instance will belong to that class). We have found it useful to also display the true class of the instance in a ring around the rim of the glyph. *Speckle* glyphs are created from several textures that, when matted together, create a fully opaque glyph. Each such texture consists

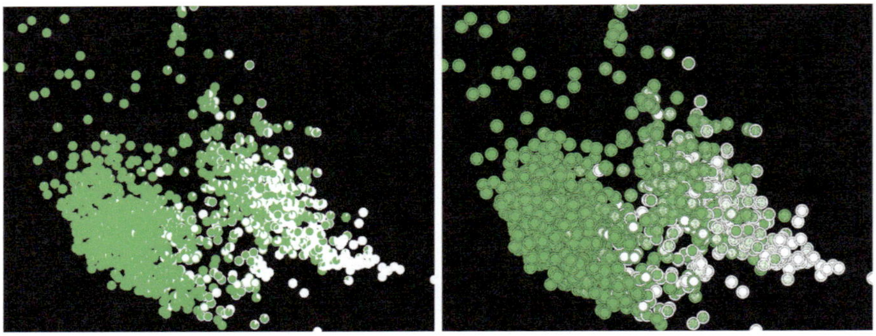

Fig. 6.2 Predictions made by model about income level in the census domain, projected using PCA and displayed with pie glyphs (*left*) and speckle glyphs (*right*)

of a set of opaque pixels with density proportional to the class likelihood, and color corresponding to the class color.

Figure 6.2 shows predictions made by a model constructed to predict income in the census domain. Based on the demographic attributes, the model predicts whether an individual will make more or less than $50K a year. The image at the left displays the predictions using a pie glyph. The size of the white segment in each glyph shows the predicted probability that the individual makes more than $50K, while the size of the green segment shows the predicted probability that they make less than $50K. The ring around the outside of the glyph shows the true class of the instance. Glyphs for which the ring matches the predominant color of the pie are accurately predicted, while those with a color mismatch are inaccurate predictions. Notice how most individuals who make less than $50K a year are predicted with high accuracy, while those making above $50K a year are much harder to correctly predict. To the upper center of the image can be seen several fully confident (yet incorrect) predictions.

The right panel of Fig. 6.2 shows the same model displayed using a speckle glyph. In both pictures, the relative likelihoods of the classes (proportion of low-income (green) and high-income (white) instances) can be seen fairly clearly. However, the relationship between the predicted and true class is somewhat more apparent than with the speckled glyphs, because the density of a class color is easier to visually interpret as a class probability than the angle width of a region in the pie chart.

Figure 6.3 shows predictions made about a multi-value nominal outcome, specifically, the general occupation of an individual in the census domain. There are 11 occupation values, including tech support, sales, executive/managerial, and machine operator/inspector. These occupations are predicted based on five input attributes: age, income, gender, race, and nationality. The images show some localization of predictions, indicated by grouping of colors, but the regions are much less well defined than in the income model. In addition, it is apparent that many instances are mispredicted (i.e., the most probable predicted class (largest "pie wedge" or dominant speckle color) is different than the true class), and that there is often high uncertainty (many different possible classes predicted with approximately equal probability).

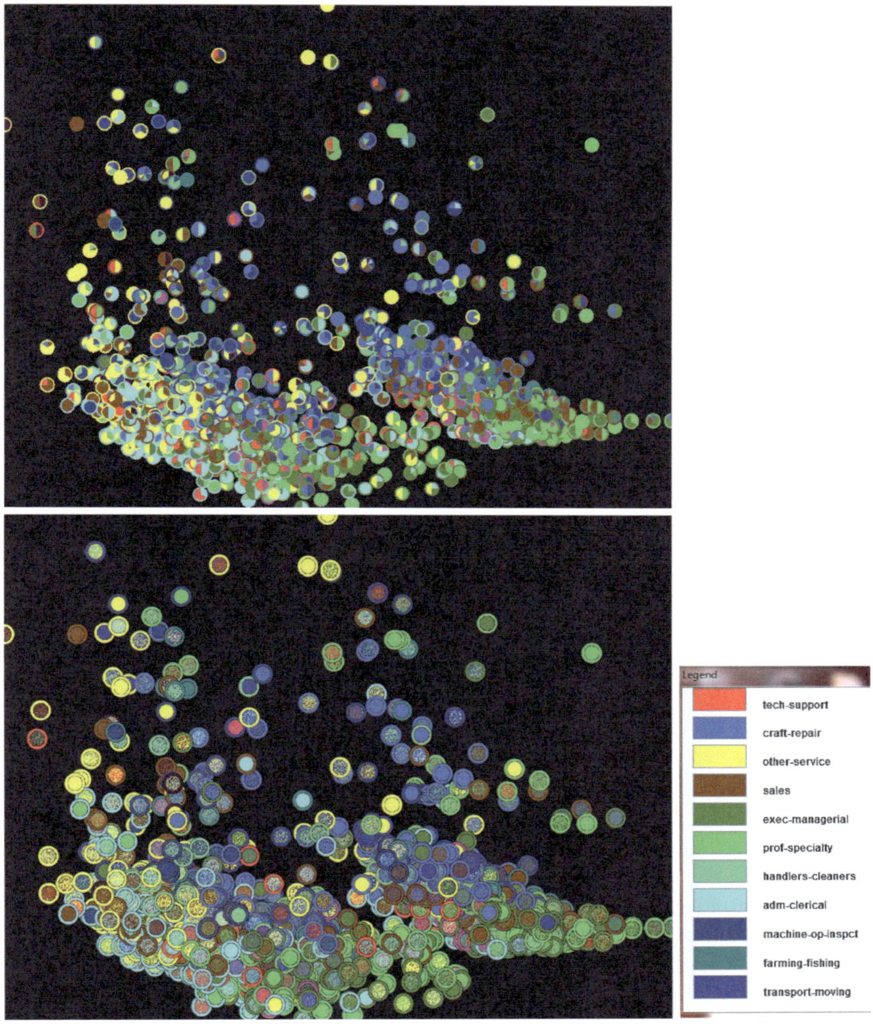

Fig. 6.3 Predictions made by model about occupation type in the census domain, projected using PCA and displayed with pie glyphs (*left*) and speckle glyphs (*right*). Legend shown on *far right*

These factors indicate that this model exhibits higher error and less confidence overall than the income model. A few specific patterns become apparent upon closer inspection: for example, in the lower right quadrant, there is a cluster of instances for which the transport-moving (royal blue) and craft-repair (pure blue) classes are conflated, and another cluster of instances for which the prof-specialty (bright green) and exec-managerial (olive green) classes are conflated. This is intuitively unsurprising (since those occupations involve similar skills and training), but the visualization also enables identifying some more surprising errors (e.g., to the far left just above the centerline of the visualization, there is an individual whose true class is tech-support (red), but who is predicted to be either in the sales (brown) or exec-managerial (dark green) classes).

It is not immediately obvious whether the speckle or pie chart glyph shapes are "better" for human understanding in the general case. It does seem to be the case that when several different classes are predicted, this pattern may be slightly easier to see using the speckle glyphs. We are currently designing a user study to investigate the validity of this hypothesis.

6.4 Future Work and Conclusions

Our research is in its early stages, but our preliminary results show promise for the use of integrated machine learning and visualization techniques to improve the representation and human understanding of uncertainty in classification models. Specific future work includes designing and carrying out a user study of the alternative visualization techniques, performing an empirical investigation of alternative dimension reduction techniques, and investigating machine learning methods for improving the confidence estimates associated with the predictive models.

Acknowledgments This preliminary work was supported by NSF EAGER #1050168 and by NSF REU Supplement #1129683. Thanks to David Mann for his contribution to the project.

References

1. Assa, J., Cohen-Or, D., Milo, T.: Displaying data in multidimensional relevance space with 2D visualization maps. In: IEEE Visualization '97, pp. 127–134. IEEE Computer Society Press, Washington, D.C. (1997)
2. Cedilnik, A., Rheingans, P.: Procedural annotation of uncertain information. In: Proceedings of IEEE Visualization '00, pp. 77–84 (2000)
3. Cox, T., Cox, M.: Multidimensional Scaling. Monographs on Statistics and Applied Probability. Chapman & Hall, London (1994)
4. Davis, T.J., Keller, C.P.: Modelling and visualizing multiple spatial uncertainties. Comput. Geosci. **23**(4), 397–408 (1997)
5. Hagh-Shenas, H., Kim, S., Interrante, V., Healey, C.: Weaving versus blending: a quantitative assessment of information carrying capacities of two alternative methods for conveying multivariate data with color. IEEE Trans. Visual. Comput. Graph. **13**(6), 1270–1277 (2007)
6. Hall, M., Frank, E., Holmes, G., Pfahringer, B., Reutemann, P., Witten, I.H.: The WEKA data mining software: an update. SIGKDD Explor. **11**(1), 10–18 (2009)
7. Johnson, C., Sanderson, A.: A next step: visualizing errors and uncertainty. IEEE Comput. Graph. Appl. **23**(5), 6–10 (2003)
8. Joshi, A.: Art-inspired techniques for visualizing time-varying data. Ph.D. thesis, University of Maryland, Baltimore County (2007)
9. Kohavi, R., Becker, B.: Adult data set (1994). Available online from UCI Machine Learning Repository: http://archive.ics.uci.edu/ml/datasets/Adult
10. Kohonen, T.: Self-organizing Maps. Springer, Berlin (1995)

11. MacEachren, A.M., Robinson, A., Hopper, S., Gardner, S., Murray, R., Gahegan, M., Hetzler, E.: Visualizing geospatial information uncertainty: what we know and what we need to know. Cartograph. Geograph. Inf. Sci. **32**, 139–160 (2005)
12. Pang, A., Wittenbrink, C., Lodha, S.: Approaches to uncertainty visualization. Visual Comput. **13**, 370–390 (1997)
13. Rushmeier, H., Lawrence, R., Almasi, G.: Case study: Visualizing customer segmentations produced by self organizing maps. In: IEEE Visualization 97, pp. 463–466. IEEE Computer Society Press, Washington, D.C. (1997)
14. Sanyal, J., Zhang, S., Bhattacharya, G., Amburn, P., Moorhead, R.J.: A user study to compare four uncertainty visualization methods for 1D and 2D datasets. IEEE Trans. Visual. Comput. Graph. **15**(6), 1209–1218 (2009)

Chapter 7
Incorporating Uncertainty in Intrusion Detection to Enhance Decision Making

Lane Harrison and Aidong Lu

Abstract Network security defense often involves uncertain data which can lead to uncertain judgments regarding the existence and extent of attacks. However, analytic uncertainty and false positive decisions can be integrated into analysis tools to facilitate the process of decision making. This paper presents an interactive method to specify and visualize uncertain decisions to assist in the detection process of network intrusions. Uncertain decisions on the degree of suspicious activity for both temporal durations and individual nodes are integrated into the analysis process to aide in revealing hidden attack patterns. Our approach has been implemented in an existing security visualization system, which is used as the baseline for comparing the effects of newly added uncertainty visualization component. The case studies and comparison results demonstrate that uncertainty visualization can significantly improve the decision making process for attack detection.

7.1 Introduction

Networking security generally deals with a large amount of false positives, which are challenging for efficient and prompt decision making. Even for simple questions like "whether an attack has occurred", security analysts often need to search for subtle traces in the data and analyze the data from different perspectives before making final decision. During this process, numerous assumptions and tests are carried out, which often overlap on dimensions like attack duration and malicious nodes. The capability to integrate partially-developed decisions and assumptions may help analysts in organizing and sorting out incomplete results in order to make more complete final decisions and reports. Closely related to decision making, uncertainty visualization can be developed for this purpose.

L. Harrison (✉) · A. Lu
The University of North Carolina at Charlotte, Charlotte, NC, USA
e-mail: ltharri1@uncc.edu

A. Lu
e-mail: aidong.lu@uncc.edu

C.D. Hansen et al. (eds.), *Scientific Visualization*, Mathematics and Visualization,
DOI 10.1007/978-1-4471-6497-5_7

In the scenario of computer network defense, security analysts are the decision makers who monitor and react to vulnerabilities, threats, and attacks. Typical problems in network attack detection include determination of whether an attack has occurred, which nodes/machines are involved, and the time range of the attack.

For example, a Sybil attack is characterized by one or more nodes assuming the identity of other legitimate nodes in the network. Sybil attacks are not limited to general computer networks. For example, sites like Amazon or Ebay, which rely on user "voting" to make recommendations to other users, are susceptible to coordinated attacks where false users promote items of their choosing. While some of these attacks can be discovered via domain-specific methods, our system is designed to allow analysts to identify attacks through exploring time-varying network connectivity data, which is generally readily available.

As the size of networks and the complexity of attacks increase, so does the time needed to accurately determine the scope of intrusions. Since infected nodes on the network must often be quarantined for repair, determining the attack scope and impact is crucial to network defense operations. Oftentimes, if an analyst cannot determine exactly which machines are affected, they must quarantine a larger section of the network. Therefore, the inability to properly determine the scope of an attack can prove costly.

In this work, we examine how integrating uncertainty views and interactions with a coordinated-views visualization and automatic classification algorithm can help analysts make accurate and timely decisions regarding the scope and duration of attacks. Based on our previous approach, a coordinated multiple views visualization, we present an integrated approach to analyze network data containing Sybil attacks. Specifically, we have designed and incorporated uncertainty management views and interactions that assist network defenders in specifying analytical uncertainty and reusing these results in future investigations. We compare two approaches with a case study and discuss the effects of uncertainty visualization for the decision making progress.

The following first describes the related work on uncertainty visualization for decision making. We then describe our approach and present the results of comparison. At the end, we discuss the effects of uncertainty visualization and conclude the paper.

7.2 Related Work

Representations of error and uncertainty have been identified as component typically missing in visualization [9]. As a good match, uncertainty visualizations deal with visual representations of error and uncertainty in the data [1, 10].

For security applications, the term "uncertainty" extends beyond the quality of the data source. For example, uncertainty can also be considered and quantified in the context of human analysis and decision making [4, 6]. This work focuses on the latter definition of uncertainty. The uncertainty interactions and views described are designed to assist analysts in quantifying and utilizing uncertainty in the human analysis process.

Network defense often requires the analysis of uncertain and deceptive data [2, 3]. Information overload is also a problem for network analysts. Recent work has found that information overload can be mitigated by mechanisms that preserve intermediate analysis results [5]. Network analysts have also shown preference to tools that help them build reports that reflect analysis processes and results [8]. Similarly, our work attempts to provide analysts a consistent and convenient method of storing intermediate analysis results regarding attack uncertainty. At any point in the analysis, these results can be exported for future analysis.

7.3 SybilVis System Overview

The SybilVis system consists of three linked views designed to allow analysts explore time-varying network connectivity data in varying levels of detail (see Fig. 7.1). The time histogram displays node activity over time. The X axis represents timesteps; The Y axis represents node ID. The node link view displays the network for a given time range. In the node link view, individual timesteps are added together to make an aggregate of network activity over time. To avoid network clutter, links are only shown between nodes if they exceed a user-defined threshold. This threshold can be changed interactively. The scatterplot view displays either the first and second eigenvectors or the non-randomness and connection-degree of the nodes.

A typical analysis session begins with the time histogram view, which shows an overview of node activity over all available timesteps. Time ranges can be explored in sequence, or the analyst may make a selection on the time histogram that corresponds to periods of increased or erratic activity. Once a time range is chosen, the node link and scatterplot views are updated accordingly. Analysts generally investigate clusters and outliers in the scatterplot view. Similarly, for the node link view, tightly linked clusters of nodes can indicate suspicious activity. Many rounds of analysis are typically required to determine attack scope. Previously, external sources (i.e. pen and paper, human memory) were required to track intermediate analysis results, which were often inconclusive until further time ranges were investigated. To address this, we developed uncertainty and annotation views and interactions, which are described in the following section.

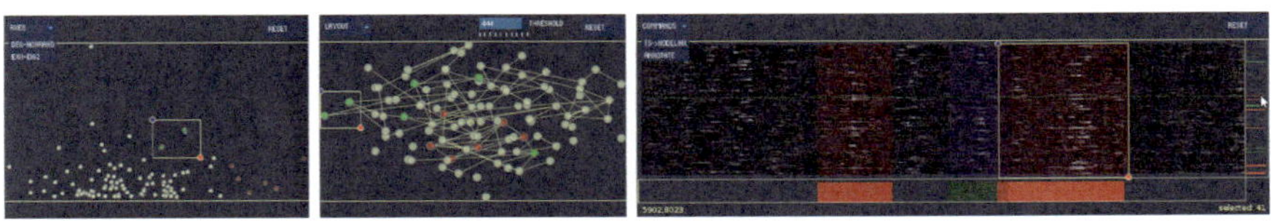

Fig. 7.1 An overview of the SybilVis system

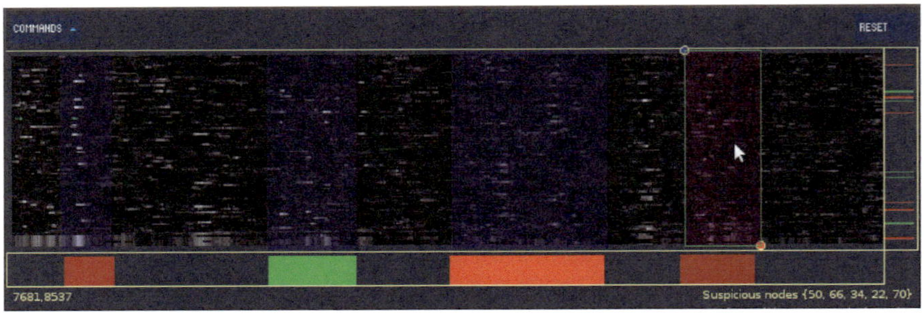

Fig. 7.2 Uncertainty views (*red* and *green bars*) and annotations (*red* and *blue overlays*) on the time histogram. The node uncertainty is to the *right* of the time histogram, and time range uncertainty is on the *bottom*

7.4 SybilVis Uncertainty

Because uncertainty is often involved in network security, tools should be designed that allow analysts to express, track, and re-use decisions during analysis. In this section, we describe how uncertainty regarding time-ranges and individual nodes is interactively specified and integrated into a visualization system.

7.4.1 Uncertainty Integration

Below and to the right of the time histogram view are the time-uncertainty and node-uncertainty views (see Fig. 7.2). As analysis progresses, a user may increase or decrease the suspiciousness-level of one or more nodes or timesteps. Changes to the uncertainty values of nodes or timesteps is then reflected as the brightness of the red/green colors in the corresponding areas in the uncertainty views. Specifically, green areas correspond to no suspected attack, while a red area denotes a likely attack period. Uncertainty colors are also then used to color nodes in the scatterplot and node-link views. This helps the analyst to identify nodes previously identified as suspicious when they appear in future views.

By having a persistent visual representation of node uncertainty, analysts can add uncertainty from separate findings in both the eigenvector and degree-nonrandomness scatterplot views. For example, a node may show up as an outlier when viewing the eigenvectors as axes but as a non-outlier with the degree-nonrandomness as axes. With the uncertainty functions, the analyst need only increase the uncertainty corresponding to the outlier to have this result recorded to inform future analyses.

The uncertainty views and interactions serve as a low-friction way to store analysis results. Once relevant nodes or timesteps are selected, the analyst need only use the mousewheel or +/− signs on the keyboard to change the uncertainty result for their selection. Furthermore, at any point a comma-delimited report can be generated with time durations of suspected attacks, the nodes involved, and their corresponding uncertainty values.

An annotation mechanism is provided in the time histogram view. Users may at any time record an annotation which is represented as a semi-transparent rectangle on top of the time histogram. Text may be added or edited as needed. Also, users may input the IDs of nodes of importance to the annotation. When a user mouses over an annotation which contains node IDs, the nodes are then highlighted in the time histogram view. Viewing suspicious nodes across time can inform users as to which time ranges may be worth investigating next. The color of the annotation can represent either different users or an analyst-defined tagging scheme.

7.4.2 Uncertainty Versus No-Uncertainty Case Study

To explore the utility of uncertainty views and interactions, we describe the analysis process of a Sybil attack dataset both with and without the uncertainty views and interactions.

While SybilVis is designed to work with general network connectivity data, we make use of datasets from a simulation system for Sybil attacks. Additionally, the use of simulation data provides both ground truth and the ability to specify variables to vary the complexity of the data. Variables of the simulation datasets include network size, number of attacking nodes, attack durations, and number of attacks. For this case study, we used a dataset that includes five attacking nodes in three attacks across 10,000 timesteps.

For a given attack duration, it is not common for all malicious nodes to be active at the same time. In fact, some malicious nodes may only be active for a less than 5 % of an attack duration, making them difficult to detect without iterative analysis. Therefore, the uncertainty views and interactions presented are designed to help capture and aggregate multiple uncertain findings into a comprehensive final result.

Suspicious time ranges are found by observing sudden drops and rises in node activity. These fluctuations may indicate the ending or beginning of an attack range. This process of segmenting and analyzing time ranges in the dataset is repeated throughout the analysis process. Once a time range is selected, it is necessary to determine if an attack has occurred, and if so, which nodes were involved. This is done by selecting outliers in the scatterplot view, and looking for tightly connected subgraphs in the node-link view. For a more in-depth discussion on this analysis process, see [7].

In the first suspicious time range, the outliers in the scatterplot view do not show up as tightly connected subgraphs in the node-link view. With uncertainty views and interactions, these nodes and time range can have their corresponding uncertainty values reduced, shown by green in the node-link views in Fig. 7.3a.

Without uncertainty techniques, this result must be either remembered or committed to some external source. Similarly, in the second suspicious timerange, several outliers in the scatterplot view do form a tightly connected subgraph in the node-link

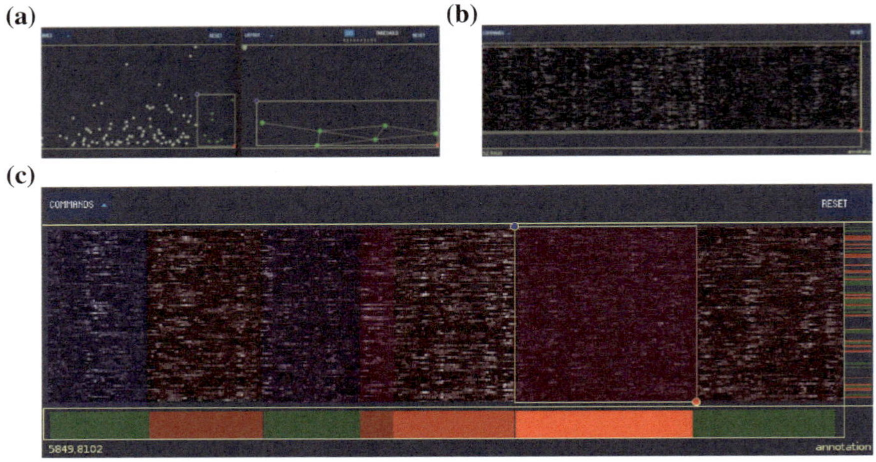

Fig. 7.3 Case study examining the analysis process of a dataset containing three Sybil attacks. **a** Scatterplot outliers and a corresponding tightly connected subgraph. **b** Initial time-histogram view. **c** Final time-histogram view showing updated time and node uncertainties

view. However, this subgraph is not *fully* connected. Therefore, the uncertainty values of these nodes can be increased according to how tightly connected the subgraph appears.

In total, six separate time ranges are analyzed for attacks (see Fig. 7.3c). Three ranges demonstrated both the presence of outliers in the scatterplot view and tightly connected subgraphs in the node-link view. One time range identified as containing an attack was previously identified as benign. With uncertainty views and interactions, time ranges can be marked during an investigation, which can guide future time range selection. Perhaps more importantly, nodes that have higher uncertainty values appear as varying shades of red in subsequent node-link and scatterplot views, which can guide selections during the iterative scatterplot/node-link investigation process. Without uncertainty techniques, previous results regarding time range and scatterplot/node-link analyses are not available for future investigation, unless they are committed to some external source and revisited/recalled at the appropriate time.

7.5 Discussion and Conclusion

In security analysis, several "soft" or "weak" decisions can be treated as evidence and combined to produce a better-informed final decision. In fact, analytic uncertainty need not be limited to security applications. For example, financial and healthcare analysis also often involve analytic uncertainty. Therefore, the capabilities of uncertainty analysis, which allows analysts to quantify and re-use uncertain decisions, should be provided by visual analysis tools rather than external sources.

Furthermore, we observe that it is useful to integrate uncertainty both into the overview visualization (the time histogram, in our case) and the detail views (scatterplot and node-link). Uncertainty in the overview visualization helps guide the analyst in selecting suspicious time-ranges for investigation, provides a summary of analysis thus far, and reduces repeated analyses. Uncertainty in the detail views directs the analyst's attention to nodes previously determined to be exhibiting likely malicious activity.

Additionally, uncertainty in the detail views can influence selections in the visualizations. For example, when selecting a loosely-defined cluster of nodes in the scatterplot view for closer examination, if there is a dark-red node (meaning it was previously determined as suspicious) near the edge of the cluster, it will likely be selected as well. It is possible that, in some cases, an analyst would want to examine nodes independent of previous decisions. In this case, analysts could simply disable the uncertainty indicators in the detail views and preserve them in the overview. However, based on our experience with analysts (below), the awareness of previous decisions on a node is crucial for ongoing security analysis.

Our approach is influenced in part by feedback from network security experts who have consistently mentioned a need to be able to raise the "threat" level of a node/machine in order to effectively track it over time to make final decisions when necessary. Threat in this sense, is essentially uncertainty regarding a true attack from a false positive. Hence, in our system, uncertainty regarding attack time ranges and malicious nodes can be easily specified and integrated into the visualizations.

To conclude, we integrate analytic uncertainty into a visualization in order to assist security analysts in identifying the time range and nodes involved in coordinated attacks. Additionally, analysis progress is documented via the annotations and time uncertainty values, which can assist in reducing repeated analyses. A case study is presented that demonstrates how integrating analytic uncertainty can inform analysis in both the overview and detail views. For future work, we will explore use of standard automatic detection methods as input to the uncertainty views. Also, we will explore ways to scale the node-link view, which is the current bottleneck for scalability in this approach. Finally, we will investigate the effectiveness of similar uncertainty management methods in other security analysis application areas, such as the VAST Challenge datasets.

References

1. Coninx, A., Bonneau, G.P., Droulez, J., Thibault, G.: Visualization of uncertain scalar data fields using color scales and perceptually adapted noise. In: Applied Perception in Graphics and Visualization (2011)
2. Conti, G., Ahamad, M., Stasko, J.: Attacking information visualization system usability overloading and deceiving the human. In: Proceedings of the 2005 Symposium on Usable privacy and security, SOUPS '05. ACM, New York (2005)
3. Cook, K.A., Thomas, J.J.: Illuminating the Path. IEEE Computer Society, Los Alamitos (2005)

4. Deitrick, S., Edsall, R.: The influence of uncertainty visualization on decision making: An empirical evaluation. In: Progress in Spatial Data Handling, pp. 719–738. Springer, Berlin Heidelberg (2006)
5. Fink, G.A., North, C.L., Endert, A., Rose, S.: Visualizing cyber security: Usable workspaces (2009)
6. Griethe, H., Schumann, H.: Visualizing uncertainty for improved decision making. In: Proceedings of the 4th International Conference on Business Informatics Research. Skövde, Sweden (2005)
7. Harrison, L., Hu, X., Ying, X., Lu, A., Wang, W., Wu, X.: Interactive detection of network anomalies via coordinated multiple views. In: Proceedings of the Seventh International Symposium on Visualization for Cyber Security, VizSec '10, ACM (2010)
8. Jaferian, P., Botta, D., Raja, F., Hawkey, K., Beznosov, K.: Guidelines for designing it security management tools. In: Proceedings of the 2nd ACM Symposium on Computer Human Interaction for Management of Information Technology, CHiMiT '08. ACM, New York (2008)
9. Johnson, C.R.: Top scientific visualization research problems. IEEE Comput. Graph. Appl. **24**(4), 13–17 (2004)
10. Johnson, C.R., Sanderson, A.R.: A next step: visualizing errors and uncertainty. IEEE Comput. Graph. Appl. **23**(5), 6–10 (2003)

Chapter 8
Fuzzy Fibers: Uncertainty in dMRI Tractography

Thomas Schultz, Anna Vilanova, Ralph Brecheisen and Gordon Kindlmann

Abstract Fiber tracking based on diffusion weighted Magnetic Resonance Imaging (dMRI) allows for noninvasive reconstruction of fiber bundles in the human brain. In this chapter, we discuss sources of error and uncertainty in this technique, and review strategies that afford a more reliable interpretation of the results. This includes methods for computing and rendering probabilistic tractograms, which estimate precision in the face of measurement noise and artifacts. However, we also address aspects that have received less attention so far, such as model selection, partial voluming, and the impact of parameters, both in preprocessing and in fiber tracking itself. We conclude by giving impulses for future research.

8.1 Introduction

Diffusion weighted MRI (dMRI) is a modern variant of Magnetic Resonance Imaging that allows for noninvasive, spatially resolved measurement of apparent self-diffusion

T. Schultz (✉)
University of Bonn, Bonn, Germany
e-mail: schultz@cs.uni-bonn.de

T. Schultz
MPI for Intelligent Systems,University of Bonn, Tübingen, Germany

A. Vilanova
TU Delft, Delft, Netherlands
e-mail: a.vilanova@tudelft.nl

A. Vilanova · R. Brecheisen
TU Eindhoven, Eindhoven, Netherlands
e-mail: r.brecheisen@tue.nl

G. Kindlmann
University of Chicago, Chicago, USA
e-mail: glk@uchicago.edu

C.D. Hansen et al. (eds.), *Scientific Visualization*, Mathematics and Visualization,
DOI 10.1007/978-1-4471-6497-5_8

coefficients. Since fibrous tissues such as nerve fiber bundles in the human brain constrain water molecules such that they diffuse more freely along fibers than orthogonal to them, the apparent diffusivity depends on the direction of measurement, and allows us to infer the main fiber direction.

Based on such data, tractography algorithms reconstruct the trajectories of major nerve fiber bundles. The most classic variant is streamline tractography, in which tracking starts at some seed point and proceeds in small steps along the inferred direction. In its simplest form, this results in one space curve per seed point. It has been observed that many of the resulting streamlines agree with known anatomy [11]. Tractography is also supported by validation studies that have used software simulations, physical and biological phantoms [25].

Tractography is currently the only technique for noninvasive reconstruction of fiber bundles in the human brain. This has created much interest among neuroscientists, who are looking for evidence of how connectivity between brain regions varies between different groups of subjects [60], as well as neurosurgeons, who would like to know the exact spatial extent of specific fiber bundles in individual patients.

However, drawing reliable inference from dMRI is challenging. Even though a randomized controlled study has shown that using dMRI in cerebral glioma surgery reduces postoperative motor deficits and increases survival times [68], neurosurgeons have observed that some methods for tractography underestimate the true size of bundles [37] and they are still unsatisfied with the degree of reproducibility that is achieved with current software packages [9].

In order to establish tractography as a reliable and widely accepted technique, it is essential to gain a full understanding of its inherent sources of error and uncertainty. It is the goal of this chapter to give an introduction to these problems, to present existing approaches that have tried to mitigate or model them, and to outline some areas where more work is still needed.

8.2 Noise and Artifacts

8.2.1 Strategies for Probabilistic Tractography

It is the goal of probabilistic tractography to estimate the variability in fiber bundle reconstructions that is due to measurement noise. This is often referred to as precision of the reconstructed bundle trajectory [33]. Due to additional types of error in data acquisition and modeling, which will be covered later in this chapter, it is not the same as accuracy (i.e., likelihood of a true anatomical connection) [35]. Current approaches do not account for factors such as repositioning of the head or variations in scanner hardware over time, which further affect repeatability in practice.

Rather than only inferring the most likely fiber direction, probabilistic approaches derive a probability distribution of fiber directions from the data. The first generation of probabilistic tractography methods has done so by fitting the diffusion tensor model

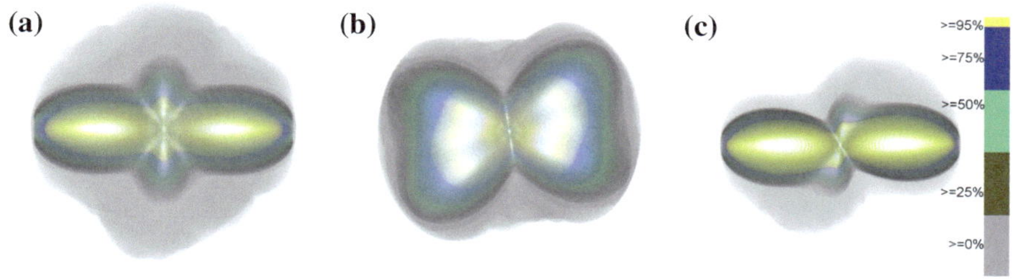

Fig. 8.1 Volume rendering ensembles of orientation distribution functions highlights regions that are included in most ensemble members. Subfigures **a**–**c** compare the uncertainty resulting from different fiber configurations and measurement setups. Images provided by the authors of [28]. **a** $b = 4{,}000$, SNR = 5, **b** $b = 7{,}000$, SNR = 10, **c** $b = 7{,}000$, SNR = 10

to the data, and using the result to parameterize a probability distribution in a heuristic manner. This often assumes that the fiber distribution is related to a sharpened version of the diffusivity profile [38], sometimes regularized by a deliberate bias towards the direction of the previous tracking step [4, 19]. Programmable graphics hardware accelerates the sampling of such models, and enables immediate visualization of the results [40]. Parker et al. [45] present two different fiber distribution models that are parameterized by measures of diffusion anisotropy. Subsequent work allows for multimodal distributions that capture fiber crossings, and uses the observed variation of principal eigenvectors in synthetic data to calibrate model parameters [44].

In contrast to these techniques, which use the model parameters from a single fit, a second generation of probabilistic tractography methods estimates the posterior distribution of fiber model parameters, based on the full information from the measurements, which includes fitting residuals. Behrens et al. [3] do so in an objective Bayesian framework, which aims at making as few assumptions as possible, by choosing noninformative priors. They have later extended the "ball-and-stick" model that underlies their framework to allow for multiple fiber compartments [2].

Bootstrapping estimates the distribution of anisotropy measures [43] or fiber directions [31, 55] by repeated model fitting, after resampling data from a limited number of repeated scans. This has been used as the foundation of another line of probabilistic tractography approaches [35, 39]. As an alternative to estimating the amount of noise from repeated measurements, wild bootstrapping takes its noise estimates from the residuals that remain when fitting a model to a single set of measurements [67]. This has been proposed as an alternative to repetition-based bootstrapping for cases where only a single acquisition is available [32]. Residual bootstrapping [12] builds on the same basic idea, but allows for resampling residuals between gradient directions, by modeling the heteroscedasticity in them. It has not only been combined with the diffusion tensor model, but also with constrained deconvolution, which allows for multiple fiber tractography [27].

To visualize the distributions estimated by bootstrapping, Jiao et al. [28] volume render ensembles of orientation distribution functions (ODFs). As shown in Fig. 8.1, this highlights regions included in most ensemble members, representing the most

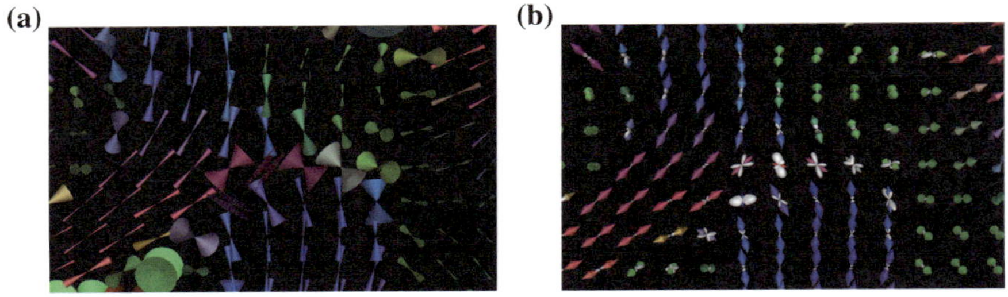

Fig. 8.2 Uncertainty in fiber directions has been visualized using cones of uncertainty, which include 95 % of all estimated directions (**a**) [31] The characteristics of the distribution are shown in greater detail by HiFiVE glyphs, which decompose it into a main direction, shown as *colored double cones*, and a residual PDF, shown as a *gray surface* (**b**) [55]

certain part of the ODF. For tractography, the main features of interest are the inferred fiber directions. The uncertainty in these directions has traditionally been visualized using cones that represent 95 % confidence intervals around a main direction [31]. Figure 8.2 compares this approach to the alternative, more recent HiFiVE glyph, which provides a more detailed impression of the distribution [55].

8.2.2 Rendering Probabilistic Tractograms

After estimating the reproducibility of white matter fiber tracts by one of the above-described methods, we can represent the results in one of two ways: voxel-centric or tract-centric. The voxel-centric representation assigns scores to individual voxels, where each voxel stores the percentage of tracts passing through it. They represent the reproducibility with which a connection from one voxel position to the seeding region is inferred from the data. The resulting 3D volume data sets are sometimes called probability or confidence maps, and are often visualized by volume rendering techniques, as in Fig. 8.3b, and 2D color maps [40].

Tract-centric techniques include the ConTrack algorithm [59], which assigns a score to each generated tract. It reflects confidence in the pathway as a whole, based on its agreement with the data and assumptions on fiber length and smoothness. Ehricke et al. [15] define a confidence score that varies along the fiber, and color code it on the streamline. Jones et al. [30, 32] use hyperstreamlines to visualize the variability of fiber tracts obtained using bootstrap or wild-bootstrap methods. They also demonstrate that using standard streamlines to render all fiber variations equally fails to give an impression of which fibers are stable and which are outliers.

Brecheisen et al. [6] propose illustrative confidence intervals where intervals are based on distances or pathway scores. Illustrative techniques, i.e., silhouette and outlines, are used to visualize these intervals. Interaction and Focus+Context widgets are used to extend the simplified illustrative renderings with more detailed information.

(a) (b)

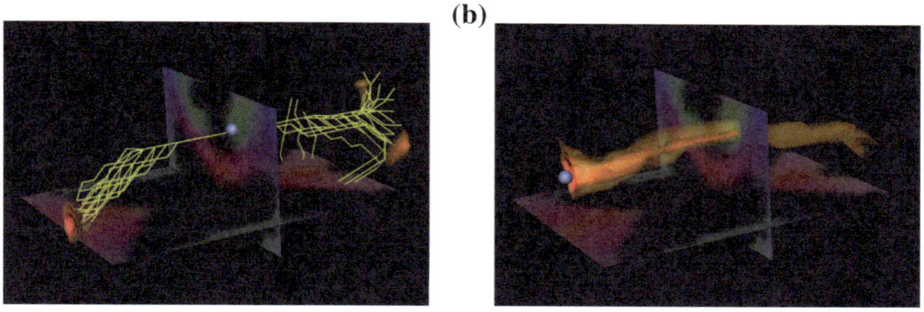

Fig. 8.3 Equally rendering all fiber variations from a seed (*blue ball*) as standard streamlines (**a**) makes it difficult to see which of them are most stable, and which are isolated outliers. Volumetric representations (**b**) are a popular alternative. Images from [53]. **a** Tract-centric visualization, showing all streamlines traced from a common seed. **b** Nested isosurfaces indicate different levels of confidence

Schultz et al. [57] cluster the voxels in which probabilistic tractography terminates, based on the seed points from which they are reached. They then derive a per-voxel score that indicates how frequently the voxel was involved in a connection between two given clusters. Fuzzy fiber bundle geometry is defined by isosurfaces of this score, with different isovalues representing different levels of precision.

8.3 Other Factors

8.3.1 Impact of Parameters

One source of uncertainty in dMRI tractography that has not received much attention is parameter sensitivity. Most tractography algorithms depend on user-defined parameters, which results in a poor reproducibility of the output results. Some reproducibility studies for concrete applications have been reported [13, 65]. However, there does not exist an automatic solution that resolves the problem in a general manner. The stability of the parameter setting is relevant information for both neuroscientists and neurosurgeons who are trying to assess whether their fiber tracking results are stable. Visualization can play an important role to help this assessment.

Brecheisen et al. [7] build a parameter space by sampling combinations of stopping criteria for DTI streamline tractography. Stopping criteria primarily affect fiber length. The investigation of parameter sensitivity is based on generating a streamline set that covers the whole parameter space of stopping criteria. Afterwards, selective culling is performed to display specific streamline collections from the parameter space. This is done by selecting parameter combinations using 2D widgets such as the feature histogram displayed in Fig. 8.4. An example feature is average fiber density per voxel. These views help the user to identify stable parameter settings, thereby improving the ability to compare groups of subjects based on quantitative tract features.

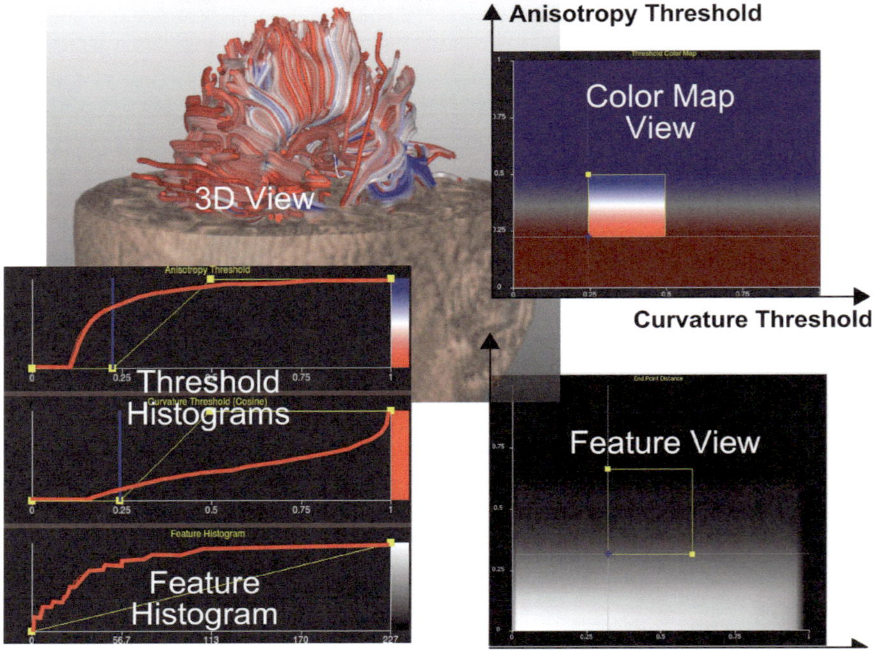

Fig. 8.4 Main viewports of Brecheisen et al. [7] exploration tool. *Top-left* 3D visualization of fiber tract together with anatomical context and axial fractional anisotropy slice. *Top-right* Color map view used for selecting individual threshold combinations and definition of color detail regions. *Bottom-right* Feature map view showing changes in quantitative tract features as a function of threshold combination at discrete sample points of the parameter space. *Bottom-left* Cumulative histograms of both threshold and feature values. © IEEE Reprinted, with permission, from IEEE Transactions on Visualization and Computer Graphics 15(6)

Jiao et al. [29] introduce a toolkit based on three streamline distances that are used to measure differences between fiber bundles. The user can vary parameters that affect the results of the fiber tractography and measure the resulting differences based on these distances. This allows them to quantify the variation and reproducibility of the fiber bundles due to different sources of uncertainty and variation in the tractography input parameters.

Although these methods provide a first step to study uncertainty due to parameter settings, it remains a time consuming exploratory task for the user. This is especially true if parameters are correlated, and their interrelation needs to be investigated.

8.3.2 Model Uncertainty and Selection

Methods for tractography that seek to recover more than a single fiber direction in a given area have to make a judgement about how many fiber directions can be meaningfully recovered from the dMRI data. The combination of measurement noise, partial voluming, and the practical constraints on how many diffusion weighted images may be acquired create uncertainty in the number of fibers present. Qualitatively different than the angular uncertainty in a single fiber direction, the traditional focus of probabilistic tractography, this uncertainty can be a viewed as a kind of *model selection uncertainty*, which is described further in Sect. 8.4.1.

Uncertainty in fiber number has been handled by different tests that either statistically sample or deterministically choose a level of model complexity (with an associated fiber number) from a nested set of models. Behrens et al. [2] use Automatic Relevance Determination (ARD) to probabilistically decide the number of "sticks" (fibers) in their ball-and-multiple-stick model. Within their probabilistic tractography, this achieves Bayesian Model Averaging [22] of the fiber orientation. For deterministic tractography, Qazi et al. [50] use a threshold on the (single, second-order) tensor planarity index c_p [66] to determine whether to fit to the diffusion weightes images a constrained two-tensor model [46] that permits tracing two crossing fibers.

Schultz et al. compare different strategies for deciding the appropriate number of fiber compartments, based on the diminishing approximation error [56], thresholding compartment fraction coefficients of a multi-fiber model [58], or by learning the number of fiber compartments using simulated data and support vector regression [54], which represents uncertainty in the form of continuous estimates of fiber number (cf. Fig. 8.5).

Much of the work on determining the number of per-voxel fiber components has been described outside of any particular tractography method, but may nonetheless inform tractographic analysis. Alexander et al. [1] use an F-Test to find an appropriate order of Spherical Harmonic (SH) representation of the ADC profile. Jeurissen et al. [26] decide the number of fibers by counting significant maxima in the fiber orientation distribution after applying the SH deconvolution (constrained by positivity) of Tournier et al. [62]. The SH deconvolution of Tournier et al. [62] in some sense involves model selection, because the deconvolution kernel is modeled by the

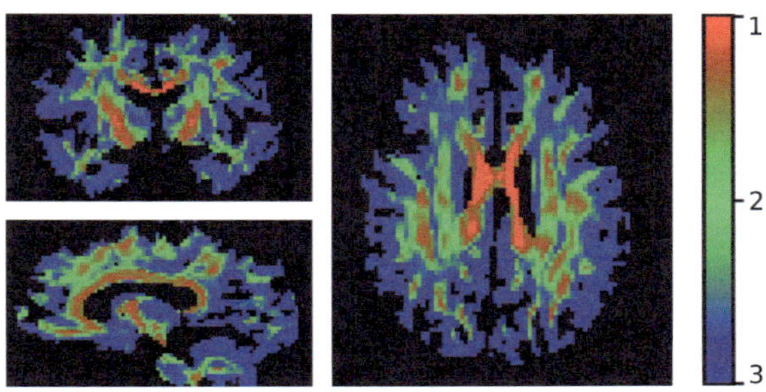

Fig. 8.5 Support vector regression estimates the number of fiber compartments per voxel as a continuous quantity, indicating regions in which a discrete number of fiber compartments can only be determined with considerable uncertainty [54]

SH coefficients of the voxels with the highest FA, presumably representing a single fiber.

Aside from the question of counting fibers, other work has examined more broadly the question of which models of the diffusion weighted signal profile are statistically supported. Bretthorst et al. [8] compute Bayesian evidence (see Sect. 8.4.1) to quantify the fitness of various models of the diffusion weighted signal, producing maps of model complexity in a fixed baboon brain, and of evidence-weighted averages of per-model anisotropy. Freidlin et al. [17] choose between the full diffusion tensor and simpler constrained tensor models according to the Bayesian information criterion (BIC) or sequential application of the F-Test and either the t-Test or another F-Test.

8.3.3 Partial Voluming

Tractography works best in voxels that contain homogeneously oriented tissue. Unfortunately, many regions of the brain exhibit more complex structures, where fibers cross, diverge, or differently oriented fibers pass through the same voxel [1]. This problem is reduced at higher magnetic field strength, which affords increased spatial resolution. However, even at the limit of what is technically possible today [20], a gap of several orders of magnitude remains to the scale of individual axons.

Super-resolution techniques combine multiple images to increase effective resolution. Most such techniques use input images that are slightly shifted with respect to each other and initial success has been reported with transferring this idea to MRI [47]. However, due to the fact that MR images are typically acquired in Fourier space, spatial shifts do not correspond to a change in the physical measurement, so it is unclear by which mechanism repeated measurements should achieve more than an improved signal-to-noise ratio [48, 51]. It is less controversial to compute images that are super-resolved in slice-select direction [18, 52] or to estimate fiber model parameters at increased resolution via smoothness constraints [42].

Track density imaging [10] uses tractography to create super-resolved images from diffusion MRI. After randomly seeding a large number of fibers, the local streamline density is visualized. It is computed by counting the number of lines that run through each element of a voxel grid whose resolution can be much higher than during MR acquisition. Visually, the results resemble those of line integral convolution, which had been applied to dMRI early on [23, 69].

8.4 Perspectives

8.4.1 Evidence for Model Selection

Many of the methods for finding per-voxel fiber count (or more generally the per-voxel signal model) described in Sect. 8.3.2 share two notable properties which may be reconsidered and relaxed in future research. First, they deterministically calculate the single best model, with hard transitions between the regions best explained by one model versus another [1, 17, 26, 50, 56, 58]. Yet we know that partial voluming (Sect. 8.3.3) creates smooth transitions between different neuroanatomic tissue regions. Though computational expensive, Markov Chain Monte Carlo (MCMC) sampling of both model parameter space and the set of models enables averaging information from more than one model [2, 8]. Second, most methods work within a particular hierarchical set of linearly ordered models (SH of different orders [1], ball and multiple sticks [2], sum of higher-order rank-1 terms [58]). One can easily imagine configurations, however, that confound such a linear ordering: an equal mix of two fibers crossing and isotropic diffusion (perhaps due to edema), or a mix of one strong fiber and two weaker equal-strength fibers. Furthermore, there is rarely objective comparison or reconciliation between disjoint sets of models.

An informative perspective on these situations may be gained by directly visualizing, either on data slices or by some form of volume rendering, the fitness of a large palette of possible models. In a Bayesian setting, the model fitness can be quantified by the marginal likelihood of the data $\mathbf{x}$ given the model M_k, or the model *evidence*, computed by integrating over the model parameter space $\boldsymbol{\theta}_k$ [36].

$$\underbrace{P(\mathbf{x}|M_k)}_{\text{evidence}} = \int \underbrace{P(\mathbf{x}|\boldsymbol{\theta}_k, M_k)}_{\text{likelihood}}\,\underbrace{P(\boldsymbol{\theta}_k|M_k)}_{\text{prior}}\,d\boldsymbol{\theta}. \tag{8.1}$$

Bretthorst et al. [8] have pioneered the calculation and visualization of model evidence for dMRI, but many possible directions are left unexplored, including the application to counting fibers, and to models that account for intra-voxel fanning or bending [41].

8.4.2 Reproducibility, Seeding, and Preprocessing

The reproducibility of tractography depends on many factors. The manual placement of seed points is an obvious concern. Detailed written instructions improve reproducibility between operators [13], especially across sites [65]. Combining multiple seed regions with logical operators makes the results more reproducible [21, 24] and seeding protocols for up to 11 major fiber bundles have been developed this way [65]. Warping individual brains to a standard template has also been reported to increase reproducibility [21, 64]. Selecting streamlines from a whole brain tractography via three-dimensional regions of interest [5] or semi-automated clustering [63] is an alternative way to reproducibly extract fiber bundles.

When the same person places the seeds on a repeated scan, the resulting variability is generally higher than when different observers follow a written protocol to place seeds in the same data [13]. Within the same session, measurement noise is the main limiting factor [14]. Between sessions, differences in exact head positioning and other subtle factors increase the variability noticably [64].

Reproducibility suffers even more when repeating the measurement on a different scanner [49]. Even a pair of nominally identical machines has produced a statistically significant bias in Fractional Anisotropy [64]. Improving calibration between sessions or scanners via software-based post-processing appears possible [64], but has not been widely explored so far.

More time consuming measurement protocols generally afford better reproducibility. Even though Heiervang et al. [21] report that the improvement when using 60 rather than 12 gradient directions was not statistically significant, Tensouti et al. [61] report a clear improvement between 6 and 15 directions, which continues—at a reduced rate—when going to 32 directions. Farrel et al. [16] use 30 directions and demonstrate a clear improvement when averaging repeated measurements.

Finally, reproducibility depends on the tractography algorithm [61], its exact implementation [9], as well as the methods used for pre-processing the data [34, 64] and their parameter settings. Given that the reproducibility of tractography will be crucial for its wider acceptance in science and medicine, more work is needed that specifically targets these problems.

8.5 Conclusion

Reproducibility of dMRI tractography is a fundamental problem that limits the acceptance of this technique in clinical practice and neuroscience research. Although some effort has been made to include uncertainty information in the tractography results, several open issues remain that need further investigation.

Probabilistic tractography is established, but visualization research has concentrated on deterministic streamline-based techniques, and few techniques have been developed to visualize the information obtained by probabilistic methods. There are several sources of uncertainty in the tractography visualization pipeline. However, only a few of them have been explored, and if at all studied, they are often considered independently with no connection to each other. Techniques that investigate the impact of parameters on the fiber tracking results and that aim to reduce the impact of user bias through parameter selection have been investigated only recently. Model selection and data preprocessing have hardly been studied with respect to their effects on tractography results.

Techniques that allow the combined analysis of uncertainty from different sources in the same framework, and that facilitate an understanding of their influence on the final tractography result are still missing. One challenge faced by visualization systems that aim to aid understanding of these uncertainties is to display this additional information efficiently and effectively, without causing visual clutter.

Ultimately, uncertainty visualization should contribute to making fiber tracking a more reliable tool for neuroscience research, and to conveying the information needed for the decision making process in clinical practice.

References

1. Alexander, D.C., Barker, G.J., Arridge, S.R.: Detection and modeling of non-gaussian apparent diffusion coefficient profiles in human brain data. Magn. Reson. Med. **48**, 331–340 (2002)
2. Behrens, T.E.J., Johansen-Berg, H., Jbabdi, S., Rushworth, M.F.S., Woolrich, M.W.: Probabilistic diffusion tractography with multiple fibre orientations: what can we gain? NeuroImage **34**, 144–155 (2007)
3. Behrens, T.E.J., Woolrich, M.W., Jenkinson, M., Johansen-Berg, H., Nunes, R.G., Clare, S., Matthews, P.M., Brady, J.M., Smith, S.M.: Characterization and propagation of uncertainty in diffusion-weighted MR imaging. Magn. Reson. Med. **50**, 1077–1088 (2003)
4. Björnemo, M., Brun, A., Kikinis, R., Westin, C.F.: Regularized stochastic white matter tractography using diffusion tensor MRI. In: Dohi, T., Kikinis, R. (eds.) Proceedings of Medical Image Computing and Computer-Assisted Intervention (MICCAI), Lecture Notes in Computer Science, vol. 2488, pp. 435–442. Springer, Berlin (2002)
5. Blaas, J., Botha, C.P., Peters, B., Vos, F.M., Post, F.H.: Fast and reproducible fiber bundle selection in DTI visualization. In: Silva, C., Gröller, E., Rushmeier, H. (eds) Proceedings of IEEE Visualization, pp. 59–64 (2005)
6. Brecheisen, R., Platel, B., ter Haar Romenij, B.M., Vilanova, A.: Illustrative uncertainty visualization for DTI fiber pathways. In: Poster Proceedings of EuroVis (2011)
7. Brecheisen, R., Vilanova, A., Platel, B., ter Haar Romenij, B.M.: Parameter sensitivity visualization for DTI fiber tracking. IEEE Trans. Vis. Comput. Graph. **15**(6), 1441–1448 (2009)
8. Bretthorst, G.L., Kroenke, C.D., Neil, J.J.: Characterizing water diffusion in fixed baboon brain. In Fischer, R., Preuss, R., von Toussaint, U. (eds.) Bayesian Inference and Maximum Entropy Methods in Science and Engineering, pp. 3–15 (2004)
9. Bürgel, U., Mädler, B., Honey, C.R., Thron, A., Gilsbach, J., Coenen, V.A.: Fiber tracking with distinct software tools results in a clear diversity in anatomical fiber tract portrayal. Cen. Eur. Neurosurg. **70**(1), 27–35 (2009)
10. Calamante, F., Tournier, J.D., Jackson, G.D., Connelly, A.: Track-density imaging (TDI): super-resolution white matter imaging using whole-brain track-density mapping. NeuroImage **53**(4), 1233–1243 (2010)
11. Catani, M., Howard, R.J., Pajevic, S., Jones, D.K.: Virtual in vivo interactive dissection of white matter fasciculi in the human brain. NeuroImage **17**, 77–94 (2002)
12. Chung, S., Ying, L.: Comparison of bootstrap approaches for estimation of uncertainties of DTI parameters. NeuroImage **33**(2), 531–541 (2006)
13. Ciccarelli, O., Parker, G.J.M., Toosy, A.T., Wheeler-Kingshott, C.A.M., Barker, G.J., Boulby, P.A., Miller, D.H., Thompson, A.J.: From diffusion tractography to quantitative white matter tract measures: a reproducibility study. NeuroImage **18**, 348–359 (2003)
14. Ding, Z., Gore, J.C., Anderson, A.W.: Classification and quantification of neuronal fiber pathways using diffusion tensor MRI. Magn. Reson. Med. **49**, 716–721 (2003)
15. Ehricke, H.-H., Klose, U., Grodd, W.: Visualizing MR diffusion tensor fields by dynamic fiber tracking and uncertainty mapping. Comput. Graph. **30**, 255–264 (2006)

16. Farrell, J.A.D., Landman, B.A., Jones, C.K., Smith, S.A., Prince, J.L., van Zijl, P.C.M., Mori, S.: Effects of signal-to-noise ratio on the accuracy and reproducibility of diffusion tensor imaging-derived fractional anisotropy, mean diffusivity, and principal eigenvector measurements at 1.5T. J. Magn. Reson. Imag. **26**, 756–767 (2007)
17. Freidlin, R.Z., Özarslan, E., Komlosh, M.E., Chang, L.-C., Koay, C.G., Jones, D.K., Basser, P.J.: Parsimonious model selection for tissue segmentation and classification applications: a study using simulated and experimental DTI data. IEEE Trans. Med. Imaging **26**(11), 1576–1584 (2007)
18. Greenspan, H., Oz, G., Kiryati, N., Peled, S.: MRI inter-slice reconstruction using super-resolution. Magn. Reson. Imaging **20**, 437–446 (2002)
19. Hagmann, P., Thiran, J.-P., Jonasson, L., Vandergheynst, P., Clarke, S., Maeder, P., Meuli, R.: DTI mapping of human brain connectivity: statistical fibre tracking and virtual dissection. NeuroImage **19**, 545–554 (2003)
20. Heidemann, R.M., Porter, D.A., Anwander, A., Feiweier, T., Calamante, F., Tournier, J.-S., Lohmann, G., Meyer, H., Knösche, T.R., Turner, R.: Whole-brain, multi-shot, diffusion-weighted imaging in humans at 7T with 1 mm isotropic resolution. In Proceedings of International Society of Magnetic Resonance in Medicine (ISMRM), p. 417 (2011)
21. Heiervang, E., Behrens, T.E.J., Mackay, C.E., Robson, M.D., Johansen-Berg, H.: Between session reproducibility and between subject variability of diffusion MRI and tractography measures. NeuroImage **33**, 867–877 (2006)
22. Hoeting, J.A., Madigan, D., Raftery, A.E., Volinsky, C.T.: Bayesian model averaging: a tutorial. Stat. Sci. **14**(4), 382–417 (1999)
23. Hsu, E.: Generalized line integral convolution rendering of diffusion tensor fields. In Proceedings of International Society of Magnetic Resonance in Medicine (ISMRM), p. 790 (2001)
24. Hao, H., Zhang, J., van Zijl, P.C.m., Mori, S.: Analysis of noise effects on DTI-based tractography using the brute-force and multi-ROI approach. Magn. Reson. Med. **52**, 559–565 (2004)
25. Hubbard, P.L., Parker, G.J.M.: Validation of tractography. In: Johansen-Berg, H., Behrens, T.E.J. (eds.) Diffusion MRI: From Quantitative Measurement to in-Vivo Neuroanatomy, pp. 353–375. Academic Press, Massachusetts (2009)
26. Jeurissen, B., Leemans, A., Tournier, J-D. Jones, D.K., Sijbers, J.: Estimating the number of fiber orientations in diffusion MRI voxels: a spherical deconvolution study. In Proceedings of International Society of Magnetic Resonance in Medicine (ISMRM) (2010)
27. Jeurissen, B., Leemans, A., Jones, D.K., Tournier, J.D., Sijbers, J.: Probabilistic fiber tracking using the residual bootstrap with constrained spherical deconvolution. Hum. Brain Mapp. **32**, 461–479 (2011)
28. Jiao, F., Phillips, J.M., Gur, Y., Johnson, C.R.: Uncertainty visualization in HARDI based on ensembles of ODFs. In: Hauser, H., Kobourov, S.G., Qu, H. (eds.) Proceedings of IEEE Pacific Visualization Symposium, pp. 193–200 (2012)
29. Jiao, F., Phillips, J.M., Stinstra, J., Krüger, J., Varma, R., Hsu, E., Korenberg, J., Johnson, C.R.: Metrics for uncertainty analysis and visualization of diffusion tensor images. In: Liao, H., Eddie Edwards, P.J., Pan, X., Fan, Y., Yang, G.-Z. (eds.) Proceedings of Medical Imaging and Augmented Reality. Lecture Notes in Computer Science, vol. 6326, pp. 179–190 (2010)
30. Jones, D.K., Travis, A.R., Eden, G., Pierpaoli, C., Basser, P.J.: Pasta: pointwise assessment of streamline tractography attributes. Magn. Reson. Med. **53**(6), 1462–1467 (2005)
31. Jones, D.K.: Determining and visualizing uncertainty in estimates of fiber orientation from diffusion tensor MRI. Magn. Reson. Med. **49**, 7–12 (2003)
32. Jones, D.K.: Tractography gone wild: probabilistic fibre tracking using the wild bootstrap with diffusion tensor MRI. IEEE Trans. Med. Imaging **27**(9), 1268–1274 (2008)
33. Jones, D.K.: Challenges and limitations of quantifying brain connectivity in vivo with diffusion MRI. Future Med. **2**(3), 341–355 (2010)
34. Jones, D.K., Cercignani, M.: Twenty-five pitfalls in the analysis of diffusion MRI data. NMR Biomed. **23**, 803–820 (2010)
35. Jones, D.K., Pierpaoli, C.: Confidence mapping in diffusion tensor magnetic resonance imaging tractography using a bootstrap approach. Magn. Reson. Med. **53**, 1143–1149 (2005)

36. Kass, R.E., Raftery, A.E.: Bayes factors. J. Am. Stat. Assoc. **90**(430), 773–795 (1995)
37. Kinoshita, M., Yamada, K., Hashimoto, N., Kato, A., Izumoto, S., Baba, T., Maruno, M., Nishimura, T., Yoshimine, T.: Fiber-tracking does not accurately estimate size of fiber bundle in pathological condition: initial neurosurgical experience using neuronavigation and subcortical white matter stimulation. NeuroImage **25**(2), 424–429 (2005)
38. Koch, M.A., Norris, D.G., Hund-Georgiadis, M.: An investigation of functional and anatomical connectivity using magnetic resonance imaging. NeuroImage **16**, 241–250 (2002)
39. Lazar, M., Alexander, A.L.: Bootstrap white matter tractography (BOOT-TRAC). NeuroImage **24**(2), 524–532 (2005)
40. McGraw, T., NadarM.S.: Stochastic DT-MRI connectivity mapping on the GPU. IEEE Trans. Visual. Comput. Graph. **13**(6), 1504–1511 (2007)
41. Nedjati-Gilani, S., Alexander, D.C.: Fanning and bending sub-voxel structures in diffusion MRI. In Proceedings International Society of Magnetic Resonance in Medicine (ISMRM), p. 1402 (2009)
42. Nedjati-Gilani, S., Alexander, D.C., Parker, G.J.M.: Regularized super-resolution for diffusion MRI. In Proceedings of IEEE International Symposium on Biomedical Imaging (ISBI), pp. 875–878 (2008)
43. Pajevic, S., Basser, P.J.: Parametric and non-parametric statistical analysis of DT-MRI data. J. Magn. Reson. **161**(1), 1–14 (2003)
44. Parker, G.J.M., Alexander, D.C.: Probabilistic monte carlo based mapping of cerebral connections utilising whole-brain crossing fibre information. In: Taylor, C.J., Noble, J.A. (eds.) Information Processing in Medical Imaging. Lecture Notes in Computer Science, vol. 2732, pp. 684–695. Springer, Berlin (2003)
45. Parker, G.J.M., Haroon, H.A., Wheeler-Kingshott, C.A.M.: A framework for a streamline-based probabilistic index of connectivity (pico) using a structural interpretation of MRI diffusion measurements. J. Magn. Reson. Imaging **18**, 242–254 (2003)
46. Peled, S., Friman, O., Jolesz, F., Westin, C.-F.: Geometrically constrained two-tensor model for crossing tracts in DWI. Magn. Reson. Imaging **24**(9), 1263–1270 (2006)
47. Peled, S., Yeshurun, Y.: Superresolution in MRI: application to human white matter fiber tract visualization by diffusion tensor imaging. Magn. Reson. Med. **45**, 29–35 (2001)
48. Peled, S., Yeshurun, Y.: Superresolution in MRI—perhaps sometimes. Magn. Reson. Med. **48**, 409 (2002)
49. Pfefferbaum, A., Adalsteinsson, E., Sullivan, E.V.: Replication of diffusion tensor imaging measurements of fractional anisotropy and trace in brain. J. Magn. Reson. Imaging **18**, 427–433 (2003)
50. Qazi, A.A., Radmanesh, A., O'Donnell, L., Kindlmann, G., Peled, S., Whalen, S., Westin, C.-F., Golby, A.J.: Resolving crossings in the corticospinal tract by two-tensor streamline tractography: method and clinical assessment using fMRI. NeuroImage, **47**(Supplement 2), T98–T106 (2009, in press)
51. Scheffler, K.: Superresolution in MRI? Magn. Reson. Med. **48**, 408 (2002)
52. Scherrer, B., Gholipour, A., Warfield, S.K.: Super-resolution in diffusion-weighted imaging. In: Fichtinger, G., Martel, A., Peters, T. (eds.) Proceedings of Medical Image Computing and Computer-Assisted Intervention (MICCAI). Lecture Notes in Computer Science, vol. 6892, pp. 124–132. Springer, Berlin (2011)
53. Schultz. T.: Feature extraction for visual analysis of DW-MRI data. Ph.D. thesis, Universität des Saarlandes (2009)
54. Schultz, T: Learning a reliable estimate of the number of fiber directions in diffusion MRI. In: Ayache, N. et al. (eds.) Proceedings of Medical Image Computing and Computer-Assisted Intervention (MICCAI), pp. 493–500 (2012) (Part III volume 7512 of LNCS)
55. Schultz, T., Schlaffke, L., Schölkopf, B., Schmidt-Wilcke, T.: HiFiVE: a Hilbert space embedding of fiber variability estimates for uncertainty modeling and visualization. Comput. Graph. Forum **32**(3), 121–130 (2013)
56. Schultz, T., Seidel, H.P.: Estimating crossing fibers: a tensor decomposition approach. IEEE Trans. Visual. Comput. Graph. **14**(6):1635–1642 (2008)

57. Schultz, T., Theisel, H., Seidel, H.P.: Topological visualization of brain diffusion MRI data. IEEE Trans. Visual. Comput. Graph. **13**(6), 1496–1503 (2007)
58. Schultz, T., Westin, C.F., Kindlmann, G.: Multi-diffusion-tensor fitting via spherical deconvolution: a unifying framework. In Jiang, T. et al. (eds.) Proceedings of Medical Image Computing and Computer-Assisted Intervention (MICCAI), pp. 673–680. Springer, Berlin (2010) (vol. 6361 of Lecture Notes in Computer Science)
59. A.J. Sherbondy, R.F. Dougherty, M. Ben-Shachar, S. Napel, and B. Wandell. Contrack: Finding the most likely pathways between brain regions using diffusion tractography. Journal of Vision, 8:1, 2008.
60. Sotiropoulos, S.N., Jbabdi, S., Xu, J., Andersson, J.L., Moeller, S., Auerbach, E.J., Glasser, M.F., Hernandez, M., Sapiro, M., Jenkinson, M., Feinberg, D.A., Yacoub, E., Lenglet, C., Van Essen, D.C., Ugurbil, K., Behrens, T.E.J.: Advances in diffusion MRI acquisition and processing in the human connectome project. NeuroImage. doi:10.1016/j.neuroimage.2013.05.057 (2013)
61. Tensaouti, F., Lahlou, I., Clarisse, P., Lotterie, J.A., Berry, I.: Quantitative and reproducibility study of four tractography alorithms used in clinical routine. Journal of Magnetic Resonance Imaging **34**, 165–172 (2011)
62. Tournier, J.D., Calamante, F., Connelly, A.: Non-negativity constrained super-resolved spherical deconvolution: robust determination of the fibre orientation distribution in diffusion MRI. NeuroImage **35**, 1459–1472 (2007)
63. Voineskos, A.N., O'Donnell, L.J., Lobaugh, N.J., Markant, D., Ameis, S.H., Niethammer, M., Mulsant, B.H., Pollock, B.G., Kennedy, J.L., Westin, C.-F., Shenton, M.E.: Quantitative examination of a novel clustering method using magnetic resonance diffusion tensor tractography. NeuroImage **45**(2), 370–376 (2009)
64. Vollmar, C., O'Muircheartaigh, J., Barker, G.J., Symms, M.R., Thompson, P., Kumari, V., Duncan, J.S., Richardson, M.P., Koepp, M.J.: Identical, but not the same: intra-site and inter-site reproducibility of fractional anisotropy measures on two 3.0 T scanners. NeuroImage **51**(4), 1384–1394 (2010)
65. Wakana, S., Caprihan, A., Panzenboeck, M.M., Fallon, J.H., Perry, M., Gollub, R.L., Hua, K., Zhang, J., Jiang, H., Dubey, P., Blitz, A., van Zijl, P., Mori, S.: Reproducibility of quantitative tractography methods applied to cerebral white matter. NeuroImage **36**, 630–644 (2007)
66. Westin, C.F., Peled, S., Gubjartsson, H., Kikinis, R., Jolesz, F.A.: Geometrical diffusion measures for MRI from tensor basis analysis. In Proceedings of International Society of Magnetic Resonance in Medicine (ISMRM) (1997)
67. Whitcher, B., Tuch, D.S., Wisco, J.J., Gregory Sorenson, A., Wang, L.: Using the wild bootstrap to quantify uncertainty in DTI. Human Brain Map. **29**(3), 346–362 (2008)
68. Wu, J.S., Mao, Y., Zhou, L.F., Tang, W.J., Hu, J., Song, Y.Y., Hong, X.N., Du, G.H.: Clinical evaluation and follow-up outcome of diffusion tensor imaging-based functional neuronavigation: a prospective, controlled study in patients with gliomas involving pyramidal tracts. Neurosurgery **61**(5), 935–949 (2007)
69. Zheng, X., Pang, A.: HyperLIC. In Proceedings of IEEE Visualization, pp. 249–256 (2003)

Chapter 9
Mathematical Foundations of Uncertain Field Visualization

Gerik Scheuermann, Mario Hlawitschka, Christoph Garth and Hans Hagen

Abstract Uncertain field visualization is currently a hot topic as can be seen by the overview in this book. This article discusses a mathematical foundation for this research. To this purpose, we define uncertain fields as stochastic processes. Since uncertain field data is usually given in the form of value distributions on a finite set of positions in the domain, we show for the popular case of Gaussian distributions that the usual interpolation functions in visualization lead to Gaussian processes in a natural way. It is our intention that these remarks stimulate visualization research by providing a solid mathematical foundation for the modeling of uncertainty.

9.1 Introduction

The visualization of uncertain field data has attracted a lot of attention in recent time. As practically no measured or simulated data is exact, visualization research attempts to incorporate uncertainty in the images presented to the user. Despite this undebated need, there has been only slow progress towards this goal. There are many field visualization methods without an extension taking uncertainty into account. We think that a major reason for this fact is a lack of knowledge regarding the necessary mathematical description of uncertainty in the case of fields. As we argue in this article, stochastic processes are a viable tool to describe uncertain functions over continuous domains. Since stochastic processes are usually not part

G. Scheuermann (✉) · M. Hlawitschka
University of Leipzig, Leipzig, Germany
e-mail: scheuermann@informatik.uni-leipzig.de

M. Hlawitschka
e-mail: hlawitschka@informatik.uni-leipzig.de

H. Hagen · C. Garth
TU Kaiserslautern, Kaiserslautern, Germany
e-mail: hagen@informatik.uni-kl.de

C. Garth
e-mail: garth@informatik.uni-kl.de

C.D. Hansen et al. (eds.), *Scientific Visualization*, Mathematics and Visualization,
DOI 10.1007/978-1-4471-6497-5_9

of the standard curriculum in computer science and sometimes even mathematics, visualization researchers are not very familiar with this non-trivial subject.

In many cases, the field visualization problem consists of a finite set of given positions where the field value is known. It shall be noted that this holds for scalar, vector, and tensor fields. Before most field visualization methods are applied, an interpolation of these values is defined creating a continuous field over the whole continuous domain. The uncertain field visualization problem is very similar: One is given a finite set of positions with a (known or estimated) distribution of the (unknown) field value at each position. We consider the prominent case of Gaussian distributions in this article and show that all the well-known interpolation methods in visualization can be used in this case to define the uncertain field over the continuous domain as a Gaussian process. This rarely known fact emphasizes the potential of stochastic processes as model for uncertain fields in visualization research.

9.2 Stochastic Processes

We want to describe a (scalar, vector or tensor) field over some closed domain $D \subset \mathbb{R}^d$, $d = 1, 2$, or 3, that depends on some unknown (typically high dimensional) parameter $\omega \in \Omega$. The whole uncertainty is contained in this parameter: If we know the parameter ω, we know the field. To keep things simple, we assume that $\Omega = \mathbb{R}^u$, but that is not necessary.[1] In addition, **we assume that Ω, is known** i.e. the number and type of parameters that determine our field.

In a first step, we need a probability measure on Ω. As Ω contains an uncountable number of elements, we use a σ-algebra $\mathbb{S}$ on Ω. Because of $\Omega = \mathbb{R}^u$, the Borelalgebra $\mathbb{B}(\mathbb{R}^u)$ is a natural choice.[2] Furthermore, we need a probability measure $\mathbb{P} : \mathbb{S} \to [0, 1]$. As usually, this means that the probability for $\omega \in A \subset \Omega$ is $\mathbb{P}(A) \in [0, 1]$. Again, **we assume that this probability measure is known**.

In our second step, we define a **random variable**

$$X : \Omega \to \mathbb{R}^v$$

as measurable[3] map where the σ-algebra on $\mathbb{R}^v$ is the Borelalgebra $\mathbb{B}(\mathbb{R}^v)$. Essentially, this is a usual (i.e. deterministic) function, assigning each (unknown) parameter

[1] In general, we only need a complete probability space, i.e. some set Ω with a σ-algebra and a probability measure on this σ-algebra. Completeness means that any subset of a set with measure zero must be in the σ-algebra. One can construct a complete probability space from an arbitrary probability space by adding elements to the σ-algebra and defining the measure on these elements accordingly [4, Suppl. 2] without any change of practical relevance.

[2] The Borelalgebra is the smallest σ-algebra that contains all open and closed subsets. This ensures in our case that we can measure the probability for all subsets of interest in practical cases.

[3] A map is measurable if each preimage of a measurable set is measurable

$\omega \in \Omega$ a value.[4] If the parameter ω is known, the random variable has a fixed value. From the probability measure $\mathbb{P}$ on Ω, we can derive a probability distribution of X on $\mathbb{R}^\nu$: For any set $A \subset \mathbb{R}^\nu$ in the Borelalgebra $\mathbb{B}(\mathbb{R}^\nu)$, we set

$$P(X \in A) := \mathbb{P}(X^{-1}(A)).$$

As final step, we will define uncertain fields now. Basically, we need to define a random variable at every position $x \in D$. However, there has to be some strong correlation between the random variables at close positions because, in visualization, we are usually dealing with continuous or even differentiable fields. Following Adler and Taylor [2], we define an **uncertain field** depending on our uncertain parameter $\omega \in \Omega$ over the domain D as a measurable, separable[5] map

$$f : \Omega \to (\mathbb{R}^\nu)^D.$$

In perfect analogy to random variables, each parameter $\omega \in \Omega$ gets assigned a deterministic function $f_\omega : D \to \mathbb{R}^\nu$, here denoted as an element of $(\mathbb{R}^\nu)^D$. Furthermore, for each position $x \in D$, we have a random variable $f_x : \Omega \to \mathbb{R}^\nu$ that assigns a fixed value at point $x \in D$ to the parameter $\omega \in \Omega$. We will use the notations

$$f_\omega(x) := f_x(\omega) := f(\omega, x) := (f(\omega))_x \in \mathbb{R}^\nu$$

for the value of the uncertain field f at position $x \in D$ given parameter $\omega \in \Omega$. The measure on $(\mathbb{R}^\nu)^D$ can be defined by a consistent description of distributions on arbitrary finite subsets of positions in D.[6]

[4] The case $\nu = 1$ means a scalar, $\nu = d, d = 2, 3$ means a vector and the case $\nu = d \times d = d^2$, $d = 2, 3$ describes a second order tensor.

[5] This condition removes subtle measurement problems without imposing restrictions of practical relevance, see Adler and Taylor [2, p. 8]. The concept was originally introduced by Doob [4] in his book on stochastic processes. In essence, it demands a dense countable subset $D \subset P$, and a fixed null set $N \in \mathbb{S}$ with $\mathbb{P}(N) = 0$ such that for any closed $B \subset \mathbb{R}^d$ and open $I \subset P$

$$\{\omega | f(x, \omega) \in B \forall x \in I\} \Delta \{\omega | f(x, \omega) \in B \forall x \in I \cap D\} \subset N$$

with symmetric set difference Δ.

[6] According to Doob [4, I.5, II.1] and going back to theorems by Kolmogorov, one needs to define probability distribution functions

$$F_{x_1,\ldots,x_n}(a_1, \ldots, a_n) = \mathbb{P}(|x_1| \leq a_1, \ldots, |x_n| \leq a_n)$$

for arbitrary finite tuples $(x_1, \ldots, x_n)$ of points in D, such that the following rather obvious two consistency conditions hold for all finite subsets of points $\{x_1, \ldots, x_n\}$ and value bounds $a_1, \ldots, a_n \in$:

$$F_{x_1,\ldots,x_n}(a_1, \ldots, a_n) = F_{x_{\alpha_1},\ldots,x_{\alpha_n}}(a_{\alpha_1}, \ldots, a_{\alpha_n}) \quad \forall \text{ permutations } \alpha$$

and

If we consider the situation at a single point $x \in D$ and a measurable subset of values $V \subset \mathbb{R}^v$, e.g. V is closed or open, we have the probability

$$P(f(\omega, x) \in V) = \mathbb{P}(\{\omega | f(\omega, x) \in V\}).$$

As an example for the probability space $(\Omega, \mathbb{S}, \mathbb{P})$, we assume that we have a set of positions $\{p^1, \ldots, p^N\} \in \mathbb{R}^2$ in the plane. At these positions, we have uncertain scalar values $\{v^1, \ldots, v^N\} \in \mathbb{R}$ with normal distributions[7] $W_i \sim N(\mu_i, \sigma_i)$. We may assume that these values are not independent with covariances

$$C_{ij} = E((v_i - \mu_i)(v_j - \mu_j)) \text{ with } \sigma_i = \sqrt{C_{ii}}.$$

Then, we have $(\Omega, \mathbb{S}, \mathbb{P}) = (\mathbb{R}^N, \mathbb{B}(\mathbb{R}^n), N(\mu, C))$. This means that our probability space is N-dimensional real space with an N-dimensional normal distribution with mean vector $\mu \in \mathbb{R}^N$ and (symmetric) covariance matrix $C \in \mathbb{R}^{N \times N}$. It shall be noted that it is possible to derive a space with independent Gaussian variables with potentially smaller dimension $M < N$ by spectral decomposition of C and using the eigenvectors with eigenvalue different from 0. In the following sections, we will see how we can define an uncertain scalar field from these data.

9.3 Gaussian Processes

The previous section introduced stochastic processes without referring to a specific type of distribution at every position. A careful look at the footnotes or intuition tells that the distributions at the different points have to be somehow consistent, and that a simple solution might be to use distributions of the same type everywhere. Looking at the literature, it can be said that Gaussian distributions are the most often used case. If one uses them, one arrives at the special topic of Gaussian processes. They have been analyzed in detail with respect to geometric properties by Adler and Taylor [1–3] in a mathematically rigorous fashion. But Gaussian processes have also been applied in other areas of computer science. A nice example is provided by machine learning as described in the book by Rasmussen

---

$$F_{x_1,\ldots,x_m}(a_1, \ldots, a_m) = \lim_{\lambda_j \to \infty, j=m+1,\ldots,n} F_{x_1,\ldots,x_n}(a_1, \ldots, a_n) \quad \forall m < n$$

We will use multivariate Gaussian distributions for this purpose in the next sections. This footnote illustrates that other distributions are possible.

[7] A normal distribution on is defined by a probability density function

$$\phi(x) = \frac{1}{\sqrt{2\pi}\sigma} \exp^{-\frac{(x-\mu)^2}{2\sigma^2}}.$$

μ is the mean of the distribution and σ the standard deviation.

and Williams [7]. This section and the rest of the article will focus on Gaussian processes.

As before, let $(\Omega, \mathbb{S}, \mathbb{P})$ be a known probability space. Let $D \subset \mathbb{R}^d, d = 1, 2,$ or 3 be the known domain of our field and let $\mathbb{R}^v$ be the set of potential values of our field, i.e. $v = 1$ means a scalar field, $v = d$ means a vector field, and $v = d \times d$ means a tensor field of second order. A measurable, separable map

$$f : \Omega \to (\mathbb{R}^v)^D$$

is called **Gaussian random field** on D if for all finite tuples $(x_1, \ldots, x_n)$ of points in D the random variable $(f_{x_1}, \ldots, f_{x_n})$ is a $v \times n$-dimensional Gaussian random variable. The function

$$\mu : D \to \mathbb{R}^v, \quad \mu(x) = E(f_x)$$

with expectation E is called **expectation function**. The map

$$C : D \times D \to \mathbb{R}^{v \times v} \quad C(x, y) := E((f_x - E(f_x))(f_y - E(f_y)))$$

is called **covariance function**. For any function $\mu : D \to \mathbb{R}^v$ and any non-negative definite function $C : D \times D \to \mathbb{R}^{v \times v}$, there is a unique Gaussian process with expectation function μ and covariance function C, see Adler and Taylor [2, p. 5]! This statement is the basis behind the design and use of Gaussian processes in machine learning as described by Rasmussen and Williams [7]. However, we think that an approach starting with interpolation is more appropriate to visualization, as this is the usual way of defining continuous fields from discrete data in our discipline.

9.4 Linear Interpolation on the Line as a Gaussian Process

This section considers a very simple example. We take the real line as domain, i.e. $D = \mathbb{R}$. We assume that we are given two uncorrelated Gaussian distributions of scalar values

$$W_1 \sim N(\mu_1, \sigma_1) \text{ and } W_2 \sim N(\mu_2, \sigma_2)$$

at the points $x_1 = 0$ and $x_2 = 1$ as data. We want to describe a simple linear interpolation. Since the two values are uncorrelated, we take $\Omega = \mathbb{R}^2$ as parameter space, the Borelalgebra $\mathbb{B}(\mathbb{R}^2)$ as σ-algebra and the 2-dimensional normal distribution $\mathbb{P} = N(\mu, C)$ with

$$\mu = \begin{pmatrix} \mu_1 \\ \mu_2 \end{pmatrix}, \quad C = \begin{pmatrix} \sigma_1 & \\ & \sigma_2 \end{pmatrix}$$

as probability distribution. This means that we assume two normally distributed, independent real parameters that will determine our uncertain field. In this simple case, the two random variables

$$W_1 : \Omega \to \mathbb{R}, W_1(\omega) = \omega_1, \quad W_2 : \Omega \to \mathbb{R}, W_2(\omega) = \omega_2$$

determine the values at the two given positions x_1 and x_2, respectively. It is natural to define the linearly interpolated uncertain field f on the real line by

$$f : \Omega \to (\mathbb{R}) \quad , \quad (f(\omega))_x := \omega_1(1-x) + \omega_2 x.$$

With the notation

$$f_\omega(x) = \omega_1(1-x) + \omega_2 x,$$

it becomes pretty clear that we are really defining a linear interpolation of the values at 0 and 1 on the real line for each given ω. However, the whole point of the chapter is that we are really defining a Gaussian process! The short argument is that this follows from slightly more abstract arguments of Adler and Taylor [3, pp. 17–19]. However, some basic computations might improve understanding of this point: At every position $x \in D$, we have the random variable

$$f_x(\omega) = \omega_1(1-x) + \omega_2 x.$$

As ω_1, ω_2 are independent Gaussian variables, this is a Gaussian variable with expectation

$$\mu(x) = E(f_x(\omega)) = \mu_1(1-x) + \mu_2 x$$

and variance

$$\sigma^2(x) = E((f_x(\omega) - \mu(x))^2) = \sigma_1^2(1-x)^2 + \sigma_2^2 x^2.$$

For the covariance function $C : D \times D \to \mathbb{R}$, we have

$$\begin{aligned} C(x, y) &= E((f_x(\omega) - \mu(x))(f_y(\omega) - \mu(y))) \\ &= E(((\omega_1 - \mu_1)(1-x) + (\omega_2 - \mu_2)x)((\omega_1 - \mu_1)(1-y) + (\omega_2 - \mu_2)y)) \\ &= (1-x)(1-y)E((\omega_1 - \mu_1)^2) + xyE((\omega_2 - \mu_2)^2) \\ &= (1-x)(1-y)\sigma_1^2 + xy\sigma_2^2 \end{aligned}$$

because of the independence of ω_1, ω_2, i.e. $E((\omega_1 - \mu_1)(\omega_2 - \mu_2)) = 0$. For $\sigma_1 = \sigma_2$, this coincides with the construction by Pöthkow and Hege [5].

9.5 General Interpolation

We turn now to a realistic interpolation scenario. We consider some closed domain $D \subset \mathbb{R}^d$. We assume that we are given N positions $p^1, \ldots, p^N \in D$. At these positions, we are given N uncertain ν-dimensional values with normal distributions, say

$$W^i \sim N_\nu(\mu^i, C^i), \quad \forall i = 1, \ldots, N$$

where $C^i \in \mathbb{R}^{(\nu\times\nu)}$ denotes the covariances between the dimensions at a single position. We still assume that the N values are independent. Our interpolation method is given by N (deterministic) weight functions

$$\phi_i : D \to \mathbb{R}, \forall i = 1, \ldots, N \text{ with } \phi_i(p^j) = \delta_{ij}$$

with Kronecker δ. This is the typical case in finite element formulations and for almost all grid based field data in visualization.

We define our probability space via $\Omega = \mathbb{R}^{N\times\nu}$, Borelalgebra $\mathbb{B}(\Omega)$ and

$$\mathbb{P} \sim N(\mu, C), \quad \mu = \begin{pmatrix} \mu^1 \\ \vdots \\ \mu^N \end{pmatrix}, \quad C = \begin{pmatrix} C^1 & & \\ & \ddots & \\ & & C^N \end{pmatrix}$$

as probability measure. Our uncertain field f is defined as

$$f : \Omega \to (\mathbb{R}^\nu)^D, \quad f(\omega, x) = f_\omega(x) = f_x(\omega) = (f(\omega))_x = \sum_{i=1}^{N} \omega^i \phi_i(x).$$

Fixing position $x \in D$, we get a random variable

$$f_x : \Omega \to \mathbb{R}^\nu$$

that describes the distribution of values at that position as a Gaussian distribution

$$f_x \sim N(\mu(x), C(x)), \quad \mu(x) = \sum_{i=1}^{N} \mu^j \phi_j(x), \quad C(x) \in \mathbb{R}^{\nu\times\nu},$$

$$C_{kl}(x) = \sum_{i=1}^{N} C^i_{kl} \phi_i^2(x).$$

Looking at the whole uncertain field again, we have the expectation function

$$\mu : D \to \mathbb{R}^v, \quad \mu(x) = \sum_{i=1}^{N} \mu^j \phi_j(x)$$

and the covariance function

$$C : D \times D \to \mathbb{R}^{v \times v}, \quad C_{kl}(x, y) = \begin{cases} 0 & k \neq l \\ \sum_{i=1}^{N} \sum_{j=1}^{N} \phi_i(x)\phi_j(y)C_{ij}^k & k = l \end{cases}$$

because of the independence of ω_k, ω_l. It should be noted that the definition of an interpolation as above and the definition of a covariance function as usually done in machine learning, see Rasmussen and Williams [7], is actually equivalent, see Adler and Taylor [3, pp. 17–19].

Finally, we describe the case of dependent data at given N positions. To simplify notation, we formulate only the scalar case. We consider a closed domain $D \subset \mathbb{R}^d$ and N positions $p^1, \dots, p^N \in D$. At these positions, we are given N uncertain scalar values with normal distributions

$$W^i \sim N(\mu_i, C_{ii}) \quad \forall i = 1, \dots, N$$

with covariances[8]

$$C_{ij} = E((W^i - E(W^i))(W^j - E(W^j))).$$

The interpolation is again given by N deterministic weight functions

$$\phi_i : D \to \mathbb{R}, \forall i = 1, \dots, N \text{ with } \phi_i(p^j) = \delta_{ij}$$

with Kronecker δ. The interesting point is that the dependence of the uncertain values typically reduces the number of independent uncertain parameters. Mathematically, this means that the (symmetric) covariance matrix C has only $M <= N$ independent rows. One can find them by principal component analysis.[9] Let $\lambda_1, \dots, \lambda_M \in \mathbb{R}$ be the non-zero eigenvalues of C, $e^1, \dots, e^M \in \mathbb{R}^N$ the corresponding eigenvectors. Let $\Lambda \in \mathbb{R}^{M \times M}$ be the diagonal matrix of the non-zero eigenvalues $\lambda_1, \dots, \lambda_M$. We model our probability space via $\Omega = \mathbb{R}^M$, Borelalgebra $\mathbb{B}(\Omega)$ and $\mathbb{P} \sim N(0, \Lambda)$ as probability measure. This probability space consists of M independent normally distributed scalar parameters with mean 0. The uncertain field f is defined as

---

[8] In praxis, the covariances are either given or have to be estimated from several given sample fields. Obviously, this estimation might be a challenge in its own right as the number of positions is almost certainly larger than the number of sample fields. Pöthkow et al. [6] made some comments in this direction.

[9] In praxis, there will be eigenvalues very close to zero in the estimated covariance matrix which one might want to set to zero. Again, this is an obvious challenge outside the scope of this article.

$$f : \Omega \to (\mathbb{R}^{\nu})^D, \quad (f(\omega))_x = \sum_{i=1}^{N} \left(\mu_i + \sum_{k=1}^{M} \omega_k e_i^k \right) \phi_i(x).$$

The Gaussian distribution at each position x, mean function and expectation function can be derived from here as before.

9.6 Conclusion

We have shown that stochastic processes provide a suitable mathematical foundation for the definition of uncertain fields in visualization. In the case of given Gaussian distributions, we have demonstrated how the well-known interpolation methods allow to define Gaussian processes from uncertain field data. We hope that these remarks will stimulate and simplify research on the visualization of uncertain field data. Of course, there is much more to say on the topic that would require more space than available here. For further reading, we recommend the cited literature below.

References

1. Adler, R.: The Geometry of Random Fields. Wiley, Chichester (1981)
2. Adler, R., Taylor, J.: Random Fields and Geometry. Springer, New York (2007)
3. Adler, R., Taylor, J.: Topological complexity of smooth random functions. Lecture Notes in Mathematics, vol. 2019. Springer, Heidelberg (2011)
4. Doob, J.L.: Stochastic Processes. Wiley, New York (1953)
5. Pöthkow, K., Hege, H.-C: Positional uncertainty of isocontours: condition analysis and probabilistic measures. IEEE Trans. Vis. Comput. Graphics **17**(10), 1393–1406 (2011)
6. Pöthkow, k., Weber, B., Hege, H.-C: Probabilistic marching cubes. Comput. Graphics Forum **30**(3), 931–940 (2011)
7. Rasmussen, C.E., Williams, C.K.I.: Gaussian Processes for Machine Learning. The MIT Press, Cambridge (2006)

Part II
Multifield Visualization

During the Dagstuhl 2011 Seminar on Scientific Visualization, members of the break-out group on *Multifield Visualization* worked extraordinarily hard in planning for this Part of the book. They were deeply immersed in evaluating several proposed definitions of the term "multifield", while cheers of a volleyball match crept into the meeting room. They showed no sign of stopping their discussions in the dining room, while everyone could taste, in the air, the delicious aroma of Dagstuhl cakes. The group was determined to produce a series of chapters that could provide visualization researchers with a coherent, and hopefully comprehensive, coverage of the subject of multifield visualization. After the Dagstuhl 2011 event, members of the group worked together through a Wiki page, and completed the 10 chapters in this Part by February 2012. As the coordinator of this Part, it has been a great joy for me to work with this group of colleagues and authors.

The structure of this Part of the book is organized following a planned flow. It starts with two introductory chapters. In the first chapter, *Definition of a Multifield*, Hotz and Peikert introduce us to the necessity of multifield visualization, and provide us with a mathematically consistent set of definitions. In the following chapter,*Categorization*, Hauser and Carr examine two taxonomic schemes for categorizing the subject of multifield visualization, and note that this Part is organized according to the categorization by *visualization approach*.

The next two chapters examine techniques for visualizing multifield data directly. In the chapter entitled Fusion of *Visual Channels*, Chen, Mueller, and Ynnerman consider methods for combining different visual channels in order to depict data from different fields. They first provide us with a large list of visual channels, then examine constructive operators on visual channels and uses of channel fusion in visualizing time-varying fields, and finally consider the techniques available for compressing multifield data. In the following chapter, *Glyph-Based Multifield Visualization*, Chung, Laramee, Kehrer, and Hauser provide us with an overview of an alternative technique that encodes multiple data attributes using glyphs. They survey the state-of-the-art glyph-based visualization, and examine design principles and guidelines that may influence the success in deploying such a technique.

Some may suggest that it is not common to encounter multifield datasets. In the next chapter, *Derived Fields*, Zhang and Natarajan show that even with a single input field, one can derive a multifield dataset, and there is a need for visualizing such derived fields. They first consider the scenario of deriving additional field data from two input fields, then discuss cases where there are more than two input fields, and finally examine the scenario of deriving a multifield dataset from a single input field. This chapter confirms the ubiquity of multifield data.

The next two chapters focus on the role of interaction in visualizing multifield data. In the chapter entitled *Interactive Visual Exploration and Analysis*, Weber and Hauser examine different aspects of *interactive visual analysis* of multifield data, ranging from the generic show-and-brush paradigm to application-specific paradigms where users may interact with application-specific features. In the following chapter, Visual Exploration of Multivariate Volume Data *Based on Clustering*, Linsen examines the visual analytics paradigm where *analysis, visualization*, and *interaction* form a closed loop. In particular, Linsen examines the attribute space in multivariate volume visualization, and the role of interactive visual exploration in cluster-based data analysis.

The following two chapters focus on the role of *feature analysis* in visualizing multifield data. In the chapter entitled *Feature-Based Visualization of Multifields*, Obermaier and Peikert begin with a general discussion on feature extraction in scientific visualization. They then provide us with definitions of multifield features and a categorization of different feature-based visualization techniques. In the following chapter, *Feature Analysis in Multifields*, Carr examines techniques for detecting features in multifield data. In particular, Carr organizes feature analysis techniques into categories of *Scalar Features in Reduced Domains, Scalar Features in the Range, Manifold Features, Overlapping Scalar Features*, and *Joint Feature Analysis*, and structure the discussions accordingly.

The last chapter, led by Laramee, captures a truly collaborative effort by this Dagstuhl break-out group, and reflects the enthusiasm and energy that made members of group so deeply immersed in the discussions and writings for this Part. The chapter, which is entitled *Future Challenges and Unsolved Problems in Multi-Field Visualization*, consists of an introduction by Laramee and contributions from Carr, Chen, Hauser, Linsen, Mueller, Natarajan, Obermaier, Peikert, and Zhang. It is a collection of proposed challenges in multifield visualization, covering a broad range of technical aspects. This collection of challenges will no doubt influence the future directions in this area of research.

Min Chen
University of Oxford, UK
e-mail: min.chen@oerc.ox.ac.uk

Chapter 10
Definition of a Multifield

Ingrid Hotz and Ronald Peikert

Abstract A challenge, visualization is often faced with, is the complex structure of scientific data. Complexity can arise in various ways, from high dimensionalities of domains and ranges, time series of measurements, ensemble simulations, to heterogeneous collections of data, such as combinations of measured and simulated data. Many of these complexities can be subsumed under a concept of multifields, and in fact, multifield visualization has been identified as one of the major current challenges in scientific visualization. In this chapter, we propose a multifield definition, which will allow us a systematic approach to discussing related research.

10.1 Motivation

In its beginnings visualization has focused on single fields, meaning data representing one specific quantity given over some domain. Thereby, the term *field* is mostly used in context with some inherently continuous domain. Typical examples are scalar, vector, tensor fields or also abstract data given over a two to four dimensional space-time. Many corresponding visualization techniques have been developed solving visualization problems related to one specific structure of the domain and data with a specific characteristic.

In many real world applications the situation often looks very different. To solve a problem often multiple fields, data from different sources, with different resolution and representing different quantities are common. E.g., in computational fluid dynamics (CFD) simulations, the result is not just a single field but a collection of

I. Hotz (✉)
Zuse Institute Berlin, Berlin, Germany
e-mail: hotz@zib.de

R. Peikert
ETH Zurich, Zurich, Switzerland
e-mail: peikert@inf.ethz.ch

C.D. Hansen et al. (eds.), *Scientific Visualization*, Mathematics and Visualization,
DOI 10.1007/978-1-4471-6497-5_10

multiple quantities, which are somehow related to each other. Other examples are imaging data acquired with different modalities or multiple runs of one simulation exploring a complex parameter space.

Applying a side-by-side visualization using established methods for the single fields might give a first insight into the data. But such an approach might be overwhelming for the user and, even worse, it ignores the information resulting from the interrelation of the different fields. The need for more integrated visualization methods for such cases have lead to the development of a scientific area focusing on "multifields", whereby the term is used in very different contexts for different applications.

Some instances of multifield visualization go back to the 20th century, e.g., visualization involving different quantities of computational fluids dynamics (CFD) results or volumetric scans using different modalities. The term *multifield visualization* has been introduced in 2001 by Johnson et al. [1] who describe it as "an area ripe for research [...] in which a scientist could visualize combinations of the above fields in such a way as to see the interactions of the fields". A recent overview of multifield visualization is given by Obermaier in his PhD thesis [2].

To start a proper analysis of the current state of this area, the first step is to agree on a basic definition for "multifields". There are many possibilities for a definition of such fields, which might be more or less appropriate for different applications.

10.2 Definitions

The purpose of this section is to define multifields in a way that is general enough to cover most of its usages, distilling the shared properties, which we consider to be the most important.

10.2.1 Fields

As a first step towards a definition of "multifields" as collections of somehow interrelated fields, it is important to agree on a formal definition of "fields".

In visualization, the term "field" has been adopted as it is used in physics where it describes a quantity which is associated to each point in space-time. Therefore, confusion with other meanings of the term, such as for algebraic structures, is unlikely. The prevalent representatives in visualization are scalar fields, vector fields and tensor fields. But the notion of fields has also been extended to functions or distributions linked to points in space-time. Essential for the understanding of fields is the assumption of some continuity of the underlying space (domain) as well as the described phenomenon (range). This is in contrast to data inherently defined over a discrete space or abstract data. The continuity assumptions together with distance measures facilitates the use of analysis methods based on derivatives.

In practice the data mostly origin from simulations and measurements and are only available on a finite subset of the domain. Due to the continuity assumptions it is principally possible to evaluate the field over the entire domain using some interpolation or approximation method. To account for the fact that data acquisition in general involves further parameters, the continuous field domain is augmented by an additional parameter space.

10.2.2 Definition of a Field

A *field* F is defined as a function

$$F : D \rightarrow R.$$

- *Domain D*
  The domain consists a Cartesian product of a finite-dimensional metrical space and some parameter space (D^P). Since for many applications time plays a distinguished role we further separate the temporal dimension (D^T) from the spatial dimensions (D^S).
  $$D = D^S \times D^T \times D^P.$$

- In the most general form the *range* R of a field F is defined as the Cartesian product of a metric space R^M and a set of categorical values Φ.
  $$R = R^M \times \Phi.$$
  The general case for R^M are finite-dimensional spaces representing scalars, vectors or tensors, which can be considered as subsets of $\mathbb{R}^n$ for some $n \in \mathbb{N}$. But this definition also allows for the more general case of function or distribution spaces. Categorical values Φ are in general of discrete nature and include classifications or binary markers.

10.2.3 Multifields

Based on this definition of fields a *multifield* $\mathscr{M}$ is defined as a set of fields

$$\begin{aligned}
\mathscr{M} &= \{F_1, F_2, \ldots, F_r\}, r \in \mathbb{N} \\
F_i &: D_i \rightarrow R_i \\
D_i &= D_i^S \times D_i^T \times D_i^P \\
R_i &= R_i^S \times \Phi
\end{aligned} \tag{10.1}$$

We further require that the various fields have a shared embedded space-time domain $D = D^S \times D^T$.

$$D_i^S \subset D^S \text{ and } D_i^T \subset D^T \text{, for all } i$$

The subsets can also be lower-dimensional, such as domain boundaries.

10.2.4 Uniqueness of Representation

It is possible that there is not one unique representation for a field or multifield according to the definition given above. E.g., it would be possible to shift the indices for the components of a vector to the parameter space. Or one could even consider a vector field already as a multifield consisting of several scalar fields.

Another example is the representation of multi-run simulations. Here two cases can be distinguished. On the one hand, a simulation can be run repeatedly for different (combinations of) parameters. In this case, a multifield representation is obtained straightforwardly by making the parameter space a part of the domain.

On the other hand, stochastic ensemble simulations, as done typically in climate research, generate a field with distributions as its values instead of scalars or vectors. Another option would be to interpret each run as a separate field.

The decision for the specific choice of the domain and range of a field should be guided by the semantics of the multifield and its fields.

10.3 Multifields and Related Concepts

The above definition of a multifield provides enough flexibility to capture most of the types of compound datasets that occur in practice. In this section we identify the types of data that can be represented by our multifield definition and describe the actual mapping to this abstract mathematical representation.

The term *multi-channel data* is commonly used to describe data having multiple variables (quantities, attributes) per point in space-time. This fits straightforwardly into our multifield concept and is in fact a standard case of it.

The same holds, conceptually, also for *multi-modal data*. Practically, a difference is that in multi-modal data the single fields are typically given in different discretization and on overlapping domains. Multi-modal data are generated in separate processes, and data from different modalities have to be registered first, before they can be represented as a multifield.

Multi-material simulations generate data for different materials. These can be represented by different fields of a multifield. Domains can overlap if materials can mix, but also at material boundaries because of discretization.

The term *multi-dimensional data* can be used to give a vague description of the complexity, but by not making a distinction between the dimension of the domain and the range, it cannot be used as a precise characterization. Similarly, the term *multivariate* should be avoided in a general discussion of multifields, because its meaning depends on the discipline. In statistics it refers to the dimension of the range and in other branches of mathematics to the dimension of the domain.

Multi-value data are a more difficult case. If there is only a bounded number of values that a point in space-time can have, the data could be represented as a multifield, although an artificial ordering of the values is introduced this way. If the data per point in space-time can be arbitrary sets, then such data cannot be expressed by multifields.

Multiphysics refers to simulations involving multiple physical models such as magnetohydrodynamics, fluid structure interaction, or fluid flow combined with chemical reactions. Multiphysics simulations are a source of multifield datasets containing an even larger number of components than are obtained from CFD simulations.

References

1. Johnson, C., Brederson, D., Hansen, C., Ikits, M., Kindlmann, G., Livnat, Y., Parker, S., Weinstein, D., Whitaker, R.: Computational field visualization. In: Mathematics Unlimited—2001 and Beyond, pp. 605–630. Springer, Berlin (2001)
2. Obermaier, H.: Multi-field visualization. Ph.D. thesis, TU Kaiserslautern (2011)

Chapter 11
Categorization

Helwig Hauser and Hamish Carr

Abstract Multifield visualization covers a range of data types that can be visualized with many different techniques. We summarize both the data types and the categories of techniques, and lay out the reasoning for dividing this Part into chapters by technique rather than by data type.

As we have seen in the previous chapter, multifield visualization covers a broad range of types of data. It is therefore possible to discuss multifield visualization according to these data types, with each type covered in a separate chapter. However, it is also possible to approach the question by considering the techniques to be applied, many of which can be applied to multiple types of multifield data. In this chapter, we therefore discuss both ways of analysing multifield visualization techniques, and why we have chosen to proceed according to technique rather than type in the subsequent chapters.

11.1 Categorization by Data Type

All multifield data shares a common attribute—that it is known or presumed that the fields are related spatially to each other. However, these relationships can arise in different ways, and this has an impact on how we analyze or visualize the data.

Broadly speaking, the individual fields in multifield data can be related in a number of ways:

1. Multi-variate data, where related properties are computed or measured,
2. Spectral data, where multiple properties are measured, but may or may not be related,

H. Hauser (✉)
University of Bergen, Bergen, Norway
e-mail: Helwig.Hauser@UiB.no

H. Carr
University of Leeds, Leeds, UK
e-mail: H.Carr@Leeds.ac.uk

C.D. Hansen et al. (eds.), *Scientific Visualization*, Mathematics and Visualization,
DOI 10.1007/978-1-4471-6497-5_11

3. Multi-run/ensemble data, where each field is a separate output of a computational or measurement process,
4. Derived fields, where new fields are generated to sharpen the understanding of existing fields,
5. Multi-scale data, where fields at different resolutions or scales are considered,
6. Other, ranging from tensor fields to time-dependent data.

We will canvass each of these types of data separately, proceeding from the types whose fields are more obviously tightly related, to those where the multifield representation is more a choice of representation than an inherent structure.

11.1.1 Multi-variate Data

Multi-variate data are common to several applications, including computational fluid dynamics (CFD), where the interaction between multiple physical quantities are modeled and computed over detailed spatial grids. In the simplest case, each location or sample in a spatial domain is assigned a coherent vector of multiple variables. The paradigm case of this type of data is CFD, where properties such as pressure and velocity are computed for each location in a grid.

Multi-variate datasets (in terms of this definition) are usually characterized by a relatively small number of variables (between two and a few dozen). Here, the visualization challenges arise from the fact that the correlation between pairs of variables is wildly heterogeneous. For example, while some variables are perfectly linearly correlated, others may be largely unrelated as, e.g., when resulting from separate solvers (say a fluid solver and a chemical reaction solver).

11.1.2 Spectral Data

Another type of multifield data is spectral data. Most commonly resulting from physical acquisition techniques (such as spectral imaging techniques), we consider datasets where data relating to different frequencies are represented as different fields. An example is spectral satellite imaging, where (concurrently) a number of images at different wavelengths are taken from the same target, resulting a multi-frequency dataset.

In comparison with multi-variate data, spectral datasets may involve much larger numbers of fields (frequencies), which leads to interesting visualization challenges. However, it is commonly the case that there is a substantial amount of coherence between all the fields. For example, the fields are often sorted in a meaningful way (usually by frequency), and responses to different frequencies tends to correlate more tightly than for example pressure and vorticity in a CFD computation.

11.1.3 Multi-run/Ensemble Data

A third type of multifield data is multi-run, multi-parameter, or ensemble data. These datasets represent multiple results from the same operation, rather than multiple related operations. Multi-run data, for example, can result from repeating a stochastical simulation a certain number of times, leading to data which can be interpreted as a statistical sample of outputs from the model. Multi-parameter or ensemble data can also result from repeating the data acquisition (simulation or measurement) while varying input parameters of either the simulated model or the measurement technique (for example regular or Monte-Carlo sampling).

Visualization of these multi-run, multi-parameter, or ensemble data usually amounts to performing a sensitivity/variability analysis of the phenomenon under consideration. In climate research, for example, the dependency of a forecast on certain model parameters can be studied. In engineering, on the other hand, the performance of a certain system component can be studied, while external driving conditions are varied.

11.1.4 Derived Fields Data

As one moves from intrinsically multifield data to data which is multifield as a result of the choice of representation, the next type to be considered is that of derived fields. In these datasets, one or more additional fields are computed directly from the known fields (as distinct from being computed at the same time as the original fields).

For example, to understand moving particles, additional descriptive quantities are often computed for each field location that—all together—explain aspects of the behaviour of the system, whether local or global.

Intrinsic to this derivation is an expectation that the derived field will depend strongly on the originating fields—thus, the derived field can either be viewed as additional information or as a reduced or simplified form of information. Even for a single scalar field, however, the opportunity of deriving fields implies that multifield visualization methods may be applicable.

11.1.5 Multi-scale Data

A further type of multifield data arises when a single field is measured at different scales or different resolutions. The effective selection of a scale can often depend on understanding the relationship between these resolutions, giving rise therefore to multifield problems. In essence, the scale axis is used to set the fields alongside each other, leading to a scale-space representation where each field represents the data at a certain scale.

Visualization questions for these data types involve the selection of an appropriate scale, or the consideration of the data through a proprietary (for example selective) reconstruction of the data (based on certain scales of interest).

11.1.6 Other Types of Multifield Data

In addition to the major types of data already listed, a multifield framework can be used as a representation for data such as tensor fields, where tensor components are interpreted as individual fields, or time-dependent data, where the time-steps are interpreted as individual fields.

Representing such data in a multifield form allows the use of existing visualization methods such as coordinated multiple views with linking and brushing, or focus+context visualization. Of course, an additional challenge is generated by the fact that an important semantic aspect of the data (that the fields actually make up a tensor or a time series) is possibly lost (or cannot be exploited).

11.1.7 Summary

If we look at the various types of multifield data, we see that nearly all of the types require similar tasks to be performed, and in particular require the detection or visualization of correlations between the fields. As a result, many of the techniques applicable to one type will tend to be applicable to other types, and a categorization by data type risks the repetitive discussion of the same techniques. We therefore consider in the next section the techniques that are applicable to multifield visualization, then return to the question of which approach to adopt.

11.2 Categorization by Visualization Approach

As we have seen above, one way to categorize multifield visualization is to focus on the type of data. A second way to categorize is to observe that many techniques cut across all of the types of data as discussed above. The advantage of this characterization is that it gives a principled context in which to discuss not only those techniques that have already been reported, but also in which to discuss classes of techniques that could be introduced in the future. A second advantage of this approach is that we can extrapolate more readily from techniques known to work for single-field data, whether scalar, vector or tensor.

Broadly speaking, we can observe that the visualization of single-field data relies on mapping the data to properties of the human visual system, on providing the user interactive tools for isolating regions of the data, and on the detection of significant features in the data. In addition to these basic categories, existing multifield visualizations often rely on the mathematical habit of reducing complex problems to simpler problems with known solutions. In the context of multifield visualization, this usually means computing a single scalar or vector field based on the input data, then visualizing that single field.

This therefore leaves four broad categories of approaches to multifield visualization, in approximate order of difficulty:

1. Visual Channel Mapping
2. Derived Fields
3. Interactive Exploration
4. Feature Detection and Analysis

Each of these will be covered in a separate section, but we start with a high-level overview of these methods first.

11.2.1 Visual Channel Mapping

For single fields or for multifields with small numbers of variables, the first set of approaches, including much of the work published to date, involves mapping data properties to visual properties. So, for example, one dependent variable may be mapped to the red channel, a second to the green channel, and a third to the blue channel. Alternately, one channel could map to hue, a second to saturation, and a third to brightness.[1] Visual channels that can be exploited this way are not, however, restricted to colour alone—as we will see in Chap. 12, texture and geometric shape are also used to represent data properties.

A core problem with visual channel mapping is that the human visual system has a limit on how many different visual channels can be perceived at once. Moreover, the amount of precision in the visual system limits the qualitative conclusions that can be drawn. However, due to the simplicity and straightforwardness of visual channel mapping, it often forms the basis for the methods to be developed in subsequent chapters.

11.2.2 Derived Fields

Once the visual channel limitations are realized, the next set of methods relies on reducing the number of visual channels by combining elements of multiple data variables in a single channel. This is usually done by computing some summative property that encapsulates a relationship between the variables, thus reducing the

[1] We note that both of these mappings are poor choices visually, but they are the easiest illustrations of the principle.

multifield to a single field, which is then visualized directly using existing techniques. These methods may include measures of complexity or correlation between variables, or derived properties such as vorticity.

While these methods can be very effective, they tend to work best at detecting relationships that are known or suspected. This is because the choice of the summative property is usually guided by a sense that a particular aspect of the data set is significant. Moreover, they presume that the phenomenon being studied is uniform throughout the domain: thus, if two properties are weakly correlated in one area, but strongly correlated elsewhere, these methods will be less successful.

11.2.3 Interactive Exploration

A third category of visualization techniques relies on the experience and intuition of the user, by providing an interactive tool for exploring the data. Inevitably, this relies on visual channel mapping and derived fields to give the user sufficient insight to identify features, and increasingly, on feature detection as well.

Interactive exploration can operate by manipulation of the visual channel mapping, by the provision of geometric tools to identify regions of the data, by selection of paradigm points or regions as seeds for similarity measures, by combination of properties through logical rules, or by reference to abstract descriptions or secondary visualizations.

While often the most effective approach, interactive exploration starts breaking down with larger data sets, as does direct visualization itself, as the volume of data outstrips the humans visual and cognitive capacity to understand the data.

11.2.4 Feature Detection and Analysis

All three categories described so far share a common difficulty: that, as the amount of data increases, less and less of it can be presented to the user. In short, the question is not "how can we visualize the data", but "what subset of the data can we visualize". As a result, visualization techniques have increasingly relied on abstract definitions of features, either specific to a domain, specific to a type of data, or common to multiple domains and data types. These features are detected computationally and presented to the user either as the answer to a question, or as the seeds to an interactive exploration.

Philosophically, these methods shade off into the disciplines of image analysis, computer vision and data analysis, all of which share a common interest in detecting features in masses of data. However, one set of methods which is distinctive in visualization is the reliance on formal mathematics such as topology to extract abstract features either for further analysis or for direct visualization.

11.2.5 Summary

Clearly, a breakdown according to techniques runs a similar risk of repetition to the risk observed for a breakdown according to types. However, as we can see from the discussion above, a breakdown according to techniques has more obvious and clearer demarcations, in addition to providing a roadmap for as yet unidentified techniques.

11.3 Conclusion

In short, while we can categorize multifield visualization either by the type of data or by the type of technique, we have chosen the latter for two reasons. First, similar techniques are visible across all types of data, and it is therefore easier to consider a single technique and, if necessary, its application to different types. Second, it allows us to extrapolate future techniques out from the accumulated experience of working with single-field data.

Chapter 12
Fusion of Visual Channels

Min Chen, Klaus Mueller and Anders Ynnerman

Abstract In this chapter, we consider the need in multifield visualization to depict information contained in two or more fields in a compositional manner. There are many different visual channels, some of which are more commonly seen in visualization than others. Channel fusion occurs when two or more visual entities have to share the same screen space. By applying appropriate constructive operations on visual channels in the composition, one may encode the integration as well as separation of the underlying information depicted by the original channels. One special situation is where multiple fields are a set of fields from different temporal steps, which imposes additional constraints on the use of visual channels. It is inevitable that the availability of visual channels will not be able to scale up to a large number of visual channels. Hence, we consider briefly several general-purpose data mapping methods that can be used to reduce the complexity of visual mapping.

12.1 Introduction

One primary goal of multifield visualization is to depict information contained in two or more fields in a compositional manner, which facilities combinational overview or visual comparison. In visualization, the most fundamental approach to achieve this goal is to make use of different visual channels available in the design space, which we refer to as *channel fusion*. This differs from a naive approach for displaying individual fields independently, typically in sequence or using juxtaposition, which usually result in a high cognitive load in gaining an insight about compositional

M. Chen (✉)
Oxford e-Research Centre, University of Oxford, Oxford, UK
e-mail: min.chen@oerc.ox.ac.uk

K. Mueller
Department of Computer Science, Stony Brook University, Stony Brook, USA
e-mail: mueller@cs.stonybrook.edu

A. Ynnerman
Norrköping Visualization Center, Linköpings Universitet, Linköpings, Sweden
e-mail: Anders.Ynnerman@liu.se

C.D. Hansen et al. (eds.), *Scientific Visualization*, Mathematics and Visualization,
DOI 10.1007/978-1-4471-6497-5_12

effects of, or making a comparison between, different fields. In this section, we first consider different visual channels that have been used in visualization (Sect. 12.2). We then discuss various constructive operations on visual channels, showing various examples of channel fusions (Sect. 12.3). This is followed by an examination of a special situation where multiple fields are a set of fields from different temporal steps (Sect. 12.4). Finally, we consider a collection of alternatives, in which data mapping plays a critical role in reducing the complexity of visual mapping (Sect. 12.5).

The fusion of visual channels is highly sensitive to the limitations of human perception as well as application-specific visual metaphors. In scientific visualization, it is often constrained by elementary properties of the data. For instance, when geometry (e.g., an isosurface) is an intrinsic property of one of the fields, the flexibility in using geometry to depict other types of properties in a composite visualization is significantly reduced. When color is an intrinsic property of one of the fields (e.g., visible spectrum in remote sensing data), it is usually very difficult to fuse different color metaphors by introducing virtual colors for other types of properties while maintaining the original color representation. Furthermore, the number of visual channels is limited; hence their use cannot be scaled to an arbitrary number of multiple fields. Because of these reasons, it is important to avoid overloading of the visual channels by enabling users to choose a subset of multiple fields or a subset of properties to be visualized (see Sect. 12.5), and by employing appropriate analytic methods for filtering out unimportant data and selecting features to be highlighted (see Chaps. 17 and 18).

12.2 Visual Channels in Multifield Visualization

There are many types of visual channels that can be utilised in visualization. Bertin provided a comprehensive study on several visual channels typically used in geo-spatial visualization in general and cartography in particular [1]. However, multifield visualization exhibits many characteristics that are untypical in geo-spatial visualization. For example, most multifield datasets represent objects or phenomena in a continuous 3D spatial domain, and many encode directional and temporal information at a much larger scale than ordinary geographical datasets. Hence it is not uncommon for multifield visualization to make use of more visual channels than those commonly-available in geo-spatial visualization. Visual channels can be roughly divided into the classes of *Geometric*, *Optical*, *Relational*, and *Semantic* channels [7]. This classification focuses on the *effect* of a visual channel rather than its *cause*. For example, the *curvature* of a surface is indubitably a geometric property, but one of the most effective ways to depict this property (i.e., to cause this effect) is *shading*, which is an optical channel. On the other hand, geometric channels can also be used to influence optical perception, hence contributing to the formation of optical channels. For example, *textures*, which are made of different geometric components, are commonly used to convey different scales of *brightness* that is an optical channel.

Geometric Channels provide means to differentiate visual entities by making use of their different geometric properties, including

- *Size/length/width/height/depth/thickness/area/volume*
- *Orientation*
- *Shape*
- *Curvature*
- *Smoothness*

Optical Channels provide means to differentiate visual entities by making use of different optical effects. Many of such effects rely extensively on perceptual interpretation. For example, different shading effects, which can be used to encode high-level concepts such as materials, are perceived through the changes of brightness and colours resulting from some rendering algorithms. In visualization, motion can be depicted explicitly as well as implicitly. The former includes both first-order and second-order motion perception, while the latter makes use of static depiction, such as motion blur patterns. Different visual effects in the class of optical channels include:

- *Intensity/brightness*
- *Colour/hue/saturation*
- *Opacity/transparency*
- *Line style/surface texture/volume texture*
- *Shading and lighting effects/halos*
- *Shadow*
- *Photographic effects such as focus, blurring, optical distortion*
- *Implicit motion/motion blur patterns*
- *Explicit motion/animation/flicker*

Relational Channels provide means to differentiate visual entities by making use of depictions of different relations. There are two main types of relations. *Spatial relationships*, such as *distance*, *depth*, and *density*, convey the difference between two visual entities through their geometric or geographic relationships with other entities in the scene, or the perceived spatial relationship with the viewer. *Topological relationships*, such as *connectivity*, *hierarchy*, and *closure*, convey the difference between two visual entities through the different structures associated with them, or their different roles in a shared structure. Some of such effects rely on perceptual interpretation, while many rely extensively on cognitive interpretation. For example, *depth* is a complex perceptual phenomenon because it may result from different visual channels. The perception of depth can be caused by explicit binocular cues (e.g., stereo vision), or implicit monocular cues (e.g., perspective and occlusion).

- *Connection/edge*
- *Node/internal node/terminator*
- *Inside/outside/enclosure/boundary*
- *Distance/displacement/offset*
- *Closure/opening*

- *Connectivity*
- *Partition/completeness*
- *Intersection/overlap*
- *Depth ordering/partial or full occlusion*
- *Hierarchy/level*
- *Density/distribution*
- *Convexity*
- *Continuity*
- *Homeomorphism/genera*
- *Similarity*
- *Translation/scaling/rotation/deformation*

Semantic Channels provide means to differentiate visual entities by making use of pre-defined visual languages that encode different semantic concepts.

- *Number*
- *Text*
- *Symbol/ideogram*
- *Sign/icon/logo/glyph/pictogram*
- *Isotype*

12.3 Constructive Operations on Visual Channels

Let $F_1, F_2, \ldots, F_n$ be a set of fields to be visualized in an amalgamated manner. Let $\mathcal{V}(F_1), \mathcal{V}(F_2), \ldots, \mathcal{V}(F_n)$ be a set of corresponding visual mappings. The composition of these visual mappings is thus a function that combines these visual mappings into a single visual representation, such that,

$$\Lambda : \mathcal{V}_1(F_1), \mathcal{V}_2(F_2), \ldots, \mathcal{V}_n(F_n) \rightarrow V$$

where V is a visual representation that is directly displayable on a 2D or 3D display, and $\mathcal{V}_i, i = 1, 2, \ldots, n$ are different visual mappings, each of which may utilize one or several visual channels.

The function Λ is usually designed to serve specific requirements of visualization tasks, which may be one or several of the followings:

- *Association and Correlation*—This is the most basic task in multifield visualization, with a goal for users to establish correspondence between features in other field and those in the other fields. The emphasis is placed on the ability to see the visual mappings of individual fields, $\mathcal{V}_i, i = 1, 2, \ldots, n$, through the composite visualization V.
- *Comparison or Contrast*—The goal of the tasks in this category is for users to examine the similarity or difference between different fields, and to estimate the changes qualitatively or quantitatively. While the requirement for the ability to

see the visual mappings of individual field remains, there is also an additional constraint, that is, in order to compare field F_a and F_b, their visual mappings, $\mathcal{V}_a$ and $\mathcal{V}_b$ have to be visually comparable.
- *Observation of Combined Effects*—In some applications, different fields can be combined meaningfully, by using, for example, logical, algebraic and statistical operators. In such cases, the emphasis has been shifted from the depiction of individual fields to that of the combined effects.

Note that these three types of tasks are not totally exclusive or performed in isolation. For example, once an observation of association is made, one may use combined effects, such as point-wise subtraction, to compare two fields and evaluate the changes from one to another.

Volume Visualization. Earlier works in volume visualization were focused primarily on logical operators on voxel occupancy in combination of multiple volume datasets (e.g., [8, 21]). At the turn of the new millennium, three groups independently developed methods for constructive operators for combining real-domain volume datasets [3, 6, 17]. Among them, Chen and Tucker [6] provided an algebraic framework, CVG, for combining discretely specified volume datasets and continuous functional specifications of scalar fields, and rendering a multi-volume scene as a CVG expression that is equivalent to a tree or scene graph. The constructive operators are defined with four separate channels for α, R, G and B, each of which can be operated on differently from others. The open-source *vlib* [22] extended this approach to allow for independent constructive operators on attribute fields for the Phong illumination model, reflection and refraction. Chen [4] later added the notion of point-based volume objects into the CVG framework to enable scalar fields defined by point clouds to be visualized with volume datasets and functional scalar fields.

It is usually difficult to evaluate the characteristics of individual fields through combined effects in multifield visualization. To support 3D visualization tasks with strong elements of association and correlation, one critical problem is occlusion. While translucency is effective for simultaneously depicting non-overlapping features in volume visualization, it is necessary to use other visual channels to deal with features that are from different fields and occupy more or less the same space. One approach is to use visual representations that exhibit a reasonable amount of empty space, such as wire frames and coarse non-photorealistic textures. For example, one method proposed in [20] is a class of pen-and-ink textures which can be used in conjunction with conventional shaded RGB surfaces and αRGB amorphous effects. The density of the pen-and-ink lines and the opacity of the empty area in the textures can be controlled using a transfer function. Another approach is to use deformation to force a field to make space for another field, provided that the users can still comprehend the actual spatial position of the deformed field. For example, Chen et al. [5] introduced deformation as a generic operator in the CVG framework and the detailed displacement as fields that can be applied onto any other scalar fields in a scene graph to facilitate focus and context views of different scalar fields.

Flow Visualization often sees the need for multifield visualization, where a vector field is depicted with other multivariate properties such as temperature and pressure [13, 16, 19]. The primary vector fields are usually visualized using the orientation and size channels, which are commonly depicted using glyphs, lines, surfaces, and textures. Different colors are commonly used for depicting one or two secondary fields. In 3D flow visualization, it is common to visualize a vector field in conjunction with an iso-surface. In this case, any volumetric data that is not on the surface is removed to reduce occlusion.

Uncertainty Visualization involves the depiction of the measurement of uncertainty in conjunction with the primary data set. When both are in field representations, it essentially becomes a problem of multifield visualization. The primary task is thus association because the foremost requirement is usually the need to observe the uncertainty measurement associated with some or all parts of the data field. The visual channels used for depicting uncertainty include color, texture, transparency, haziness, blurring, uncertainty glyph, and geometric transformation (for details, see discussions in [11, 18]).

One common dilemma in uncertainty visualization is that the underlying data fields often require the use of several visual channels. For example, in surface and volume rendering, many geometrical channels and optical channels are used explicitly or implicitly. The continuous spatial usage prevents substantial use of relational and semantic channels. If the visual channels for depicting uncertainty were confused with those for the primary data fields, it would introduce additional and undesirable visual uncertainty. One effective means for addressing this dilemma is to use repetitively-animated glyphs [15]. The dynamic nature of the uncertainty depiction makes a clear distinction from the static depiction of the primary data fields.

12.4 Composition of Time-Varying Fields

Time-varying fields are often visualized as an animated sequence of images, each of which depicts a single field or multiple fields at a particular time step. While this form of visualization is intuitive and commonly deployed in practice, it has several shortcomings. For example, viewing animations requires full attention, and is prone to change blindness. Because of the limitation in short-term memory, an animation is often watched over and over again in order to make comparison between different time steps.

One alternative approach is to compose a static visualization to depict several time steps. Hence, even when we consider only a signal field, F, at two different time steps t_1 and t_2, the composition of the two fields, F_{t_1} and F_{t_2} transforms the problem to that of a conventional multifield visualization as discussed in the previous sections. The requirements for such visualizations approaches often place a great emphasis on spatial association, comparison and differentiation, posing a non-trivial challenge to the selection of visual channels. On one hand visually disparate channels (e.g., color, outline, symbol) would create difficulties for comparison and differentiation. On the other hand, similar visual channels (e.g., red and green hues) might not deal with occlusion effectively.

Botchen et al. [2] conducted a small study on visual channels in the context of video visualization. They conducted a survey that rated the suitability of six visual channels (i.e., color, luminance, opacity, thickness, symbols and textures) for three data types (category, uncertainty and size), and optimized their selection of multiple channels based on the rating. The proposed visual design, VideoPerpetuoGram, is scalable to an arbitrary number of time steps.

Woodring et al. considered a more challenging problem of time-varying volume visualization, and used constructive operations to combine different color channels corresponding with different time steps [24] and to create combined effects for visualizing set and numerical relationships between different time steps [23]. Hsu et al. used color and outline channels for different time steps, in conjunction with spatial layouts that separate different time steps [10].

One approach is to have one time step as a reference, and illustrate the relative changes of the succeeding (or occasionally preceding) time steps using a different visual channel. The illustration-inspired techniques proposed by Joshi and Rheingans [12] exemplifies this approach.

12.5 Compression of Multifields

A main obstacle with multifield data is that they can exceed the number of visual channels available, at least in a general sense. This is very reminiscent to the problem of multivariate data when the aim is to reduce the dimensionality of the data for display. A number of strategies have been devised to achieve this, what is called *low-dimensional embedding* through *dimension reduction*. Note that in the following it is assumed that the multifield data "live" in a spatial context where this spatial context does not need to be a regular lattice. While spatial context is not a strict requirement, the spatial coordinates could also be integrated into the analysis.

1. **Principal component analysis (PCA)**: Methods based on this technique would first determine the covariance matrix of the multifield data. An eigenvector analysis would then determine the k most significant axes onto which the multifield data would be projected and then mapped to the available visual channels.
2. **Multi-dimensional scaling (MDS)**: These methods perform a linear or non-linear projection of the data into a lower-dimensional space. Since this is an optimization problem, many different techniques with different strategies are available. Typically the quality of an embedding is measured by a stress metric which is the RMS error of the point-wise distances in data space and the respective distance in embedded space. For multifield data, one would perform MDS on the data and set k to the number of visual channels.
3. **Linear Discriminant Analysis (LDA)**: LDA is somewhat related to PCA but unlike PC is does maximize to identify projections of highest variation, but instead

it seeks to model the difference between the classes of data. This assumes that some clustering or classification of the data exists. An inherent constraint of LDA is that the number of remaining dimensions cannot be below the number of classes else overlap in the embedded space will occur. Thus, for the multifield data case one would first perform data clustering, say via k-means setting k to the total number of available visual channels and then perform LDA.

4. **Information-Theoretical Analysis**: This approach seeks to identify spatial regions where different fields may or may not have similar information or a similar amount of information. Information measures, such as mutual information can be used to quantify the levels of similarity. Based on such a measurement, different fields can be fused together, for example, by computing a weighted average of their values at each point. Haidacher et al. used this approach to design a joint transfer function in visualizing multiple volume datasets from different imaging modalities [9].
5. **Expert-guided dimension reduction**: PCA, MDS, LDA can be guided by experts via suitable visual interfaces. This has come to be known as visual cluster analysis, where experts use information visualization tools to guide the clustering, the selection of the principal components, the weighting of the dimensions, and the selection of influential data points. For multifield data the expert would use a bi-modal interface consisting of an information display and a suitable scientific visualization display to interactively and iteratively steer the insight gained.

Lawrence et al. [14] present a technique that can fuse an arbitrary number of aligned images into a single color or intensity image. Their method falls into the second grouping of above, i.e., MDS-based compression. They specifically target multi-spectral imagery as obtained from remote sensing. They use an iterative stress majorization method in conjunction with clustering to determine the low-dimensional subspace into which the solution is embedded. A very useful feature of their algorithm is that it allows users to incorporate direct constraints onto the mapping process. This, for example, allows for better preservation of object colors that a user may want to maintain.

References

1. Bertin, J.: Semiology of Graphics: Diagrams, Networks. ESRI Press, Maps (2008)
2. Botchen, R.P., Bachthaler, S., Schick, F., Chen, M., Mori, G., Weiskopf, D., Ertl, T.: Action-based multifield video visualization. IEEE Trans. Vis. Comput. Graphics **14**(4), 885–899 (2008)
3. Cai, W., Sakas, G.: Data intermixing and multivolume rendering. Comput. Graphics Forum **18**(3), 359–368 (1999)
4. Chen, M.: Combining point clouds and volume objects in volume scene graphs. In: Proceedings of Volume Graphics pp. 127–135 (2005)
5. Chen, M., Silver, D., Winter, A.S., Singh, V., Cornea, N.: Spatial transfer functions—a unified approach to specifying deformation in volume modeling and animation. In: Proceedings of Eurographics/ACM Volume Graphics, pp. 35–44 (2003)
6. Chen, M., Tucker, J.V.: Constructive volume geometry. Comput. Graphics Forum **19**(4), 281–293 (2000). A short version of the paper was presented in VG99 (July 1999)

7. Chen, M., Floridi, L.: An Analysis of Information in Visualisation. Synthese **190**(16), 3421-3438 (2013)
8. Fang, S., Dai Srinivasan, R.: Volumetric CSG—a model-based volume visualisation approach. In: Proceedings of 6th International Conference in Central Europe on Computer Graphics and Visualisation, pp. 88–95 (1998)
9. Haidacher, M., Bruckner, S., Kanitsar, A., Gröller, M.E.: Information-based transfer functions for multimodal visualization. In: Proceedings of Eurographics Workshop on Visual Computing for Biomedicine (2008)
10. Hsu, W.H., Mei, J., Correa, C.D., Ma, K.L.: Depicting time evolving flow with illustrative visualization techniques. In: Proceedings of International Conference on Arts and Technology (ArtsIT2009) (2009)
11. Johnson, C., Sanderson, A.: A next step: visualizing errors and uncertainty. IEEE Comput. Graphics Appl. **23**(5), 6–10 (2003)
12. Joshi, A., Rheingans, P.: Illustration-inspired techniques for visualizing time-varying data. In: Proceedings of IEEE Visualization, pp. 679–686 (2005)
13. Laramee, R.S., Hauser, H., Doleisch, H., Post, F.H., Vrolijk, B., Weiskopf, D.: The state of the art in flow visualization: dense and texture-based techniques. Comput. Graphics Forum **23**(2), 203–221 (2004)
14. Lawrence, J., Arietta, S., Kazhdan, M., Lepagne, D., Ogan, C.: A user-assisted approach to visualizing multidimensional images. IEEE Trans. Vis. Comput. Graphics **17**(10), 1487–1498 (2011)
15. Lundstrom, C., Ljung, P., Persson, A., Ynnerman, A.: Uncertainty visualization in medical volume rendering using probabilistic animation. IEEE Trans. Vis. Comput. Graphics **13**(6), 1648–1655 (2007)
16. McLoughlin, A., Laramee, R.S., Peikert, R., Post, F.H., Chen, M.: Over two decades of integration-based, geometric flow visualization. Comput. Graphics Forum **29**(6), 1807–1829 (2010)
17. Nadeau, D.R.: Volume scene graphs. In: Proceedings of IEEE Symposium on Volume Visualization (2000)
18. Pang, A.T., Wittenbrink, C.M., Lodha, S.K.: Approaches to uncertainty visualization. Vis. Comput. **13**, 370–390 (1997)
19. Post, F.H., Vrolijk, B., Hauser, H., Laramee, R.S., Doleisch, H.: The state of the art in flow visualization: feature extraction and tracking. Comput. Graphics Forum **22**(4), 775–792 (2003)
20. Treavett, S., Chen, M.: Pen-and-ink rendering in volume visualization. In: Proceedings of IEEE Visualization, pp. 203–209 (2000)
21. Wang, S.M., Kaufman, A.: Volume sampled voxelization of geometric primitives. In: Proceedings of IEEE Symposium on Volume Visualization, pp. 78–84 (1993)
22. Winter, A.S., Chen, M.: vlib: a volume graphics API. In: Proceedings of Volume Graphics, pp. 133–147. Springer Wien New York (2001)
23. Woodring, J., Shen, H.W.: Multi-variate, time varying, and comparative visualization with contextual cues. IEEE Trans. Vis. Comput. Graphics **12**(5), 909–916 (2006)
24. Woodring, J., Wang, C., Shen, H.W.: High dimensional direct rendering of time-varying volumetric data. In: Proceedings of IEEE Visualization, pp. 417–424 (2003)

Chapter 13
Glyph-Based Multi-field Visualization

David H.S. Chung, Robert S. Laramee, Johannes Kehrer and Helwig Hauser

Abstract In this chapter, we present a state of the art on glyph-based visualization techniques that address the complex challenges of multi-field visualization. Glyphs are discrete parametrized visualization objects that encode multiple data values based on appearance (i.e., visual channels) such as size, shape, color, and opacity, and are effective for conveying multiple fields of data simultaneously. We provide a categorization of these techniques with the aim for an informative overview of recent literature. Our categorization is based on visual channels utilized by the glyph for mapping each data attribute, and the spatial dimensionality of the glyph-based visualization. We also discuss critical design aspects of glyph-based visualization to deal with the perceptual challenges inherent with this approach.

13.1 Introduction

The visualization of data that are given as fields of values is a classical topic in visualization research. A substantial amount of relevant work has been done, offering a wealth of well-proven techniques for revealing insight into such data fields. When visualizing multiple fields of data that co-exist with respect to a joint domain of reference, additional challenges are faced. On the one hand, there is a *technological challenge* of how to realize a visualization mapping that can reveal multiple fields of data at a time. On the other hand, there is a *perceptual challenge* of how easy it is to understand and correctly interpret such a visualization.

D.H.S. Chung (✉) · R.S. Laramee
Swansea University, Swansea, UK
e-mail: d.h.s.Chung@swansea.ac.uk

J. Kehrer · H. Hauser
University of Bergen, Bergen, Norway
e-mail: helwig.hauser@uib.no

C.D. Hansen et al. (eds.), *Scientific Visualization*, Mathematics and Visualization,
DOI 10.1007/978-1-4471-6497-5_13

Glyph-based visualization is one possible approach to realize such a visualization of multi-field data (and other chapters of this book part describe alternative approaches). A parameterized visualization object is considered—called a *glyph* (or sometimes also an icon)—such that certain specifics with respect to its form, e.g., its shape, color, size/orientation, texture, etc., are given according to data values which this glyph should represent. A glyph-based visualization is then created by arranging a certain number of these glyphs across the domain of reference (these could be just a few, or just one, or many, even so many that they merge into a dense visualization) such that every glyph becomes a visualization of the data at (or nearby) the location where the glyph is placed.

Glyph-based visualization approaches span a certain spectrum from, for example, dense arrangements of relatively simple shapes (stick figures [17] would be an example) to individual instances of complex glyphs that reveal a lot of information (but only for few, selected places)—the local flow probe [13] would be an example for this type of a glyph-based data visualization. Glyph-based visualization approaches also vary with respect to whether they are constructed in a 2D or 3D visualization space. We think that it also makes sense to consider glyph-based visualization approaches, which are based on the placement of glyphs on surfaces within 3D (called 2.5D in the following). Additionally, we can differentiate visualization solutions according to which form aspects are varied according to the data, and how many different values a glyph eventually represents (usually this number is not too large, often 2–4, but then also examples exist where dozens of values are represented).

A property of all glyph-based visualization approaches is that a discrete visualization is created (instead of a continuous representation like a color map)—only at certain locations across the domain individual glyphs are instantiated to represent the data. This means that this approach is only suitable, when it is possible to assume a certain minimal degree of continuity of the data such that a mental reconstruction of the data, in particular also in the space between the glyphs, is at least principally possible. In scientific visualization, this often is possible, making glyph-based visualization particularly interesting for this particular field of application. Alternatively, a glyph-based visualization also makes sense for discrete data, if a one-to-one relation between every instance of the data and the glyphs is established.

In the following, we first review a selection of techniques that have been proposed for glyph-based data visualization. Then, we continue with a discussion of critical design aspects of glyph-based visualization, not at the least oriented at opportunities to deal with the perceptual challenge that is inherently associated with this form of visualization approach.

13.2 State-of-the-Art

This section presents a selection of important papers with a focus on glyph-based multi-field visualization. A categorization is given based on the visual channels such as color, shape, size, texture and opacity occupied by the glyph in requirement for

Table 13.1 Table illustrating a classification of multi-variate glyph-based visualization techniques based on the visualization dimensionality and the visual channels required to depict the data set

Visual channel	Visualization dimensionality		
	2D	2.5D	3D
Color	[5] [11] [22]	[3] [6] [16]	[21] [12] [9] [2] [10] [15] [8]
Shape	[8]		[1] [13] [21] [9] [7] [10] [15]
Size	[25] [22] [20]	[3] [16]	[21] [9] [2] [15]
Texture	[22]	[3] [6]	
Opacity	[11] [22]		[15]

mapping each data attribute. We further cluster the techniques with respect to the spatial dimensionality of the visualization e.g., 2D, 2.5D and 3D. Texture can be subjective in terms of glyph-based classification, however, we find that it is very relevant in the research of multi-field. The following work can be acknowledged without the use of this classification, but we include this in the table for completeness.

13.2.1 Spatial Dimensionality: 2D

A common technique for representing multi-field data is to overlay multiple visualizations onto a single image. Kirby et al. [11] stochastically arrange multiple visualization layers to minimize overlap. Given a permutation of layers, a user-specified importance value is attached to each visualization of increasing weights in order to provide greater emphasis to higher layers. Visual cues such as color and opacity indicate regions and layers of importance (e.g., Rate of strain tensor example emphasized the velocity more by using black arrows). This method enables the simultaneous depiction of 6–9 data attributes, in which the authors apply to a simulated 2D flow field past a cylinder at different reynolds number. The example shows the visualization of velocity, vorticity, rate of strain tensor, turbulent charge and turbulent current.

Visualizing Multiple Fields on the Same Surface by Taylor [22] provides an overview of successful and unsuccessful techniques for visualizing multiple scalar fields on the same surface. The author first hypothesizes that the largest number of data sets that can be displayed by mapping each field to the following: a unique surface characteristic, applying a different visualization technique to each scalar field or by using textures/glyphs whose features depend on the data sets. This framework is limited to visualizing up to four scalar fields. The author then describes two techniques that prove effective for visualizing multiple scalar fields, (1) *data-driven spots (DDS)*—using different spots of various intensities and heights to visualize each data set, and (2) *oriented slivers*—using sliver like glyphs of different orientations that are unique to each data set along with various blending.

13.2.2 Spatial Dimensionality: 2.5D

A Scientific Visualization Synthesizer by Crawfis and Allison [3] introduces a novel approach for visualizing multiple scientific data sets using texture mapping and raster operations. The authors present an interactive programming framework that enables users to overlay different data sets by defining raster functions/operations. Using a generated synthetic data, the author presents a method for reducing the visual clutter by mapping color to a height field and using a bump map to represent the vector plots and contour plots. The final texture is mapped onto a 3D surface.

Peng et al. [16] describes an automatic vector field clustering algorithm and presents visualization techniques that incorporate statistical-based multi-variate glyphs. In summary, the authors clustering algorithm is given by: (1) derive a mesh resolution value for each vertex, (2) encode vector and mesh resolution values into R, G, B and α in image space. Clusters naturally form in this space based on pixel values. (3) The clusters are merged depending on a similarity value derived using Euclidean distance, mesh resolution, average velocity magnitude and velocity direction. Several clustering visualizations are given, using $|v|$-range glyph that depicts the local minimum and maximum vector, and a θ-range glyph that shows the variance of vector field direction along with the average velocity direction and magnitude. Other visualization options include streamlets that are traced from the cluster centre, and color coding with mean velocity. The authors demonstrate their clustering results on a series of synthetic and complex, real-world CFD meshes.

13.2.3 Spatial Dimensionality: 3D

Geometric shapes are often used to represent multiple data values. Superquadrics and Angle-Preserving Transformations by Barr [1] introduces such an approach for creating and simulating three-dimensional scenes. The author defines a mathematical framework used to explicitly define a family of geometric primitives (superquadrics) from which their position, size, and surface curvature can be altered by modifying a set of different parameters. Example glyphs include a torus, star-shape, ellipsoid, hyperboloid or toroid. In addition, the paper describes a group of invertible transforms developed to bend and twist mathematical objects in three dimensions into a new form where shape properties such as volume, surface area and arc length is conserved.

De Leeuw and van Wijk [13] present an interactive probe-glyph for visualizing multiple flow characteristics in a small region. In particular, the authors focus on visualizing six components: velocity, curvature, shear, acceleration, torsion and convergence. The construction of the glyph is given by, (1) a curved vector arrow where the length and direction represents the velocity, and the arc shape is mapped to the curvature, (2) a membrane perpendicular to the flow where its displacement to the center is mapped to acceleration, (3) candy stripes on the surface of the velocity arrow illustrates the amount of torsion, (4) a ring describes the plane perpendicular to the flow over time (shear-plane), and finally (5) the convergence and divergence of the flow is mapped to a "lens" or osculating paraboloid. Placement of such probes are interactively placed by users along a streamline to show local features in more detail.

Data Visualization Using Automatic, Perceptually-Motivated Shapes by Shaw et al. [21] describes an interactive glyph-based framework for visualizing multi-dimensional data through the use of superquadrics. The author uses the set of superquadrics defined by Barr [1] and describes a method for mapping data attributes appropriately to shape properties such that visual cues effectively convey data dimensionality without depreciating the cognition of global data patterns. They map in decreasing order of data importance, values to location, size, color and shape (of which two dimensions are encoded by shape). Using superellipsoids as an example, the authors applied their framework on two different data sets.

Superquadric Tensor Glyphs by Kindlmann [9] introduces a novel approach of visualizing tensor fields using superquadric glyphs. Superquadric tensor glyphs address the problems of asymmetry and ambiguity prone in previous techniques (e.g. cuboids and ellipsoids). The author provides an explicit and implicit parameterization of the primitives defined by Barr [1] that uses geometric anisotropy metrics c_l, c_p, c_s to quantify the certainty of a tensor based on shape, and a user-controlled edge sharpness parameter γ. The parametrization forms a barycentric triangular domain of tensor glyphs that change in shape, flatness and orientation under different tensor eigen vectors. A subset of the family of superquadrics is chosen and applied towards visualizing a DT-MRI tensor field which is then compared against an equivalent ellipsoid visualization.

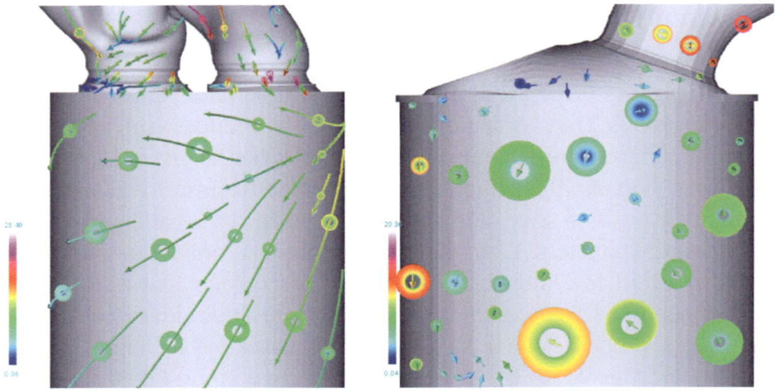

Fig. 13.1 Visualization of the flow in an engine using composite glyphs that depict the range of vector magnitude and direction in each cluster by Peng et al. [16]

13.3 Critical Design Aspects of Glyph-Based Visualization

It was a wide-spread opinion for a long time that "just" knowing the basic principles of glyph-based visualization would suffice to its successful usage. More recently, however, it has been understood that only well designed glyphs, where different glyph properties are carefully chosen and combined, are actually useful. In this section, we discuss critical design aspects and guidelines for glyph-based visualization.

In the context of information visualization, Ward [24] discusses glyph placement strategies such as data- or structure-driven placement. Ropinski and Preim [19] propose a perception-based glyph taxonomy for medical visualization. The authors categorize glyphs according to (1) preattentive visual stimuli such as glyph shape, color and placement, and (2) attentive visual processing, which is mainly related to the interactive exploration phase (e.g., changing the position or parameter mapping of a glyph). Additional usage guidelines are proposed, for instance, that parameter mappings should focus the user's attention and emphasize important variates in the visualization. Also, glyph shapes should be unambiguous when viewed from different viewing directions. Kindlmann [9], for example, uses superquadric glyph shapes that fulfill the latter criterion.

Inspired by the work of Ropinski and Preim, Lie et al. [14] propose further guidelines for glyph-based 3D visualization. Aligned with the visualization pipeline [4], the task of creating a glyph-based 3D visualization is divided into three stages as shown in Fig. 13.2: (1) during *data mapping*, the data variates are remapped (to achieve, for example, some contrast enhancement) and mapped to the different glyph properties; (2) *glyph instantiation* creates the individual glyphs, properly arranged across the domain; and (3) during *rendering*, the glyphs are placed in the visualization, where one has to cope with issues such as visual cluttering or occlusion. In the following, we discuss critical design aspects for each of these steps.

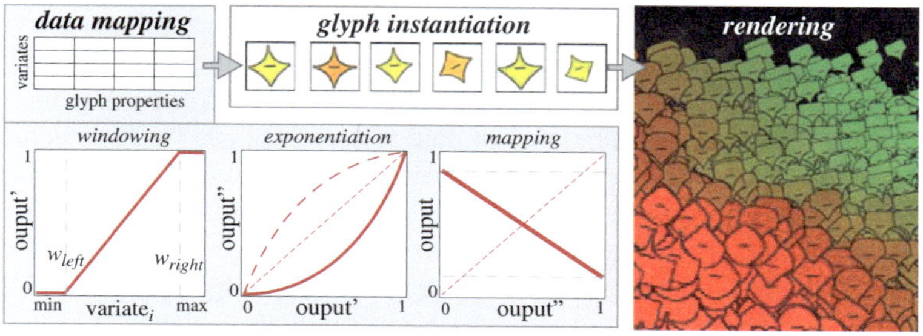

Fig. 13.2 Each data variate is subject to three stages of data mapping: windowing, exponentiation and mapping. The values are mapped to different glyph properties and used to instantiate the individual glyphs. Finally, the glyphs are rendered in their spatial context

Similar to Ward [24], Lie et al. consider it useful that the glyphs expect normalized input from the depicted data variates such as values in the range [0, 1]. During data mapping, the authors identify three consecutive steps. First, the data values within a user-selected range $[w_{left}, w_{right}]$ are mapped to the unit interval. Values outside this range are clamped to 0 or 1, respectively. This allows to enhance the contrast of the visualization with respect to a range of interest (sometimes called windowing). A natural default choice for this step would be a linear map between $[w_{left}, w_{right}]$ and [0, 1], but also other forms of mapping could be considered (for example, a ranking-based or discontinuous mapping). After the windowing, an optional exponential mapping $e(x) = x^{\gamma}$ can be applied in order to further enhance the contrast on the one or the other end of the spectrum. Finally, a third mapping step enables the user to restrict or transform the output range that should be depicted by a glyph property. Here, also semantics of the data variates can be considered (compare to the usage guidelines of Ropinski and Preim [19]). Using a reverse mapping, for instance, smaller data values that are possibly more important can be represented in an enhanced style while larger values are deemphasized.

Several considerations are important for the instantiation of individual glyphs. When using a 3D glyph shape, one has to account for possible distortions introduced when viewing the glyph from a different point of view [9]. In order to avoid this problem, Lie et al. suggest to use 2D billboard glyphs instead.[1] In certain scenarios, however, it makes sense to use 3D glyphs, for example, when depicting a flow field via arrow glyphs. Another challenge in glyph design is the *orthogonality* of the different glyph components, meaning that it should be possible to perceive each visual cue individually (or to mentally reconstruct them as suggested by Preim and Ropinski [19]). When representing a data variate by glyph shape, for example, this affects the area (size) of the glyph as well. Accordingly, such effects should be *normalized* against each other, for instance, by altering the overall glyph size in order to compensate for implicit changes of the glyph shape.

[1] A billboard is a planar structure placed in a 3D scene, which automatically adjusts its orientation such that it always faces the observer.

However, it is not always easy to design a glyph-based visualization such that the different data-to-property mappings are independent and do not influence each other (the interpretation of shape details, for example, is usually influenced by the size of the glyph). In this context, the number of data variates that can be depicted must be seen in relation to the available screen resolution. Large and complex glyphs such as the local probe [13] can be used when only a few data points need to be visualized. If many glyphs should be displayed in a dense manner, however, a more simple glyph may be desirable [10]. Another design guideline is the usage of *redundancies*, for instance, to use symmetries that ease the reconstruction of occluded parts of the glyph. Important properties can, moreover, be mapped to multiple glyph properties in order to reduce the risk of information loss.

Important aspects when rendering many glyphs in a dense 3D context are depth perception, occlusion, and visual cluttering. In cases where many glyphs overlap, *halos* can help to enhance the depth perception and to distinguish individual glyphs (compare to Piringer et al. [18]). For improving the depth perception for non-overlapping glyphs a special color map (called *chroma depth* [23]) can be used to represent depth. Finally, appropriate glyph placement [19, 24], interactive slicing, or filtering via brushing are strategies for dealing with occlusion and cluttering issues.

References

1. Barr, A.H.: Superquadrics and angle-preserving transformations. IEEE Comput. Graph. Appl. **1**(1), 11–23 (1981)
2. Chlan, E.B., Rheingans, P.: Multivariate glyphs for multi-object clusters. In: INFOVIS, p. 19 (2005)
3. Crawfis, R., Allison, M.J.: A scientific visualization synthesizer. In: Proceedings of the IEEE visualization, pp. 262–267 (1991)
4. Hauser, H., Schumann, H.: Visualization pipeline. In: Liu, L., Özsu, M.T. (eds.) Encyclopedia of database systems, pp. 3414–3416. Springer, Berlin (2009)
5. Healey, C., Booth, K., Enns, J.: High-speed visual estimation using preattentive processing. ACM Trans. Comput.-Human Interact. **3**, 107–135 (1996)
6. Healey, C.G., Enns, J.T.: Large datasets at a glance: combining textures and colors in scientific visualization. IEEE Trans. Vis. Comput. Graph. **5**(2), 145–167 (1999)
7. Jankun-Kelly, T., Mehta, K.: Superellipsoid-based, real symmetric traceless tensor glyphs motivated by nematic liquid crystal alignment visualization. IEEE Trans. Vis. Comput. Graph. **12**(5), 1197–1204 (2006)
8. Kehrer, J., Muigg, P., Doleisch, H., Hauser, H.: Interactive visual analysis of heterogeneous scientific data across an interface. IEEE Trans. Vis. Comput. Graph. **17**(7), 934–946 (2011)
9. Kindlmann, G.: Superquadric tensor glyphs. In: Deussen, O., Hansen, C., Keim, D., Saupe, D. (eds.) Joint Eurographics—IEEE TCVG symposium on visualization, Konstanz, Germany, pp. 147–154 (2004)
10. Kindlmann, G., Westin, C.F.: Diffusion tensor visualization with glyph packing. IEEE Trans. Vis. Comput. Graph. **12**(5), 1329–1336 (2006)
11. Kirby, R.M., Marmanis, H., Laidlaw, D.H.: Visualizing multivalued data from 2D incompressible flows using concepts from painting. In: Proceedings of the IEEE visualization, pp. 333–340 (1999)
12. Kraus, M., Ertl, T.: Interactive data exploration with customized glyphs. In: WSCG (Posters), pp. 20–23 (2001)

13. de Leeuw, W.C., van Wijk, J.J.: A probe for local flow field visualization. In: Proceedings of the IEEE visualization conference (Vis'93), pp. 39–45 (1993)
14. Lie, A.E., Kehrer, J., Hauser, H.: Critical design and realization aspects of glyph-based 3D data visualization. In: Proceedings of the spring conference on computer graphics (SCCG 2009), pp. 27–34 (2009)
15. Meyer-Spradow, J., Stegger, L., Döring, C., Ropinski, T., Hinrichs, K.: Glyph-based SPECT visualization for the diagnosis of coronary artery disease. IEEE Trans. Vis. Comput. Graph. **14**(6), 1499–1506 (2008)
16. Peng, Z., Grundy, E., Laramee, R., Chen, G., Croft, N.: Mesh-driven vector field clustering and visualization: an image-based approach. IEEE Trans. Vis. Comput. Graph. (TVCG), 283–298 (2011)
17. Pickett, R.M., Grinstein, G.G.: Iconographic displays for visualizing multidimensional data. In: IEEE International Conference on Systems, Man, and Cybernetics, vol. 1, pp. 514–519 (1988)
18. Piringer, H., Kosara, R., Hauser, H.: Interactive focus+context visualization with linked 2D/3D scatterplots. In: Proceedings on coordinated and multiple views in exploratory visualization (CMV 2004), pp. 49–60 (2004)
19. Ropinski, T., Preim, B.: Taxonomy and usage guidelines for glyph-based medical visualization. In: Proceedings on simulation and visualization, pp. 121–138 (2008)
20. Sanyal, J., Zhang, S., Dyer, J., Mercer, A., Amburn, P., Moorhead, R.: Noodles: a tool for visualization of numerical weather model ensemble uncertainty. IEEE Trans. Vis. Comput. Graph. **16**(6), 1421–1430 (2010)
21. Shaw, C.D., Ebert, D.S., Kukla, J.M., Zwa, A., Soboroff, I., Roberts, D.A.: Data visualization using automatic, perceptually-motivated shapes. In: SPIE 3298, visual data exploration and analysis (1998)
22. Taylor, R.: Visualizing multiple fields on the same surface. IEEE Comput. Graph. Appl. **22**(3), 6–10 (2002)
23. Toutin, T.: Qualitative aspects of chromo-stereoscopy for depth-perception. Photogram. Eng. Remote Sens. **63**(2), 193–203 (1997)
24. Ward, M.O.: A taxonomy of glyph placement strategies for multidimensional data visualization. Info. Vis. **1**(3–4), 194–210 (2002)
25. Wittenbrink, C.M., Pang, A., Lodha, S.K.: Glyphs for visualizing uncertainty in vector fields. IEEE Trans. Vis. Comput. Graph. **2**(3), 266–279 (1996)

Chapter 14
Derived Fields

Eugene Zhang and Vijay Natarajan

Abstract This chapter reviews various methods for multifield visualization that are based on the notion of derived fields. The derived fields are categorized based on properties like the number and type of input fields. Mathematical properties, algorithms, and applications are discussed for each derived field. Correlation and alignment measures are examined for a set of homogeneous fields, including pairwise similarity/dissimilarity measurements. Multifield analysis is also discussed in the context of input fields being the components of the decomposition of another field, possibly of a different type. Finally, research challenges are discussed in the context of the design of multifield analysis and visualization methods based on the concept of derived fields.

14.1 Introduction

In this chapter we consider the notion of *derived fields* in the context of multifield analysis and visualization. We discuss a categorization based on the number of fields studied, their homogeneity, and the type of relationship between the input fields that is captured by the derived field.

First, given a set of at least two fields of the same type, it is possible to define pairwise similarity and dissimilarity for any two of the fields as well as the global alignment and dependency of the fields considered as a whole. These quantities,

E. Zhang (✉)
School of Electrical Engineering and Computer Science,2111 Kelley Engineering Center, Oregon State University, Corvallis OR 97331, USA
e-mail: zhange@eecs.oregonstate.edu

V. Natarajan
Department of Computer Science and Automation, Supercomputer Education and Research Center, Indian Institute of Science, Bangalore 560012, India
e-mail: vijayn@csa.iisc.ernet.in

C.D. Hansen et al. (eds.), *Scientific Visualization*, Mathematics and Visualization,
DOI 10.1007/978-1-4471-6497-5_14

namely pairwise similarity/dissimilarity, global alignment, and dependency, are derived quantities that can provide critical information on the input fields. For example, when tracking features in fluid flow datasets, it is often desirable to measure the alignment of a sequence of consecutive time-slices in the data. In Sect. 14.2 we will review existing work on pairwise derived fields, i.e., the number of input fields is two. In Sect. 14.3 we will consider global alignment and dependency measures for the case when there are more than two input fields.

Another scenario of derived fields in the context of multifield visualization is referred to as decomposition and componentization. In this case, this input may be considered as a single field. However, its key characteristics are revealed by a decomposition into multiple derived fields. The behavior of the input field can be better understood by studying each derived field in the decomposition as well as the interplay among them. An example of this is the well-known *Hodge-Helmholtz decomposition*, where an input vector field is decomposed into the sum of three vector fields: (1) *divergence-free*, (2) *curl-free*, and (3) *harmonic* vector fields. We will review techniques corresponding to this category in Sect. 14.4.

14.2 Pairwise Distances and Correlation Measures

A first step towards capturing the relationships between fields in multifield data is to consider pairwise interactions. In this context we discuss the use of distance measures, similarity measures, and local correlations between two fields.

14.2.1 Correlation Measures

The correlation coefficient is a standard and popular statistical measure used to determine if two sets of real values are linearly related by comparing their deviations from the respective mean values [4, Chap. 8]. When two scalar functions are sampled at discrete points, the correlation coefficient is computed as

$$\rho = \frac{\sum_{i=1}^{n}(x_i - \bar{x}) \cdot (y_i - \bar{y})}{\sqrt{\left(\sum_{i=1}^{n}(x_i - \bar{x})^2\right) \cdot \left(\sum_{i=1}^{n}(y_i - \bar{y})^2\right)}}$$

where x_i, y_i are the corresponding values of the two functions and $\bar{x}$, $\bar{y}$ are the mean values of the two functions. Two scalar functions have a high correlation coefficient if they deviate consistently from their respective mean values i.e., if one function takes a value close to its mean then so does the other function at the same point in the domain. Note that the correlation coefficient as defined above is a global measure. However, in the context of two time-varying fields, the correlation coefficient can be computed at each point resulting in a derived field over the domain. This field captures the linear relationship between the two time-series data at each point.

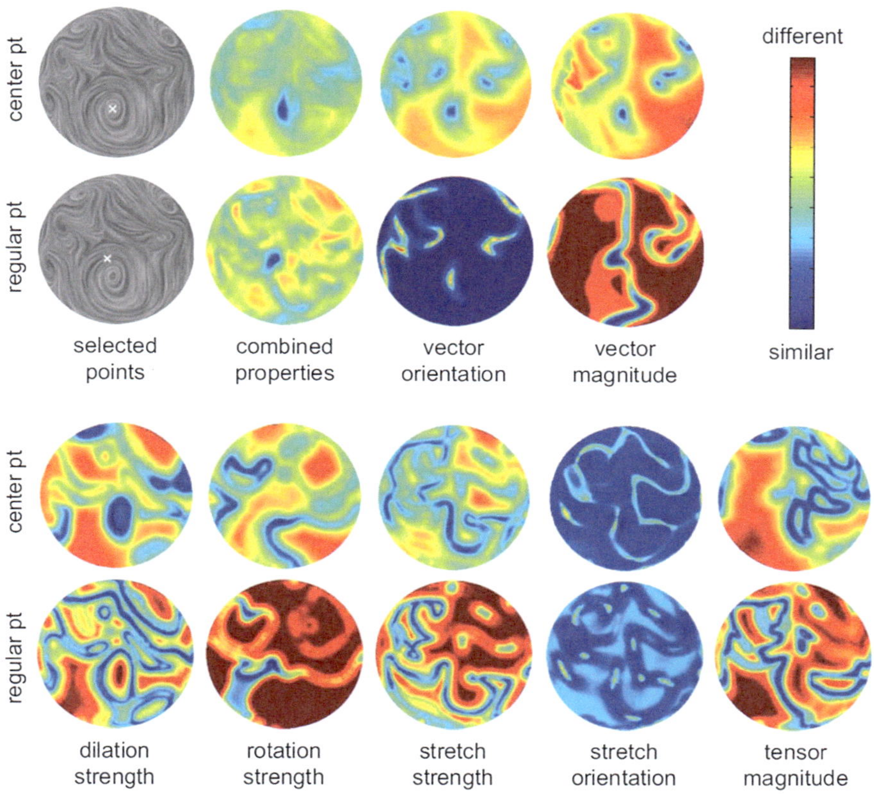

Fig. 14.1 Tracking flow features by treating a stack of vector fields as spin images and using vector and tensor field measurements [27]. © IEEE Reprinted, with permission, from IEEE Conference on Computer Vision and Pattern Recognition, 2009

Xu et al. [27] track features in a flow dataset by treating the 3D flow field as a stack of 2D vector fields. Given a point in one slice in the stack, points in other slices can be correlated to this point by treating the 2D vector fields as spin images. This idea is then extended to using the velocity gradient tensor fields of the vector fields, leading to more efficient feature matching (Fig. 14.1).

14.2.2 Gradient Comparison

Correlation between a pair of scalar fields has also been defined based on the gradients. The use of the gradients allows the incorporation of the spatial locality into the correlation computation. We now describe two derived fields that compare gradients and discuss their relative merits.

14.2.2.1 Definitions and Properties

Sauber et al. [22] introduce a gradient similarity measure (GSIM) between two gradient fields ∇f_i and ∇f_j that assumes high values when the gradients have similar magnitude and direction. The measure is defined at each point as

$$s(\nabla f_i, \nabla f_j) = (s_d(\nabla f_i, \nabla f_j) \cdot s_m(\nabla f_i, \nabla f_j))^r, \text{ where}$$

$$s_d(\nabla f_i, \nabla f_j) = \left(\frac{\nabla f_i^T \nabla f_j}{\| \nabla f_i \| \cdot \| \nabla f_j \|} \right)^2, \text{ and}$$

$$s_m(\nabla f_i, \nabla f_j) = 4 \frac{\| \nabla f_i \| \cdot \| \nabla f_j \|}{(\| \nabla f_i \| + \| \nabla f_j \|)^2}.$$

In the above expression, s_d represents the similarity in gradient direction, s_m represents the similarity in gradient magnitude, and the exponent r is a parameter that determines the sensitivity of the measure. The fields are normalized to have a common range before computing gradients.

Edelsbrunner et al. [3] define a derived field that assumes high values when the gradients are orthogonal to each other. The derived field, denoted by κ, is essentially the length of the cross product between the two gradients vectors.

While the two fields are different in the sense that GSIM measures similarity whereas κ measures dissimilarity, both derived fields have many similarities besides the fact that both are based on gradient comparison. Both GSIM and κ depend on the scale and length of the gradients, are pointwise comparisons, and do not distinguish between positive and negative correlation. The similarities imply that both techniques are applicable to the same data sets. Gosink et al. [5] also compute correlation between gradient fields to study the interactions between the different pairs of scalar fields in multifield data. The inner product of the gradients of two fields of interest is computed over principle level sets of a third field. They employ this approach to study combustion in methane and hydrogen. The correlation field proposed by Gosink et al. is similar to GSIM described above with the difference being the domain over which the correlation field is computed.

14.2.2.2 Applications

Figure 14.2 shows the derived field GSIM for two pairs of quantities measured in the simulation of hurricane Isabel. The transfer function assigns non-zero opacity to regions with values of GSIM greater than 0.9. Patterns in the derived field can help in the analysis of various phenomena like fronts in the hurricane.

A visualization of κ helps in the study of the different phases in a combustion simulation as shown in Fig. 14.3. Three time steps are shown: the ignition, burning, and the final phase. The flame front is tracked by regions with large values of κ computed for the scalar field pair $prog$ and H_2, which represent the progress of

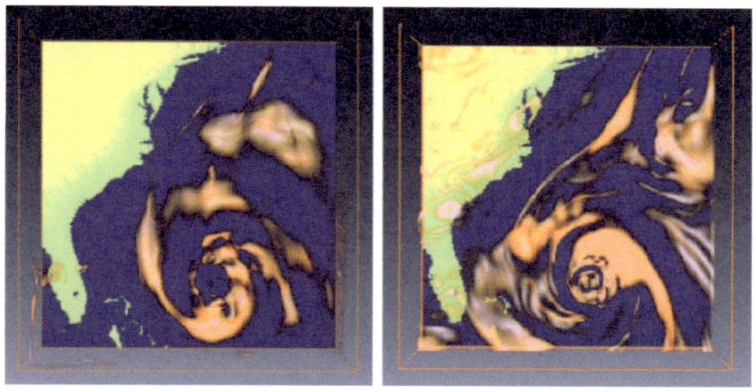

Fig. 14.2 Gradient similarity measure (GSIM) computed for two pairs of scale fields: precipitation versus vapour (*left*) and vapour versus temperature. Image courtesy of Sauber et al. [22].

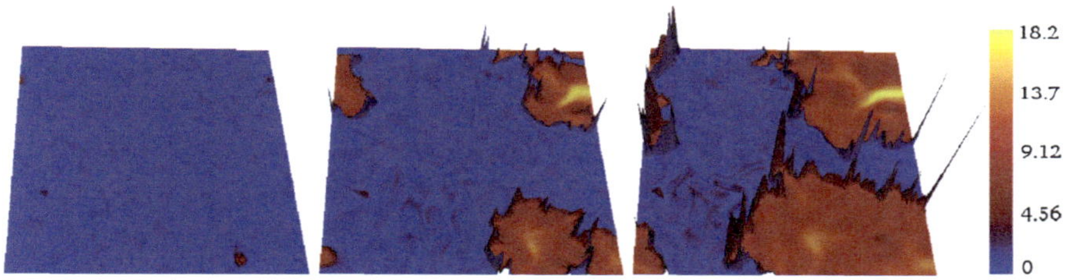

Fig. 14.3 Local comparison of two scalar fields $prog$ and H_2 from a combustion simulation. The derived field κ, which compares the gradients of the two scalar fields, is shown using a terrain map and $prog$ is mapped to color. From *left* to *right*: ignition phase, burning phase, and the end of combustion. The fronts of the flames are tracked by a region with higher values of κ. This region is represented by the peaks that enclose the burnt region. Image courtesy of Edelsbrunner et al. [3]

combustion and fuel concentration, respectively. The higher peaks in the terrain correspond to sections of the flame front that are progressing faster. They also discuss an application to the study of a protein–protein complex in structural biology. A protein–protein complex consists of two or more proteins docked in a stable conformation. For example, the barnase-barstar complex (1BRS) consists of two proteins. The electrostatic potential defined by barnase (N) and barstar (S) individually in their docked conformation and the potential defined by the complex are available as scalar fields sampled over the space, namely f_N, f_S, and f_{1BRS}. Regions where κ between f_N and f_S is high correspond to salt bridges/strong hydrogen bonds. Figure 14.4 shows a visualization of κ. The colored dots in the figure indicate high values of κ values, namely those in the range [0.002, 0.0207] and are mapped from blue to red. The dots with values lower than 0.002 are not displayed. The gold lines indicate the hydrogen bonds corresponding to those regions of space.

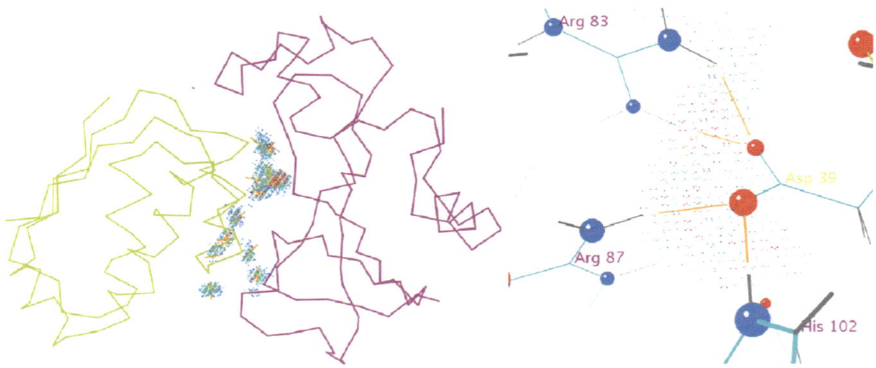

Fig. 14.4 Visualization of the derived field κ between electrostatic potentials defined by barnase and barstar in the complex 1BRS. *Left* an overview of the regions with high values of κ in the complex. The proteins are shown as alpha-carbon traces, with barnase in magenta and barstar in yellow. *Right* a closeup of a hydrogen bond cluster. `Asp 39` of barstar hydrogen bonds with `Arg 87`, `Arg 83`, and `His 102` of barnase. All four residues are highly important in the interaction between barnase and barstar. Image courtesy of Edelsbrunner et al. [3]

14.3 Alignment and Dependency Measures

We now discuss derived fields that capture the variation or dependency between multiple (greater than two) fields.

14.3.1 Local Gradient-Based Comparison Measures

The gradient comparison measures discussed in the previous section also extends to multiple scalar fields. We now describe these extensions, their properties and applications.

14.3.1.1 Definitions and Properties

The gradient similarity measure GSIM is extended to k gradient fields by computing the minimum gradient pair similarity

$$C_k = \min\{s(\nabla f_i, \nabla f_j) \mid 1 \leq i < j \leq k\}.$$

The measure assumes low values if the gradient directions are equally distributed in the domain. Given k fields, GSIM can be computed for all possible subsets of fields. The size of this set grows exponentially with the number of fields and hence it is impractical to compute and analyze GSIM for all subsets of fields.

Sauber et al. [22] address this issue by introducing the multifield-graph. Nodes of this graph correspond to each subset of input fields and are displayed with icons that graphically represent the similarity between the fields. Nodes are laid out in layers corresponding to the number of fields in the subset. Two nodes in adjacent layers are connected by an edge if the fields in lower layer node are also compared in the upper layer node. The correlation/similarity and the size of domain with high correlation/similarity are represented by the size and color of a disk displayed within each node. A selective display of nodes enables focusing on nodes that represent high correlations.

The derived field κ also extends to multiple fields. Given k scalar fields, κ is defined as the norm of the wedge product between the 1-forms $\mathrm{d}f_i$,

$$\kappa = \| \mathrm{d}f_1 \wedge \mathrm{d}f_2 \wedge \cdots \wedge \mathrm{d}f_k \| .$$

Each one form $\mathrm{d}f_i$ corresponds to the gradient ∇f_i. The wedge product is a natural extension of the cross product of two gradient vectors and represents the k-dimensional volume of the parallelpiped spanned by the k gradients [3]. While the comparison measure κ does not satisfy the triangle inequality, it satisfies a number of useful algebraic properties.

1. Symmetry: $\kappa(\ldots, f_i, \ldots, f_j, \ldots) = \kappa(\ldots, f_j, \ldots, f_i, \ldots)$.
2. Degeneracy: $\kappa(F) = 0$ if $\mathrm{d}f_i = \mathrm{d}f_j$ for $1 \leq i \neq j \leq k$.
3. Scaling: $\kappa(\alpha f_1 + \beta, f_2, \ldots, f_k) = |\alpha| \cdot \kappa(f_1, f_2, \ldots, f_k)$, with $\alpha, \beta \in \mathbb{R}$.
4. Sub-additivity: $\kappa(f_1 + g_1, f_2, \ldots, f_k) \leq \kappa(f_1, f_2, \ldots, f_k) + \kappa(g_1, f_2, \ldots, f_k)$.
5. Sub-multiplicativity: $\frac{\kappa(f_1, \ldots, f_i, f_{i+1}, \ldots, f_k)}{\mathrm{vol}(\mathbb{M})} \leq \kappa(f_1, \ldots, f_i) \cdot \kappa(f_{i+1}, \ldots, f_k)$.

14.3.1.2 Computation and Applications

In practice, the scalar fields are measured at discrete points in the domain and linearly interpolated within elements in a triangulation of the manifold. In such a setting, GSIM and κ can be computed in a loop over the d-simplices in the triangulation. Since all functions are linear over a d-simplex, their gradients/differentials are constant within each mesh element. The norm of the k-form is evaluated at a point within the d-simplex directly from the formula and weighted by the volume of the d-simplex.

In a typical application of the multifield-graph, the user selects a particular node using a visual interface and analyzes the derived field corresponding to that particular node. Figure 14.5 shows the multifield-graph for the hurricane Isabel data set with six scalar fields. Selected nodes are displayed within three of the five possible layers. The extension of κ to k fields is directly visualized for two- and three-dimensional domains to study the relationship between the fields.

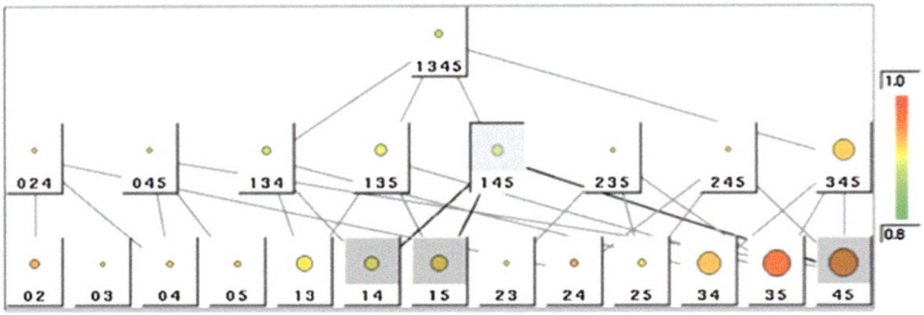

Fig. 14.5 Multifield-graph computed for the hurricane Isabel data set. The size and color of the disks represent the degree of correlation/similarity between the fields. Image courtesy of Sauber et al. [22]. © IEEE Reprinted, with permission, from IEEE Transactions on Visualization and Computer Graphics 12(5)

14.3.2 Local Statistical Complexity

Multifield data have also been studied using statistical and information theoretic methods. Jänicke et al. [10] adapt the notion of local statistical complexity to the context of time-varying fields and apply it to study data available from PDE simulations [11]. The local statistical complexity is a measure of the amount of information required from the past to predict the field in the current time step a specific point. It is computed using the notion of entropy and mutual information as a time-varying scalar field. Consider a time-varying field. All points that could possibly influence the value of the field at a point p are arranged into a light cone. The size of the region of influence increases by one for each time step away from p. A light cone (l^+) into the future time steps is also considered, similar to the light cone in the past (l^-), for the computation. A conditional distribution $P(l^+|l^-)$ can be defined on the light cones. The local statistical complexity is computed as the mutual information between the distribution represented by a particular light cone and the equivalence class of past light cones that have similar conditional distribution. Jänicke et al. [10, 11] describe efficient algorithms to compute the local statistical complexity. Features are identified as complex if the probability that they occur again is low. They demonstrate applications of this derived field to a wide range of data from diffusion, flow, and weather simulations.

14.3.3 Multifield Comparison Measure

Nagaraj et al. introduced a gradient-based comparison measure for multiple scalar fields [14]. The measure is defined as the norm of a matrix comprising the gradient vectors of the different functions. Let A be a $m \times n$ matrix of real numbers. The *norm* of the matrix A, denoted as $\|A\|$, is defined as

$$\|A\| = \max_{\|x\|=1,\, x \in \mathbb{R}^n} \|Ax\|,$$

where $\|x\|$ represents the Euclidean norm of vector x [9]. Let $F = \{f_1, f_2, f_3, \ldots, f_m\}$ be a set of smooth scalar fields defined on a manifold $\mathbb{M}$. The derivative at a point $p \in \mathbb{M}$ is written as a matrix of partial derivatives,

$$dF(p) = \begin{bmatrix} \frac{\partial f_1}{\partial x_1}(p) & \ldots & \frac{\partial f_1}{\partial x_n}(p) \\ \vdots & \ddots & \vdots \\ \frac{\partial f_m}{\partial x_1}(p) & \ldots & \frac{\partial f_m}{\partial x_n}(p) \end{bmatrix}$$

The *multifield comparison measure* η_p^F at point p is defined as the norm of the matrix $dF(p)$, $\eta_p^F = \|dF(p)\|$.

14.3.3.1 Properties and Computation

The measure η_p^F satisfies three important properties: symmetry, coordinate system independence and stability.

- **Symmetry**. The measure is independent of the permutation of the functions in F.
- **Coordinate system independence**. The norm of the matrix dF at a point p does not depend on the coordinate system used to represent p.
- **Stability**. A finite change in the functions results in a bounded change in the multifield comparison measure. The amount of change additionally depends on the size of the triangle.

Evaluating the multifield comparison measure at a point requires the solution to a maximization problem. Nagaraj et al. show that this computation can be reduced to the faster evaluation of the maximum eigenvalue of a positive semi-definite matrix $\Lambda = (dF(p))^T(dF(p))$

$$\begin{aligned} \eta_p^F &= \left(\max_{x \in \mathbb{R}^n, \|x\|=1} x^T \Lambda x \right)^{\frac{1}{2}} \\ &= \max\{\sqrt{\lambda} : \lambda \text{ is a diagonal element of } \Lambda\} \\ &= \max\{\sqrt{\lambda} : \lambda \text{ is an eigenvalue of } (dF(p))^T(dF(p))\}. \end{aligned}$$

The derivative matrix $dF(p)$ is constant within each mesh element if the scalar field is available as a sample and linearly interpolated within elements of a triangulation.

14.3.3.2 Applications

The multifield comparison measure has been applied to study various real-world data from weather modeling, climate simulations, and combustion simulations. In particular, it was used to study a simulation of the hurricane Isabel and the analysis of a global wind pattern data set.

Rainbands and fronts. A simulation of the hurricane Isabel that struck the Atlantic region in USA was performed on a physical area of 2139 km $\times$ 2004 km $\times$ 19.8 km over 48 simulated hours [26]. The data is available over a $600 \times 600 \times 600$ grid over 48 time steps. Among the multiple quantities computed, the scalar fields corresponding to pressure (Pf) and the horizontal wind velocity components (Uf and Vf) were considered in this experiment. Cloud structures associated with an area of rainfall, called rainbands, occur mainly at boundaries separating two masses of air of different densities and temperatures, called fronts. The leading edge of the cooler mass of air is called the cold front and the leading edge of a warm air mass is called the warm front. The turbulence of the horizontal wind velocity is high near rain bands. The fronts can be analyzed by computing the multifield comparison measure for the pair of 3D scalar fields Uf and Vf, where the 3D domain corresponds to the volume in the altitude range 1,500–5,800 m.

First, the multifield comparison measure is computed for the fields Uf and Vf in the 10th time step. A visualization of the measure clearly shows the two warm fronts and a cold front [14]. The warm front leads the cold front. This information about fronts cannot be extracted from the two functions individually. The multifield comparison measure is computed next for the fields Uf and Vf in the 40th hour of simulation. The warm front at the north disappears, see Fig. 14.6c, d. The previously leading warm front is overtaken by the cold front resulting in an occlusion.

Wind patterns. Prevailing winds blow in a dominant direction at a particular point and are affected by movements in the Earth's atmosphere. In regions of mid-latitudes, the winds blow from west to the east and are known as westerlies. The winds found in the tropics near the equator are easterlies or trade winds. Data from a climate simulation for a 50 year period between 1960 and 2009 is available over 600 time steps corresponding to each month [21]. Data within each time step is available on a 3D grid with resolution corresponding to $1° \times 1° \times 16$ plev (pressure elevations) on earth.

The wind velocity on the grid is a vector field. The matrix norm for 600 vector fields is computed by replacing the rows with the wind velocities. The norm η^F measures the variation of the wind velocities over the time period of the simulation, see Fig. 14.7b. Comparing with the wind patterns in Fig. 14.7a, we see that high values of η^F

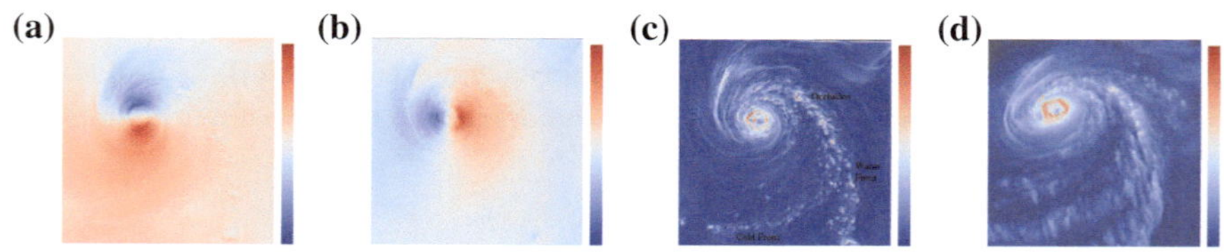

Fig. 14.6 Fronts in Hurricane Isabel at hour 40. **a** Volume rendering (*top view*) of horizontal wind speed Uf. **b** Volume rendering (*top view*) of horizontal wind speed Vf. **c** Volume rendering (*top view*) of multifield comparison measure η^F computed for Uf and Vf showing the rainbands at different fronts. The cold front leads the warm front resulting in an occlusion. **d** Volume rendering from a different viewpoint. Image courtesy of Nagaraj et al. [14]

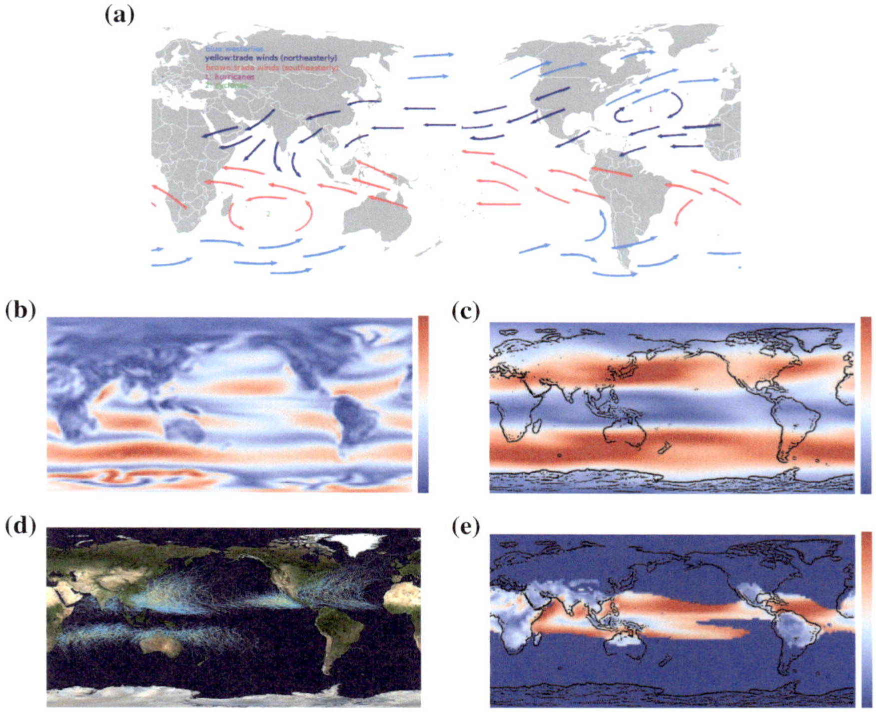

Fig. 14.7 Multifield comparison measure η^F computed for wind velocities over the years 1960–2009, where the comparison is over a set of six hundred 3D vector fields. **a** Map of world showing wind patterns (*source* Wikipedia). **b** Distribution of η^F over surface corresponding to pressure elevation 925 hPa. The dark red regions correspond to the wind patterns. **c** Distribution of η^F over surface corresponding to pressure elevation 300 hPa. The temperate regions exhibit higher values. **d** Storm track for the years 1985–2005 (*source* Wikipedia). **e** Distribution of η^F after removing regions with low mean temperature ($<27\,^{\circ}$C). *Red* regions correspond to the storm tracks. The world map is overlaid for clarity. Image courtesy of Nagaraj et al. [14]

correspond to the prevailing winds, particularly the westerlies found in the regions surrounding Antarctica, the region of hurricanes in Atlantic, the cyclone prone region between Madagascar and Australia, and the trade winds across the Atlantic sea traveling towards the Caribbean sea. The distribution of the comparison measure over the isobar for pressure level 300 hPa, which corresponds to approximately 30,000 feet above sea level, is shown in Fig. 14.7c. The comparison measure assumes high values over the temperate regions corresponding to the westerly jet. This is a semi-permanent feature of the mid-latitudes. Many regions in the tropics undergo a seasonal reversal of wind (called the monsoons). Lower values of the comparison measure over the tropics indicates unsteadiness and corresponds to a seasonal reversal in wind pattern over this part of the world.

Storm tracks. The regions over the ocean with warm temperatures (>27 °C) are susceptible to storms. Filtering out regions with lower temperatures and restricting the analysis to the months from June to November helps locate storm tracks. Regions shown in blue in Fig. 14.7e have been filtered out. The red regions match closely with the storm tracks shown in Fig. 14.7d. Even though the west coast of South America has trade winds, storms are particularly absent due to lower temperatures. The storm prevalent regions in the Indian, Atlantic, and Pacific oceans have high values of the comparison measure.

14.4 Decomposition and Componentization

In this section we examine a different situation in which multiple fields can arise as the components of a decomposed field.

14.4.1 Hodge Decomposition

A classical example of this is the Hodge-Helmholtz decomposition [18, 25] of a vector field V as follows:

$$V = V_c + V_d + V_h \tag{14.1}$$

where V_c is *curl-free* ($\bigtriangledown \times V_c = 0$), V_d is *divergence-free* ($\bigtriangledown \cdot V_d = 0$), and V_h is *harmonic* ($\bigtriangledown \cdot V_h = 0$ and $\bigtriangledown \times V_h = 0$).

Such a decomposition can have applications in many scientific and engineering domains such as fluid simulation and modeling, electromagnetism, weather prediction, engine design, scientific visualization, and computer graphics. In these applications, one often needs to analyze an input vector field such as the velocity of fluid particles and the direction of the magnetic field. One of the most important aspects of a vector field is *singularities*, which are points in the domain that satisfy $V(\mathbf{p}) = 0$. A singularity can be classified by its *Jacobian* (gradient tensor) as follows [6]:

1. source: both eigenvalues of the Jacobian are positive.
2. sink: both eigenvalues are negative.
3. center: both eigenvalues are imaginary numbers.
4. saddle: one of the eigenvalues is positive and the other negative.

Through the decomposition, the sources, sinks, and some saddles can be captured by the curl-free component, while the centers and some other saddles are captured by the divergence-free component. The harmonic component is often seemingly featureless in the planar case. However, on hyperbolic surfaces, the harmonic component can capture the saddles that arise as a result of surface topology. For example, any

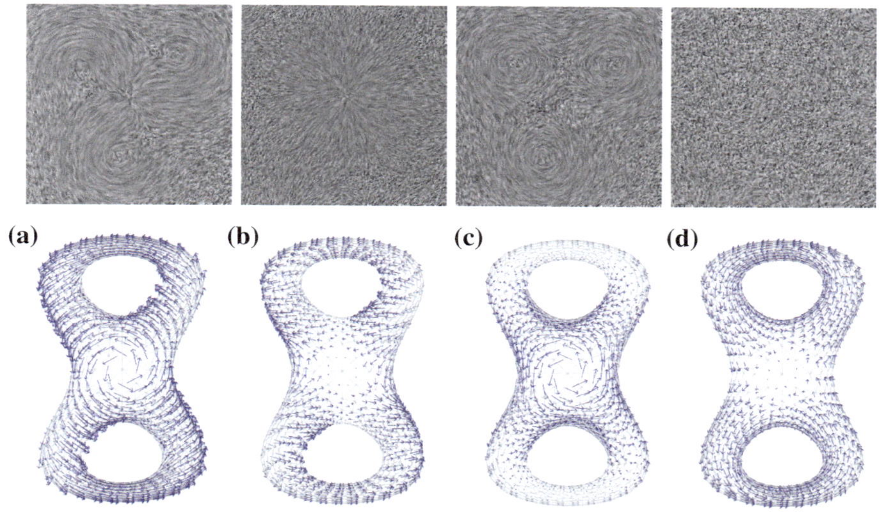

Fig. 14.8 Two examples of Hodge-Helmholtz decomposition: (*top*) a planar vector field, and (*bottom*) a vector field defined on a torus. From *left* to *right* are: (**a**) the original field V, (**b**) the curl-free component V_c, (**c**) the divergence-free component V_d, and the harmonic component V_h. Notice that singularities in the original field can be captured effectively by the decomposition. Moreover, the harmonic component is more prominent for fields defined on a hyperbolic manifold. Image courtesy of Polthier and Preuss [18, 19]

smooth vector field on a genus-two surface must contain at least four saddles or some higher-order saddles.

Polthier and Preuss develop techniques to efficiently perform the Hodge-Helmholtz decomposition on a triangular mesh with a piecewise constant vector field [18, 19] (Fig. 14.8). Such techniques are later extended to volumes [25].

Another important application of the Hodge-Helmholtz decomposition is in fluid simulation. In this case the fluids are assumed to divergence-free. However, numerical solvers often introduce errors which lead to flow fields with a non-zero divergence, thus causing unrealistic fluid behaviors. This is corrected by a *projection* step, for which the Hodge-Helmholtz decomposition is performed on the vector field, and the curl-free part is removed [23, 24].

14.4.2 Components of Tensor Field

There has been some recent trend in studying asymmetric tensor fields [1, 13, 28, 29], with applications in flow visualization and earthquake engineering. Given a vector field V such as the velocity of fluid particles or the deformation of land, the gradient $T = \bigtriangledown V$ is an asymmetric tensor field which can be used to describe the deformation of particles in both fluid and solid movements. This can be explained by the following decomposition of the gradient tensor:

$$T = D + R + S \tag{14.2}$$

where $D = \gamma_d Id$ is a multiple of the identity matrix, R is an anti-symmetric matrix, and S is a *traceless*, symmetric matrix. There are three fundamental fluid motions besides translation, and they are *isotropic scaling*, *rotation*, and *anisotropic stretching* or *pure shearing*. Interestingly, these motions correspond to the three components described in Eq. (14.2). D describes the isotropic stretching. When $\gamma_d > 0$, the particle's volume will increase, while when $\gamma_d < 0$, the particle will lose volume when it travels. R represents rotations, i.e., spinning around the center of the particle. This is related to the *vortices* in the flow. S corresponds to the anisotropic stretching. In this case the particle is under *pure shearing*, which refers to simultaneous expansion along some axis or axes and contraction along perpendicular directions without changing the volume. Pure shearing is linked to the rate of angular deformation, rate of mixing of multiple interacting fluid materials, and energy dissipation.

While these fields can be studied independently, in this context it is often important to study their interaction. For example, for two-dimensional cases, i.e., T is a 2×2 matrix, Zhang et al. [28] introduce the notion of *eigenvalue manifold* and *eigenvector manifold*. We will examine these concepts in detail.

In 2D, the components in Eq. (14.2) can be written as follows:

$$D = \gamma_d \begin{pmatrix} 1 & 0 \\ 0 & 1 \end{pmatrix}, \quad R = \gamma_r \begin{pmatrix} 0 & -1 \\ 1 & 0 \end{pmatrix}, \quad S = \gamma_s \begin{pmatrix} \cos\theta & \sin\theta \\ \sin\theta & -\cos\theta \end{pmatrix} \tag{14.3}$$

where γ_d, γ_r, and $\gamma_s \geq 0$ are the strengths of isotropic scaling, rotation, and pure shearing, respectively. θ decodes the orientation of the shearing. Note that the eigenvalues of T are purely decided by γ_d, γ_r, and γ_s. Zhang et al. [28] treat the triple $(\gamma_d, \gamma_r, \gamma_s)$ as a vector and consider the configurations corresponding to unit vectors. Such vectors form a hemisphere which they refer to as the eigenvalue manifold (Fig. 14.9: left). There are five canonical points on this manifold (Fig. 14.9: colored dots), corresponding to $(\gamma_d = 1, \gamma_r = 0, \gamma_s = 0)$ (pure expansion), $(\gamma_d = -1, \gamma_r = 0, \gamma_s = 0)$ (pure contraction), $(\gamma_d = 0, \gamma_r = 1, \gamma_s = 0)$ (pure counterclockwise rotation), $(\gamma_d = 0, \gamma_r = -1, \gamma_s = 0)$ (pure clockwise rotation), and $(\gamma_d = 0, \gamma_r = 0, \gamma_s = 1)$ (pure shearing). A configuration is said to be dominated by one of these five canonical motions, μ, if the point corresponding to the this configuration has the smallest geodesic distance to the canonical motion μ. The partition of the eigenvalue manifold in turn leads to a partition of the domain of tensor field T, although the map is not bijective.

Figure 14.10 illustrates this with an example vector field that is generated by combining two counter-rotating Sullivan vortices. Notice that the flow is predominantly expanding in the middle (yellow), contracting on the outside (blue), rotating (red and green), and pure shearing (white) elsewhere. Note that a region of predominant expansion motion cannot be directly adjacent to a region of predominant contraction. Similarly, a counterclockwise rotation region cannot be adjacent to a region domi-

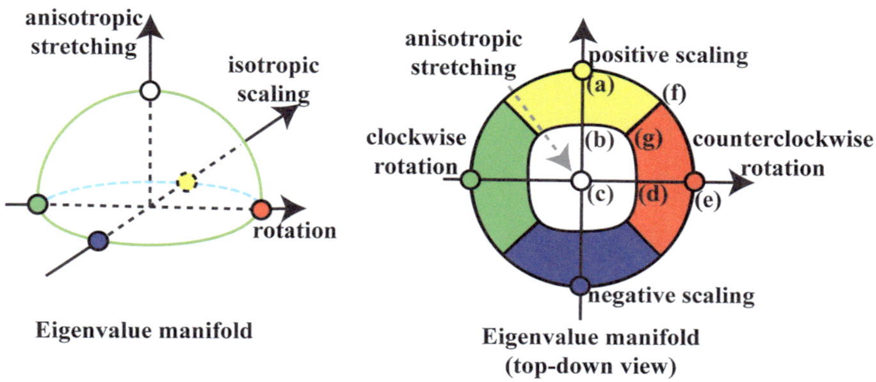

Fig. 14.9 Eigenvalue Manifold: there are five special points on the manifold, which are positive and negative scaling, counterclockwise and clockwise rotation, and anisotropic stretching. The Voronoi decomposition with respect to these five special points partitions the manifold into five cells where the flow is dominated by different characteristics [28]. © IEEE Reprinted, with permission, from IEEE Transactions on Visualization and Computer Graphics 15(1)

Fig. 14.10 Color-coding based on the eigenvalue manifold (*left*) and the combined eigenvalue and eigenvector manifold (*right*) [28]. © IEEE Reprinted, with permission, from IEEE Transactions on Visualization and Computer Graphics 15(1)

nated by clockwise rotation. Such results have led Lin et al. [13] to define asymmetric tensor field topology in terms of graphs whose nodes correspond to the regions in the partition and whose edges encode adjacency relationships between the regions.

The eigenvector information in the tensor field is determined purely by γ_r, γ_s, and θ from Eq. (14.3). Note that asymmetric tensors may have real eigenvalues (real domains) or complex eigenvalues (complex domains). In the latter case no real-valued eigenvectors exist. Zheng and Pang [29] introduce the notion of *dual-eigenvectors* which they show are the continuous extension of major and minor eigenvectors from the real domains into complex domains. Dual-eigenvectors are also the semi-axes of the elliptical flow patterns inside the complex domains.

Zhang et al. [28] realize that the decomposition from Eq. (14.3) can be simplified and reparameterized as follows:

$$T(\rho, \theta, \varphi) = \rho \cos\varphi \begin{pmatrix} \cos\theta & \sin\theta \\ \sin\theta & -\cos\theta \end{pmatrix} + \rho \sin\varphi \begin{pmatrix} 0 & -1 \\ 1 & 0 \end{pmatrix} \qquad (14.4)$$

where $\rho = \sqrt{\gamma_r^2 + \gamma_s^2}$ is the tensor magnitude. Notice that D, the isotropic scaling component, does not impact the directional information in a tensor field and can be dropped when considering eigenvectors. Furthermore, (γ_s, γ_r) is considered as a vector since it is their respective strength that determines whether a tensor is in the real domain or complex domain, the angle between the major and minor eigenvectors in the real domain, and the eccentricity of the elliptical flow patterns in the complex domain. Similar to the definition of eigenvalue manifold, Zhang et al. define the eigenvector manifold by considering unit vectors, i.e., $\gamma_r^2 + \gamma_s^2 = 1$. Such tensors can be parametrized using spherical coordinates shown in Eq. (14.4). Zhang et al. [28] demonstrate that a tensor is in the real domain if $-\frac{\pi}{4} < \phi < \frac{\pi}{4}$ and complex domain if $\phi < -\frac{\pi}{4}$ or $\phi > \frac{\pi}{4}$. The equator ($\phi = 0$) corresponds to the pure shearing tensors while the poles ($\phi = \pm\frac{\pi}{2}$) correspond to pure rotations (degenerate points in the tensor). The boundary between the real and complex domains ($\phi = \pm\frac{\pi}{4}$) is referred to *degenerate curves*. Points on these curves correspond to *simple shears* which are different from pure shears. Notice that equator serves as the boundary between counterclockwise rotating flows and clockwise rotating flows. Figure 14.11 illustrates these facts, while Fig. 14.12 demonstrates some special configurations.

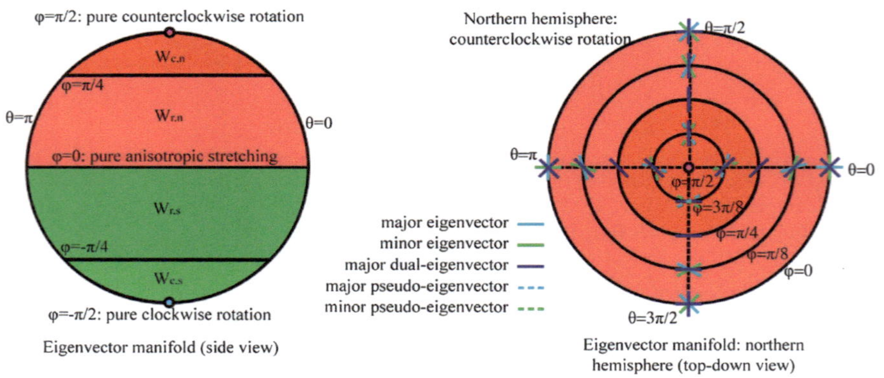

Fig. 14.11 Eigenvector manifold: the orientation of the rotational component is counterclockwise in the northern hemisphere and clockwise in the southern hemisphere. Each hemisphere is partitioned into real domains and complex domains. The equator represents pure symmetric tensors (irrotational flows), while the poles represent pure rotations. The directions of expansions and contraction in the real domain as well as the orientations of the elliptical patterns are determined by the relative stretches between the rotation and stretching components in the decomposition [28]. © IEEE Reprinted, with permission, from IEEE Transactions on Visualization and Computer Graphics 15(1)

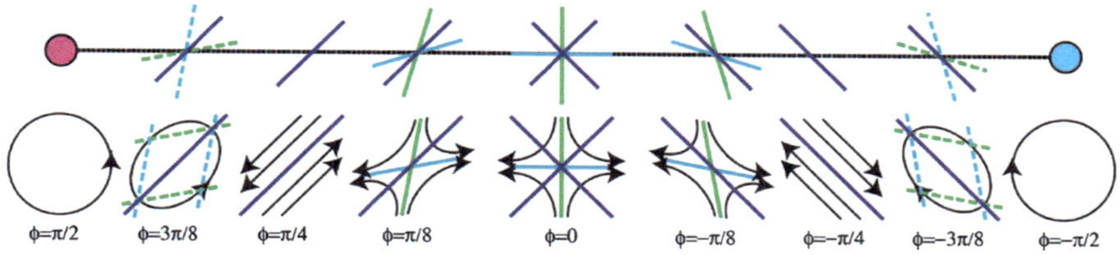

Fig. 14.12 Example tensors and their corresponding vector field patterns [28]. © IEEE Reprinted, with permission, from IEEE Transactions on Visualization and Computer Graphics 15(1)

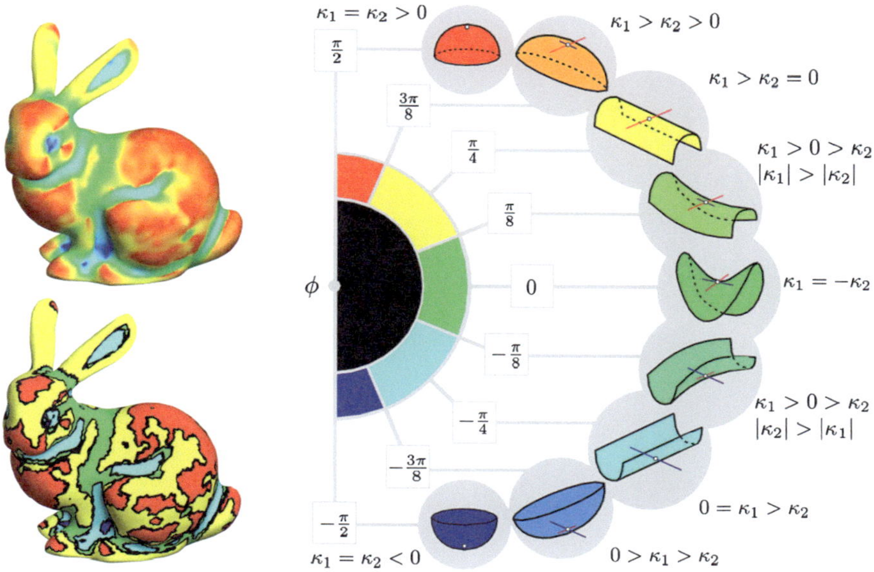

Fig. 14.13 The tensor decomposition in Eq. (14.4) can be adapted to symmetric tensors. In this example the symmetric tensor is the curvature tensor in the surface. Note that this tensor decomposition can lead to surface classification and feature extraction [15]. © IEEE Reprinted, with permission, from IEEE Transactions on Visualization and Computer Graphics 18(6)

The decomposition in Eq. (14.4) can also be used to *symmetric* tensor fields. In this case $\gamma_r = 0$ and the tensor can be rewritten as:

$$\rho \sin\phi \begin{pmatrix} 1 & 0 \\ 0 & 1 \end{pmatrix} + \rho \cos\phi \begin{pmatrix} \cos\theta & \sin\theta \\ \sin\theta & -\cos\theta \end{pmatrix} \tag{14.5}$$

where $\rho = \sqrt{\gamma_d^2 + \gamma_s^2}$ again is the tensor magnitude. Like Eq. (14.4), this equation is a special case of Eq. (14.2) where one of the three components disappears (the anti-symmetric component). Nieser et al. [15] have applied this to the curvature tensor to extract surface features for remeshing purposes. Figure 14.13 illustrates this

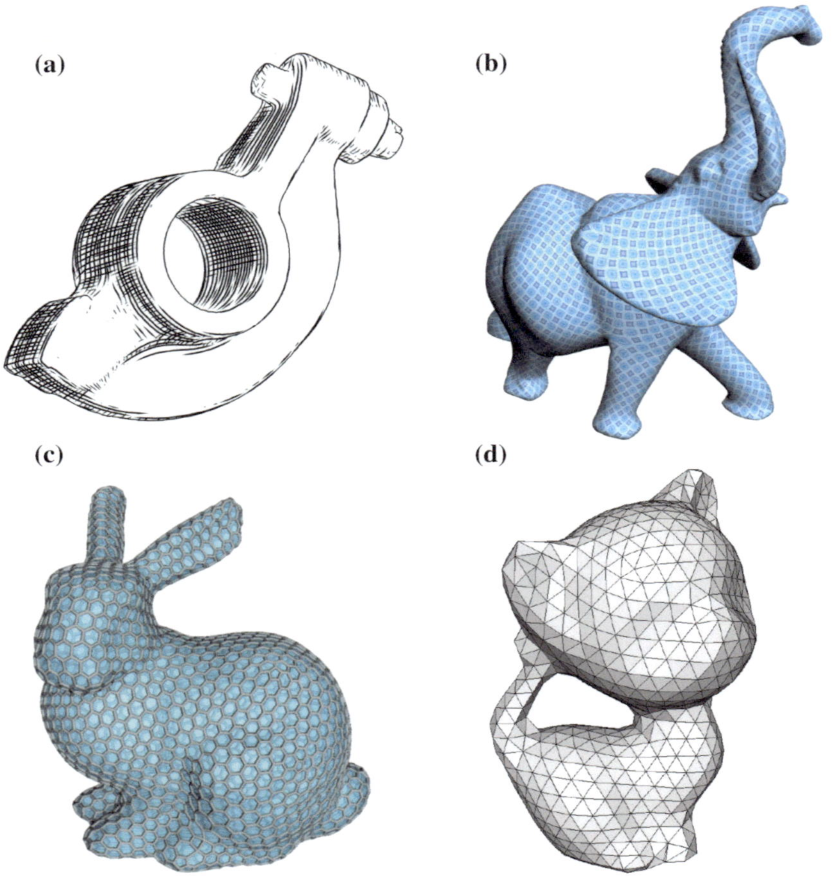

Fig. 14.14 Higher-order tensors have been used to present N-way rotational symmetries (N-RoSy [16]), with applications in pen-and-ink sketching (**a**), regular pattern synthesis (**b**), architectural modeling (**c**), and geometry remeshing (**d**). Image (**a**) and (**c**) are courtesy of [17]. © IEEE Reprinted, with permission, from IEEE Transactions on Visualization and Computer Graphics 17(7)

classification with a bunny surface. have applied this to the curvature tensor to extract surface features for remeshing purposes. Figure 14.13 illustrates this classification with a bunny surface.

14.4.3 Higher Order Tensor Fields

The decomposition of the asymmetric tensor field is intrinsically linked to the Hodge-Helmholtz decomposition. D, R, and S from Eq. 14.2 correspond to the curl-free, divergence-free, and harmonic component in the Hodge-Helmholtz decomposition (Fig. 14.14).

Higher-order tensors, i.e., tensors of a rank larger than two, are of great interests to scientists and engineers in many application domains. For example, general relativity deals with higher-order tensors. Elasticity tensor, a fourth-order tensor, relates the strain tensor (deformation) to the stress tensor (force). The spatial gradient of an N-th order tensor is an $N + 1$-th order tensor. This has been used by Delmarcelle and Hesselink to classify degenerate points for symmetric second-order tensors [2]. A special class of higher-order tensors have also been used to describe rotational symmetries on surfaces [16], with applications in pen-and-ink sketching [7], remeshing [15, 20], and regular texture and geometry synthesis on surfaces [15].

There have been a number of decomposition methods [12]. However, physical interpretation of these decompositions as well as effective analysis and visualization is still lacking. The only prominent work available at this point is [8].

14.5 Conclusions

In this chapter we have examined applications and existing techniques on multifield visualization based on the notion of derived fields. The derived fields play an important role in understanding relationships between multiple input fields. In addition, for a single input field, multiple derived fields can be generated as a result of decomposition which have the potential of providing insights on the input field.

There are a number of future research directions that we believe are important and can have major impact on multifield visualization:

1. Adaptation of pairwise similarity/dissimilarity as well as global alignment and dependency measures to a set of heterogeneous fields.
2. A detailed study of the sensitivity of the derived fields to the mesh that represents the domain. This study will be particularly useful if the input fields are specified on different meshes representing a common domain.
3. An effective interface using derived fields that supports identification of (a) important fields that can further analyzed in detail and (b) redundant fields that can be discarded from further studies.
4. Integrating the derived fields with existing techniques for interactive exploration like query-based visualization framework [5], focus+context, and show and brush for visual analysis.

Acknowledgments Natarajan acknowledges support from the Department of Science and Technology, India under grants SR/S3/EECE/048/2007 and SR/S3/EECE/0086/2012. Zhang is partially supported by National Science Foundation awards IIS-0546881, CCF-0830808 and IIS-0917308.

References

1. Chen, G., Palke, D., Lin, Z., Yeh, H., Vincent, P., Laramee, R.S., Zhang, E.: Asymmetric tensor field visualization for surfaces. In: IEEE Proceedings on Visualization (2011)
2. Delmarcelle, T., Hesselink, L.: The topology of symmetric, second-order tensor fields. In: IEEE Computer Graphics and Applications, pp. 140–147 (1994)
3. Edelsbrunner, H., Harer, J., Natarajan, V., Pascucci, V.: Local and global comparison of continuous functions. In: Proceedings of IEEE Conference on Visualization, pp. 275–280 (2004)
4. Feller, W.: An Introduction to Probability Theory and Its Applications, vol. I. Wiley, New York (1968)
5. Gosink, L.J., Anderson, J.C., Wes Bethel, E., Joy, K. I.: Variable interactions in query-driven visualization. IEEE Trans. Visual. Comput. Graphics **13**(6), 1400–1407 (2007)
6. Helman, J.L., Hesselink, L.: Representation and display of vector field topology in fluid flow data sets. IEEE Comput. **22**(8), 27–36 (1989)
7. Hertzmann, A., Zorin, D.: Illustrating smooth surfaces. In: Proceedings of Annual Conference Series on Computer Graphics (SIGGRAPH 2000) pp. 517–526 (2000)
8. Hlawitschka, M., Scheuermann, G.: HOT lines: tracking lines in higher order tensor fields. Proc. IEEE Visual. **2005**, 27–34 (2005)
9. Horn, R., Johnson, C.: Matrix Analysis. Cambridge University Press, Cambridge (1985)
10. Jänicke, H., Bottinger, M., Tricoche, X., Scheuermann, G.: Automatic detection and visualization of distinctive structures in 3D unsteady multi-fields. Comput. Graphics Forum **27**(3), 767–774 (2008)
11. Jänicke, H., Wiebel, A., Scheuermann, G., Kollmann, W.: Multifield visualization using local statistical complexity. IEEE Trans. Visual. Comput. Graphics **13**, 1384–1391 (2007)
12. Kolda, T.G., Bader, B.W.: Tensor decompositions and applications. SIAM Rev. **51**(3), 455–500 (2009)
13. Lin, Z., Yeh, H., Laramee, R.S., Zhang, E.: 2D asymmetric tensor field topology. In: Proceedings of Topological Methods in Data Analysis and Visualization II (Topo-In-Vis 2011). Springer, Berlin (2012)
14. Nagaraj, S., Natarajan, V., Nanjundiah, R.S.: A gradient-based comparison measure for visual analysis of multifield data. Comput. Graphics Forum **30**(3), 1101–1110 (2011)
15. Nieser, M., Palacios, J., Polthier, K., Zhang, E.: Hexagonal global parameterization of arbitrary surfaces. In: IEEE Transactions on Visualization and Computer Graphics (2012)
16. Palacios, J., Zhang, E.: Rotational symmetry field design on surfaces. ACM Trans. Graph. **26**(3), 55:1–55:10 (2007)
17. Palacios, J., Zhang, E.: Interactive visualization of rotational symmetry fields on surfaces. IEEE Trans. Visual. Comput. Graphics **17**(7), 947–955 (2011)
18. Polthier, K., Preuss, E.: Variational approach to vector field decomposition. In: Proceedings of Eurographics Workshop on Scientific Visualization. Springer, Berlin (2000)
19. Polthier, K., Preuss, E.: Identifying vector fields singularities using a discrete hodge decomposition. In: Hege, H.C., Polthier, K. (eds.) Visualization and Mathematics III. Springer, Berlin (2002)
20. Ray, N., Li, W.C., Lévy, B., Sheffer, A., Alliez, P.: Periodic global parameterization. ACM Trans. Graphics **25**(4), 1460–1485 (2006)
21. Roeckner, E., Bäuml, G., Bonaventura, L., Brokopf, R., Esch, M., Giorgetta, M., Hagemann, S., Kirchner, I., Kornblueh, L., Manzini, E., Rhodin, A., Schlese, U., Schulzweida, U., Tompkins, A.: The atmospheric general circulation model ECHAM5. Part 1: model description. Technical Report 349, Max Planck Institute of Meteorology (2003)
22. Sauber, N., Theisel, H., Seidel, H.P.: Multifield-graphs: an approach to visualizing correlations in multifield scalar data. IEEE Trans. Visual. Comput. Graphics **12**(5), 917–924 (2006)
23. Stam, J.: Stable fluids. In: Proceedings of the 26th Annual Conference on Computer Graphics and Interactive Techniques. SIGGRAPH '99, pp. 121–128. ACM Press/Addison-Wesley Publishing Co., New York (1999)

24. Stam, J.: Flows on surfaces of arbitrary topology. ACM Trans. Graphics (SIGGRAPH 2003) **22**(3), 724–731 (2003)
25. Tong, Y., Lombeyda, S., Hirani, A.N., Desbrun, M.: Discrete multiscale vector field decomposition. ACM Trans. Graph. **22**(3), 445–452 (2003)
26. Wang, W., Bruyere, C., Kuo, B.: Competition data set and description in 2004 IEEE visualization design contest. http://vis.computer.org/vis2004contest/data.html (2004)
27. Xu, L., Dinh, H.Q., Zhang, E., Lin, Z., Laramee, R.S. A distribution-based approach to tracking points in velocity vector fields. IEEE Conf. Comput. Vis. Pattern. Recogn
28. Zhang, E., Yeh, H., Lin, Z., Laramee, R.S.: Asymmetric tensor analysis for flow visualization. IEEE Trans. Visual. Comput. Graphics **15**(1), 106–122 (2009)
29. Zheng, X., Pang, A.: 2D asymmetric tensor analysis. In: IEEE Proceedings on Visualization, pp. 3–10 (2005)

Chapter 15
Interactive Visual Exploration and Analysis

Gunther H. Weber and Helwig Hauser

Abstract Interactive exploration and analysis of multi-field data utilizes a tight feedback loop of computation/visualization and user interaction to facilitate knowledge discovery in complex datasets. It does so by providing both overview visualizations, as well as support for focusing on features utilizing iterative drill-down operations. When exploring multi-field data, interactive exploration and analysis relies on a combination of the following concepts: (i) *physical views* that show information in the context of the spatiotemporal domain (domain perspective), (ii) *range views* show relationships between multiple fields (range perspective), and (iii) selecting/marking data subsets in one view (e.g., regions in a physical view) leading to a consistent highlighting of this subset in all other views (brushing and linking). Based on these principles, interactive exploration and analysis supports building complex feature definitions, e.g., using Boolean operations to combine multiple selections. Utilizing derived fields, statistical methods, etc., adds a further layer of flexibility to this approach. Using these concepts, it is also possible to integrate feature detection methods from the other chapters of this part, as well as application-specific feature extraction methods into an joint framework. This methodology of interactive visual data exploration and analysis has proven its potential in a larger number of successful applications. It has been implemented in a larger number of systems and is already available for a wide spectrum of different application domains.

G.H. Weber (✉)
Computational Research Division, Lawrence Berkeley National Laboratory, Berkeley, USA
e-mail: ghweber@lbl.gov

G.H. Weber
Department of Computer Science, University of California, Davis, USA
e-mail: ghweber@ucdavis.edu

H. Hauser
Department of Informatics, University of Bergen, Bergen, Norway
e-mail: helwig.hauser@UiB.no

C.D. Hansen et al. (eds.), *Scientific Visualization*, Mathematics and Visualization,
DOI 10.1007/978-1-4471-6497-5_15

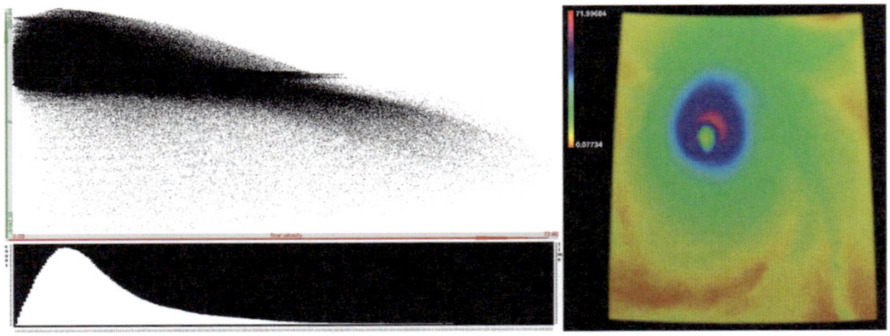

Fig. 15.1 Interactive visual analysis (IVA) uses two types of views: *Spatiotemporal views* (*like the image on the right*), e.g., false color plots, show the distribution of a quantity within the domain. Range views (*like the images on the left*), including scatter plots (*top left*) and histograms (*bottom left*), show the correlation between multiple fields or additional information about a single field, respectively

15.1 Basic Concepts

At its basis, interactive visual analysis (IVA) builds on the combination of different views on data with the ability to emphasize data subsets interactively (most commonly features of interest). In the context of multi-field data exploration and analysis, two aspects of data are of primary interest: (i) the spatiotemporal distribution of one or more fields, and (ii) the relationship of one or multiple fields with respect to each other. For example, examining multiple fields in a simulation of a hurricane, one may be interested in the spatial location of regions of high velocity, but also in learning how velocity correlates with pressure. To provide this information, IVA utilizes two types of views displaying complementary information. (i) *Spatiotemporal views*, such as false color plots or volume rendered images provide a domain-centric perspective on the data. For example, in the hurricane example we can map velocity to color and display a false color plot that shows the spatial distribution of velocity in the simulation domain [Fig. 15.1 (right)]. (ii) *Range views*, such as scatter plots [3] or parallel coordinate plots [4, 13], show the correlation between two or more fields and show the data from a range perspective. For example, for the hurricane example, a scatter plot of pressure and velocity shows their correlation [Fig. 15.1 (top left)]. Individually, the use of these types of views has a long history in science and statistics.

Considering only one aspect at a time limits data analysis capabilities. The fundamental idea underlying IVA is to combine different views on the same data in such a way that a user can correlate the different views. One way to achieve this correlation is to enable the interactive selection of data subsets, and highlight such a data subset in other views in a consistent manner, i.e., ensuring the same data items are visually emphasized over their context in all views. Selection is often performed directly on a view by interactive visual means, similar to those in a drawing program, and therefore

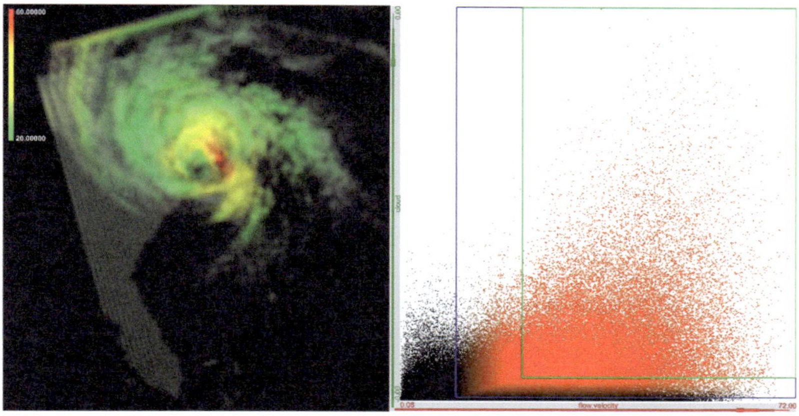

Fig. 15.2 *Brushing-and-linking* correlates multiple views by making it possible to select a data subset in one view, e.g., selected ranges in attribute/field space, and consistently highlighting this subset in all other views. Changing color (e.g., showing the selected points in a different color (*left*) or using saturation (*right*)) is a common way to achieve this effect of highlighting the emphasized data subset over its context (commonly referred to as focus plus context (F + C)

are usually called *brushing*. Highlighting in this instance serves as means of *linking* views together, and this technique is referred to as *brushing-and-linking*[1] [1, 19].

For example, in the hurricane case, one might be interested in spatial regions corresponding to fast moving clouds. Using brushing-and-linking it is possible to select such a feature in the scatter plot [Fig. 15.2 (right)] and highlight the corresponding regions in other views, such as in a physical view of the hurricane [Fig. 15.2 (left)]. Using brushing-and-linking, it is possible to formulate simple queries interactively, such as "where are regions of high temperature and low velocity" and visualize the results in a physical view. While there are many instances, where such features of interest are known a priori and analysis is driven by known queries [29], the full power of IVA lies in the fact that a user can discover features of interest during interaction and pose or refine queries during interactive analysis.

As a consequence, IVA often defines "features" as data subsets of interest to the user, be it due to prior knowledge or because a data subset has caught the user's attention. Common user interactions include brushing for outliers (e.g., to determine why a subset is behaving differently from the rest), regions of strong correlation (e.g., to verify if this correlation holds in the entire data set or only in particular subsets), or a spatial region of interest, like an inlet or outlet of a flow simulation (e.g., to determine if a correlation exists in that region). In general, we distinguish three patterns of explorative/analytical procedures:

---

[1] We note that the visualization community uses the order "linking-and-brushing" more commonly, while the database community uses the order "brushing-and-linking". We use the term "brushing-and-linking" here as brushing is usually performed before linking.

1. From the domain to the range perspective, we select a subset of data items in a physical view and examine the selection in range views. This type of analysis serves to *localize the investigation* to a region of interest such as an inlet or outlet.
2. From the range to the domain perspective, we select a subset of data items in a range view, such as in a scatter plot, and examine the result in a physical view. This type of analysis enables *localizing features*. In this case brushing defines a feature, usually as a set of thresholds, and highlighting in a spatiotemporal view shows whether the selection corresponds to a localized feature.
3. Within the range perspective, we select subset of data items in a range view and observe the selection in another range view. This type of analysis provides a means of performing an interactive *multivariate analysis*, e.g., by brushing in one scatter plot and examining the selection in another scatter plot of different variables. This pattern was originally introduced in the field of information visualization [1, 31].

Using one or more of these patterns is the simplest form of IVA, more recently referred to as "Show & Brush." It utilizes multiple views, usually at least one range view for visually correlating multiple fields and one domain view to show properties in a physical domain context. Though being the simplest form of IVA, this method already covers a large percentage of use cases in multi-field analysis and serves as powerful basis for more advanced types of exploration and analysis. This type of IVA has proven valuable in many application areas, including aeronautical design [12], climate research [15, 20], biomedical visualization [8, 25], the analysis of gene expression data [32], the analysis of combustion engines [7, 22], and the analysis of simulations of particle accelerators [27].

15.2 Additional Concepts

Based on the simple "Show & Brush" paradigm, a few extensions can greatly enhance the expressiveness of IVA. First, in many cases it is useful to define brushes not as binary classifiers into two categories "of interest" and "not of interest" but as a means to map each data item to a *degree of interest* [6]. It is possible to define this degree by specifying two selections (e.g., regions in a scatter plot). All items inside an *inner range* have a degree of interest of 100 % (i.e., are definitely of interest), and all items outside an *outer range* have a degree of interest of 0 % (i.e., not of interest). Between those regions, a transfer function maps the distance of a sample from inner and outer range to a degree of interest between 100 and 0 %. A linear ramp is a common choice for this transfer function. More generally, we can utilize fuzzy logic operators to combine multiple smooth brushes.

This smooth drop-off of a degree of interest makes it possible to transition seamlessly between data items of interest and those of not interest and use generalized focus plus context (F+C) methods [5, 9, 26] to reduce cluttering in resulting visualizations and draw a user's attention to the most important details. Traditionally, focus and context methods use space distortion such as a fish-eye lens to assign more space to data of interest while presenting the remainder as context for orientation. However, in a generalized setting, various visual attributes can serve to emphasize or deemphasize data items, including, for example:

- Color (hue, saturation, brightness, or an alternative representation) and opacity: a typical example would be to present the data subset in focus in color and its context in gray scale by mapping the degree of interest to saturation [8]. Alternatively, the degree of interest can be mapped to opacity, rendering the focus opaque and the context semi-transparent [21].
- Style: different visualization modalities (isosurfaces, volume rendering, etc.) can be used to discriminate focus and context. Alternatively, rendering styles, in particular non-photorealistic/illustrative styles (halos, outlines, cross-hatched/dotted lines/polygonal primitives) can serve this purpose (for example in a *two-level volume rendering* approach [11]).
- Frequency: Only use the full spectrum of spatial frequencies for the data subsets in focus and render the context band-limited. This approach is called *Semantic Depth of Field* [18] and results in a blurred style for the context, directing the user's attention to the sharply rendered data subsets in focus.
- Space: This approach refers to the traditional notion of F+C visualization, i.e., that the visualization space is distorted in order to give more space (or time) to the visualization of data subsets in focus.

15.3 Levels of IVA

So far, with "Show & Brush", we have seen the base level of IVA. Based on the complexity of feature definitions, we distinguish additional, more complex (and thereby also more powerful) levels of IVA. It is our experience, however, that in many cases—if not in most cases—the simple Show & Brush technique already provides sufficient functionality to enable an effective data analysis; the more complex levels of IVA, as introduced below, are only advanced solutions for more complicated cases which cannot be served with the base-level IVA.

1. **Show & Brush (level 1)**: This level captures the analysis as described so far. It utilizes at least two linked views, usually one physical and one range view. The interactive selection of features of interest is accomplished by brushing in one view, leading to a focus plus context visualization in the linked view(s).
2. **Relational analysis (level 2):** This level supports the combination of brushes using logical operations and a simple feature definitions language.
3. **Complex analysis (level 3):** This level integrates computational analysis, e.g., derived fields, statistical methods, machine learning [28], etc., into the interactive visual approach, thus adding a new dimension of possible procedures. A typical scenario would be that, prompted by insights gained during visual exploration and analysis, the user decides to initiate a certain computational analysis procedure, such as clustering. This procedure results in at least one additional (synthetic)

data attribute, such as membership in a cluster, which can subsequently be used together with all other data to improve the analysis.
4. **Proprietary analysis (level 4):** This class is a container for everything beyond complex analysis and includes, e.g., the integration of application-specific feature definitions (such as flow feature detectors [2]) or could entail the integration of higher-level feature definition languages. Identifying common concepts and refining IVA beyond this level is a subject of future research.

We note that this terminology is potentially controversial and that "relational analysis" and "complex analysis" have other possible meanings. Consequently, we present this classification as a starting point that can evolve as research in IVA progresses. In the following, we describe the higher levels of IVA in greater detail.

15.4 Relational Analysis

Relational analysis takes the selections in form of brushes and provides means to combine these brushes (or selections) into more complex feature definitions. A simple feature definition language uses Boolean expressions, for example, to combine brushes into more complex feature definitions. Figure 15.3 shows an example from the analysis of three-dimensional gene expression data. Here, positions correspond to the locations of cells in an organism, and the multiple fields represent expression values of genes, i.e., they specify whether a certain gene is expressed in a given cell. Individual brushes select expression patterns based on single genes. Combining these brushes using Boolean operations, it is possible to define complex selections. The example in the figure uses this capability, to combine patterns based on a priori knowledge about how genes interact, and verify whether the genes involved completely explain the arising pattern.

It is possible to generalize logical operations to smooth brushes [5, 6] and enable F+C visualization in relational analysis. One associated challenge is to extend the visual means, which discriminate data subsets in focus from their context, in such a way that takes this more complex form of feature definition into account. Within each view, an appropriate F+C visualization is necessary to reflect the brush(es) applied to this view. Another level of F+C visualization must reflect the overall feature specification, possibly also involving multiple features. One possible solution to this problem is a four-level F+C visualization approach proposed by Muigg et al. [23], which, as one particular aspect, is based on an intelligent color combination scheme.

Combining brushes usually defines a relation between multiple fields. Early work on query-driven visualization (QDV) [29] used similar concepts in that it defined features as a Boolean combination of relational expressions. However, in this QDV work, the features and expressions were known a priori and not refined during analysis. An important aspect of QDV visualization is the use of indices, such as FastBit [33], to accelerate data selection based on queries. However, there is also work on combining QDV concepts with IVA, e.g., using parallel coordinates [27].

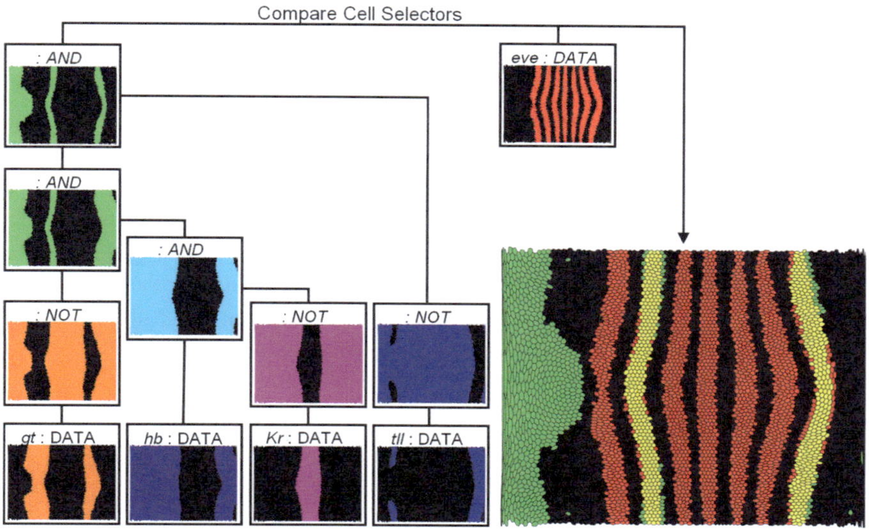

Fig. 15.3 Building complex feature definitions from individual brushes using Boolean operations in an example from three-dimensional gene expression. Genes are expressed in spatial patterns that control specialization of cells into different tissue types. More complex patterns, such as the seven stripes of the gene *even skipped (eve)* (*red pattern in the image*), arise from simpler expression patterns when expression of one gene controls (enhances or suppresses) the expression of other genes. The image shows the use of brushes to verify known relations that create *eve* stripes two and seven. The expression patterns of the genes *giant (gt)*, *hunchback (hb)*, *Krüppel (Kr)*, and *tailless (tll)* are first classified by defining an independent brushes in scatter plots. Subsequently, the brushes defining the *gt*, *Kr*, and *tll* patterns are inverted using a NOT operation (to model suppression of gene expression). Afterwards these brushes as well the brush defining the *hb* pattern are combined using a sequence of AND operations. In this way the overlap of the hb expression pattern, and the inverted *gt*, *Kr*, and *tll* expression patterns can be determined. The result (green) is compared to the *eve* expression pattern (red) identified by another brush. ©IEEE/ACM Reprinted, with permission, from IEEE/ACM Transactions on Computational Biology and Bioinformatics 6(2)

15.5 Complex Analysis

The levels of IVA described so far are an extremely versatile and powerful framework for enabling effective and efficient visual data analysis. Certain aspects of complex datasets, however, cannot be captured with these mechanisms. In such situations, the integration of computational data analysis tools, like those known from statistics, data mining, or machine learning, can help, leading to a solution which is tightly aligned with the currently modern *visual analytics* methodology [17, 30]. Alternatively, the implementation of extended interaction mechanisms, such as brushes that are capable of grasping aspects of the data that are not explicitly represented in a visualization, can also help in these situations. In the following, we exemplify both approaches to achieve *complex analysis* in the context of IVA.

A very powerful extension of the IVA methodology as described up to here is adding the capability to interactively derive new, user-defined data attributes, based on computational data analysis procedures. While in principle there are no limits to the set of potentially useful data derivation mechanisms, it is the authors' opinion that it is worthwhile to emphasize a few more general examples:

Interactive Spatiotemporal Data Derivation: Interactive estimation of gradients with respect to the usually spatiotemporal domain is a generally useful data derivation mechanisms. Spatial and temporal derivatives, including higher order derivatives obtained by repeated application of an interactive derivation operator, are often useful for defining features definition since features are often based on some notion of change. Using temporal derivatives, for example, supports a more advanced analysis of time-dependent aspects of such datasets, where the consideration of first- and second-order derivatives (wrt. time) leads to a massively parallel data analysis similar to how curve sketching is performed for individual time series.

Interactive and Targeted Data Normalization: Data analysis commonly adopts two types of perspective: an *absolute* perspective that considers absolute data values (or derived attribute values), and a *relative perspective* that examines relative values. One mechanism that enables a relative perspective in IVA is to support interactive data normalization. A powerful aspect of performing this normalization as part of IVA is that it not only allows for global normalization procedures, which usually do not add too much in terms of opportunities to understand data aspects that otherwise would not be accessible, but to also enables more localized normalization operations. Examples are normalization per time step, normalization per height-level, etc. Useful normalization operators include the scaling to the unit interval, z-standardization, or the normalization against other data statistics like the median and the MAD.

Interactive Derivation of Data Statistics: Statistics are powerful means to summarize and characterize data. Having data statistics, in particular localized data statistics, available for subsequent computations and interactive feature specifications, enriches the spectrum of possibilities in IVA substantially. A very good starting point are the standard descriptive statistics *mean*, *standard deviation*, *skewness*, and *kurtosis*. Interesting complements include more robust estimates such as the *median*, *MAD*, etc., as well as ranking-based statistics (e.g., based on quartiles or octiles). Interesting applications for IVA have been demonstrated, for example, in the context of multi-run data analysis for climatology [14].

Considering correlation information, data clustering, etc.: Data analysis techniqu es from statistics, data mining, machine learning, etc., are very rich in terms of history and available related work, and the potential set of useful mechanisms that are promising candidates for integration into IVA is almost unlimited. Particularly interesting candidates for extending the power of IVA are: the interactive derivation of *correlation* information between data attributes (e.g., based on the standard Pearson correlation, or Spearman's correlation measure), techniques for *attribute selection* or *dimension reduction* (such as PCA or LDA, for example), the consideration of data *clustering* (e.g., based on supervised or unsupervised clustering techniques), the integration of measures of *outlyingness* (e.g., based on the Mahalanobis distance of the data points, or derived from normality tests such as Shapiro's p-test), etc. One example for combining IVA with clusterings is the analysis of three-dimensional gene expression data with integrated clustering [28].

Advanced brushing mechanisms can be integrated in IVA as an alternative or in addition to these data derivation approaches. Brushes developed for special purposes include *angular brushing* [10] of parallel coordinates to access the slopes of the lines, or *similarity brushing* [23, 24], which utilizes a more advanced similarity measure between data and brush to determine the data items that are selected by a certain brushing interaction.

In principle, it is possible to design advanced brushes for any of the data aspects that otherwise could be made accessible (to standard brushing) via the further above described data derivation mechanism. The more indirections, however, in terms of implicitly considered data derivations, are built into an advanced brush, the more challenging the additional cognitive load becomes when using such a brush. It therefore stands to reason that highly complicated relations in the data, which only can be accessed through a number of concepts as described above (some statistics, some dimension reduction, some outlyingness measure, etc.), are better made available to interactive feature specification in a step-by-step procedure (a certain sequence of data derivation steps, for example) than packing too much into a single advanced brushing tool.

Figure 15.4 shows an example of a Complex Analysis—in this case an outlier analysis in a multi-run climate simulation dataset. As part of a coupled atmosphere–ocean–biosphere simulation model, temperature values in the world's big oceans, represented by three 2D cross-sections (longitude vs. depth), are analyzed, which are given over a 500 year period at about 6000 BC. The goal of this analysis was to identify spatiotemporal locations where the simulated temperature values exhibit large differences (as compared to the main trend) in some simulation runs. Using the interactive data derivation mechanism, first the overall number of outliers per space-time location was computed (this step uses a mild univariate outlyingness measure, i.e., all values which lie more than $3 \cdot \mathrm{IQR}/2$ above q_3 (the 3rd quartile) or below q_1 (with IQR being the interquartile range $q_3 - q_1$). The scatter plot in Fig. 15.4a identifies all locations according to how many such outliers exist (x-axis) and to which degree they are large- or small-value outliers (y-axis). A smooth brush was then used to highlight all locations with a substantial number of outliers, and the glyph-based visualization in Fig. 15.4b shows these locations emphasized (larger, less transparent glyphs). In a next step, the analysis was confined to lower-value outliers. This restriction was achieved by first using the data derivation mechanism, again, to "normalize" the y-axis wrt. its vertical extent per x-location. This step enables a selection—with a standard rectangular brush—of those outliers, which are mainly lower-value outliers. The scatter plot after loading this new attribute and the according brush are illustrated in Fig. 15.4c.

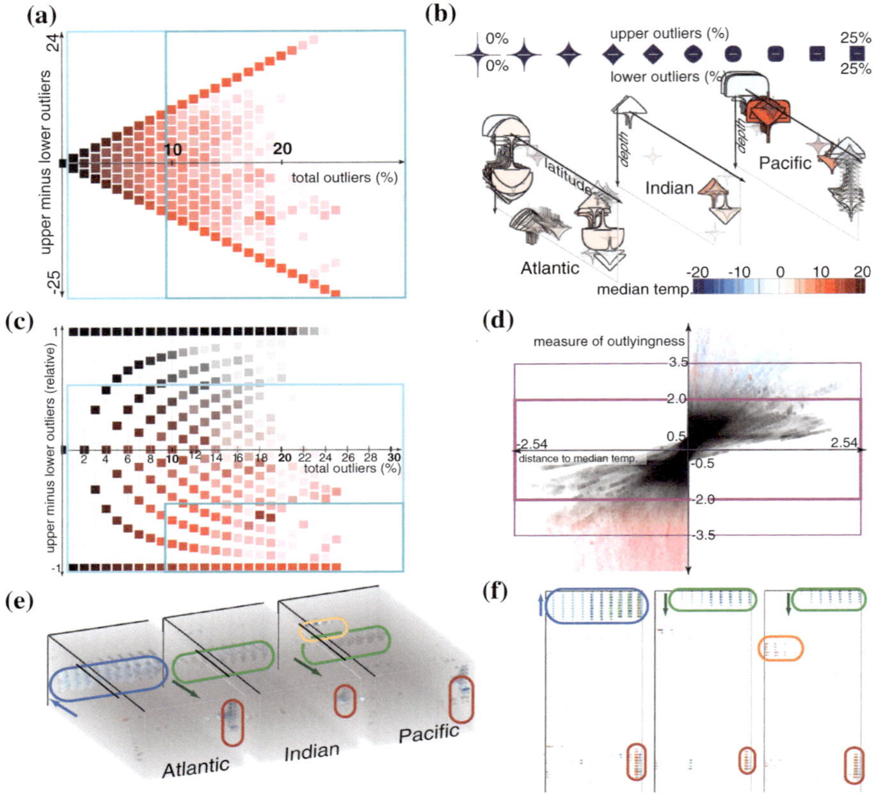

Fig. 15.4 Selected steps within a complex analysis example (outlier analysis in a multi-run climate simulation dataset—more details in the main text). After having used the interactive data derivation mechanism to compute an IQR-based outlyingness measure, spatiotemporal locations are identified with outliers. This identification is achieved by brushing scatter plot (**a**) and observing the selected locations in the linked visualization (**b**). In a next step, the analysis was confined to lower-value outliers (using the data derivation mechanism, again) by brushing scatterplot (**c**). Subsequently, to see the actual outliers themselves, a new scatter plot was used, with detrended and accordingly normalized temperature values on y, to focus on the actual outliers, then observed in views (**e**) and (**f**). More details about this study are available in the main text and in a paper by Kehrer et al. [16]

Up to this point, the entire analysis was solely focused on delimiting locations that have outliers of a particular characteristic. In the next step, the focus was directed to the outliers themselves. To select them, another data derivation steps was performed, computing detrended and normalized temperature values per location (the performed operation was to first subtract the median temperature wrt. all simulation runs, per location, and then divide by IQR). A new scatter plot, shown in Fig. 15.4 (d), was used to show all data points wrt. their distance to the median (x-axis) and this detrended and normalized temperature measure (y-axis). Consistent with Fig. 15.4a, b, all points with y-values beyond ± 2 are also considered as outliers (and brushed accordingly). This brushing leads to their identification in the views Fig. 15.4e, f, where each ocean section is repeated 100 times (once for every computed simulation run). This analysis resulted in an interesting deep-water pattern of some "outliers" in the north of the simulation, translating from the Atlantic slice into the Arctic basin (which actually look much more like a distinct pattern than just outliers) as well as some surface-water outliers (warm water, half-way north in the Pacific, marked orange) and some other outliers near Antarctica (circled red). More details about this study have been presented by Kehrer et al. [16].

15.6 Conclusions and Future Directions

IVA has already proven valuable in a wide range of application areas, including engineering, climate research, biomedical research and economy. The ability to define features interactively and refine feature definitions based on insights gained during visual and exploration and analysis provides an extremely powerful and versatile tool for knowledge discovery. Future challenges lie in the integration of alternate feature detection methods and their utilization in intelligent brushes. Furthermore, integrating IVA and simulations, thus supporting computational steering, offers a wide range of new possibilities for knowledge discovery.

Acknowledgments This work was supported by the Director, Office of Advanced Scientific Computing Research, Office of Science, of the U.S. Department of Energy under Contract No. DE-AC02-05CH11231. We thank the members of the LBNL visualization group and the Berkeley *Drosophila* Transcription Network. Special thanks are extended to Helmut Doleisch and colleagues from SimVis GmbH in Vienna, Austria, for the cooperation over many years. We thank Johannes Kehrer and Peter Filzmoser as well as Thomas Nocke, Michael Flechsig, and colleagues from the Potsdam Institute for Climate Impact Research in Germany for the collaboration on the climate data analysis. We also thank Krešimir Matković and colleagues from the VRVis Research Center in Vienna, Austria, for many years of fruitful collaboration on IVA research, as well as many others from Vienna, Bratislava, Magdeburg, Zürich, and Bergen.

References

1. Becker, R.A., Cleveland, W.S.: Brushing scatterplots. Technometrics **29**(2), 127–142 (1987)
2. Buerger, R., Muigg, P., Ilčík, M., Doleisch, H., Hauser, H.: Integrating local feature detectors in the interactive visual analysis of flow simulation data. In: Museth, K., Möller, T., Ynnerman, A. (eds.) Proceedings of EuroVis 2007, pp. 171–178. Eurographics Association, Norrköping, Sweden (2007)
3. Chambers, J.M., Cleveland, S,W., Kleiner, B., Tukey, P.A.: Graphical Methods for Data Analysis. Wadsworth/Brooks Cole, Monterey (1983)
4. d'Ocagne, M.: Coordonnées Parallèles et Axiales: Méthode de transformation géométrique et procédé nouveau de calcul graphique déduits de la considération des coordonnées parallèlles. Gauthier-Villars, Paris (1885)

5. Doleisch, H., Gasser, M., Hauser, H.: Interactive feature specification for focus+context visualization of complex simulation data. In: Bonneau, G.P., Hahmann, S., Hansen, C.D. (eds.) Data Visualization 2003, Proceedings of the Eurographics/IEEETCVG Symposim on Visualization (2003)
6. Doleisch, H., Hauser, H.: Smooth brushing for focus+context visualization of simulation data in 3d. In: Proceedings WSCG'02, pp. 147–154 (2002)
7. Doleisch, H., Mayer, M., Gasser, M., Wanker, R., Hauser, H.: Case study: Visual analysis of complex, time-dependent simulation results of a diesel exhaust system. EuroGraphics Symposium on Visualization 2004, pp. 91–96 (2004)
8. Gresh, D.L., Rogowitz, B.E., Winslow, R.L., Scollan, D.F., Yung, C.K.: WEAVE: a system for visually linking 3-d and statistical visualizations, applied to cardiac simulation and measurement data. In: Ertl, T., Hamann, B., Varshney, A. (eds.) Proceedings IEEE Visualization 2000, pp. 489–492. IEEE Computer Society Press, Los Alamitos (2000)
9. Hauser, H.: Generalizing Focus+Context Visualization, pp. 305–327. Springer, Berlin (2006)
10. Hauser, H., Ledermann, F., Doleisch, H.: Angular brushing for extended parallel coordinates. In: Proceedings of the IEEE Symposium on Information Visualization 2002 (InfoVis 2002), pp. 127–130 (2002)
11. Hauser, H., Mroz, L., Bischi, G.I., Gröller, M.E.: Two-level volume rendering. IEEE Trans. Vis. Comput. Graph. **7**, 242–252 (2001)
12. Henze, C.: Feature detection in linked derived spaces. In: Ebert, D., Rushmeier, H., Hagen, H. (eds.) Proceedings IEEE Visualization '98, pp. 87–94. IEEE Computer Society Press, Los Alamitos (1998)
13. Inselberg, A.: The plane with parallel coordinates. Vis. Comput. **1**(4), 69–91 (1985)
14. Kehrer, J., Filzmoser, P., Hauser, H.: Brushing moments in interactive visual analysis. Comput Graph Forum (Proceedings of EuroVis) **29**(3), 813–822 (2010)
15. Kehrer, J., Ladstädter, F., Muigg, P., Doleisch, H., Steiner, A., Hauser, H.: Hypothesis generation in climate research with interactive visual data exploration. IEEE Trans. Visu. Comput. Graph. **14**, 1579–1586 (2008)
16. Kehrer, J., Muigg, P., Doleisch, H., Hauser, H.: Interactive visual analysis of heterogeneous scientific data across an interface. IEEE Trans. Vis. Comput. Graph. **17**(7), 934–946 (2011)
17. Keim, D., Kohlhammer, J., Ellis, G., Mansmann, F. (eds.): Mastering the Information Age-Solving Problems with Visual Analytics. Eurographics (2010)
18. Kosara, R., Miksch, S., Hauser, H.: Semantic depth of field. In: IEEE Symposium on Information Visualization pp. 97–104 (2001)
19. Kosara, R., Sahling, G.N., Hauser, H.: Linking scientific and information visualization with interactive 3D scatterplots. In: Short Communication Papers Proceedings of the 12th International Conference in Central Europe on Computer Graphics, Visualization, and Computer Vision (WSCG), pp. 133–140 (2004)
20. Ladstädter, F., Steiner, A.K., Lackner, B.C., Pirscher, B., Kirchengast, G., Kehrer, J., Hauser, H., Muigg, P., Doleisch, H.: Exploration of climate data using interactive visualization. J. Atmos. Oceanic Technol. **27**(4), 667–679 (2010)
21. Lokuge, I., Ishizaki, S.: Geospace: an interactive visualization system for exploring complex information spaces. In: Proceedings of the ACM CHI '95 Conference on Human Factors in Computing Systems, pp. 409–414 (1995)
22. Matković, K., Gračanin, D., Jelović, M., Ammer, A., Lež, A., Hauser, H.: Interactive visual analysis of multiple simulation runs using the simulation model view: Understanding and tuning of an electronic unit injector. IEEE Trans. Vis. Comput. Graph. **16**(6), 1449–1457 (2010)
23. Muigg, P., Kehrer, J., Oeltze, S., Piringer, H., Doleisch, H., Preim, B., Hauser, H.: A four-level focus+context approach to interactive visual analysis of temporal features in large scientific data. Comput. Graph. Forum (Proceedings of EuroVis 2008) **27**, 775–782 (2008)
24. Novotný, M., Hauser, H.: Similarity brushing for exploring multidimensional relations. J. WSCG **14**(1–3), 105–112 (2006)
25. Oeltze, S., Doleisch, H., Hauser, H., Muigg, P., Preim, B.: Interactive visual analysis of perfusion data. IEEE Trans. Vis. Comput. Graph. **13**, 1392–1399 (2007). doi: http://doi.ieeecomputersociety.org/10.1109/TVCG.2007.70569

26. Piringer, H., Kosara, R., Hauser, H.: Interactive focus+context visualization with linked 2D/3D scatterplots. In: Proceedings of the Second International Conference on Coordinated & Multiple Views in Exploratory Visualization (CMV'04), pp. 49–60. IEEE Computer Society, Washington, DC, (2004). doi: http://dx.doi.org/10.1109/CMV.2004.11
27. Rübel, O., Prabhat, Wu, K., Childs, H., Meredith, J., Geddes, C.G.R., Cormier-Michel, E., Ahern, S., weber, G.H., Messmer, P., Hagen, H., Hamann, B., Bethel, E.W.: High performance multivariate visual data exploration for extemely large data. In: SuperComputing 2008 (SC08) (2008)
28. Rübel, O., Weber, G.H., Huang, M.Y., Bethel, E.W., Biggin, M.D., Fowlkes, C.C., Hendriks, C.L., Keränen, S.V.E., Eisen, M.B., Knowles, D.W., Malik, J., Hagen, H., Hamann, B.: Integrating data clustering and visualization for the analysis of 3d gene expression data. IEEE Trans. Comput. Biol. Bioinform. **7**(1), 64–79 (2010)
29. Stockinger, K., Shalf, J., Wu, K., Bethel, E.W.: Query-driven visualization of large data sets. In: Proceedings of IEEE Visualization 2005, pp. 167–174. IEEE Computer Society Press (2005)
30. Thomas, J.J., Cook, K.A. (eds.): Illuminating the path: the research and development agenda for visual analytics (2005). Available at URL http://nvac.pnl.gov/agenda.stm
31. Wang Baldonado, M.Q., Woodruff, A., Kuchinsky, A.: Guidelines for using multiple views in information visualization. In: AVI '00: Proceedings of the Working Conference on Advanced Visual Interfaces, pp. 110–119. ACM Press, New York (2000). doi: http://doi.acm.org/10.1145/345513.345271
32. Weber, G.H., Rübel, O., Huang, M.Y., DePace, A.H., Fowlkes, C.C., Keränen, S.V., Luengo Hendriks, C.L., Hagen, H., Knowles, D.W., Malik, J., Biggin, M.D., Hamann, B.: Visual exploration of three-dimensional gene expression using physical views and linked abstract views. IEEE Trans. Computat. Biol. Bioinform. **6**(2), 296–309 (2009). doi: 10.1109/TCBB.2007.70249.
33. Wu, K., Otoo, E., Shoshani, A.: Optimizing bitmap indices with efficient compression. ACM Trans. Database Syst. **31**, 1–38 (2006)

Chapter 16
Visual Exploration of Multivariate Volume Data Based on Clustering

Lars Linsen

Abstract The attribute space of a multivariate volume data set is hard to handle interactively in the context of volume visualization when more than three attributes are involved. Automatic or semi-automatic approaches such as involving clustering help to reduce the complexity of the problem. Clustering methods segment the attribute space, and the segmentation can be exploited for visual exploration of the volume data. We discuss user-guided and automatic clustering approaches of the multi-dimensional attribute space and visual representations of the results. Coordinated views of object-space volume visualization with attribute-space clustering results can be applied for interactive visual exploration of the multivariate volume data and even for interactive modification of the clustering results. Respective methods are presented and discussed and future directions are outlined.

16.1 Introduction

Volume visualizations rely on some segmentation of the given volumetric domain. Typical examples are the choice of an isovalue for isosurface extraction from a scalar field or the application of a one-dimensional transfer function to the range of a scalar field for direct volume rendering. Such a segmentation of the volumetric domain is implicitly given by segmenting the range of the scalar field. Some early approaches extended this idea to the segmentation of 2D or even 3D spaces formed by the range and some derived properties such as magnitude of first- and second-order derivatives. The segmentation in these spaces are performed interactively by providing respective interaction methods and widgets. This is possible as long as the interaction takes place in 2D or, with some limitations, in 3D visual spaces. These visual spaces are often obtained by histogram computations and their visualization by using color coding. Interaction mechanisms, then, allow the user to select regions of interest and assign parameters for the volume visualization methods to them. For example, the idea of using multi-dimensional transfer functions to direct volume rendering goes back to

L. Linsen (✉)
Jacobs University, Bremen, Germany
e-mail: l.linsen@jacobs-university.de

C.D. Hansen et al. (eds.), *Scientific Visualization*, Mathematics and Visualization,
DOI 10.1007/978-1-4471-6497-5_16

Kindlmann and Durkin [21]. They proposed using 2D histograms of the value and the magnitudes of first- or second-order directional derivatives of a scalar field to define transfer functions. Kniss et al. [22] extend their work by introducing a set of direct manipulation widgets for multi-dimensional transfer functions. For vector fields, Daniels II et al. [9] presented an approach for interactive vector field exploration by brushing on a 2D scatterplot of two derived scalar properties of the vector field.

When dealing with multivariate volume data, the number of attributes per sample is typically (significantly) larger than two or three. Hence, purely interactive methods are of limited use for selecting regions of interest. Automatic components help to alleviate the problem. The segmentation of the multi-dimensional attribute space can be achieved by employing a clustering method. Automatic clustering methods of multi-dimensional spaces is an intensively researched topic and many different methods can be applied, see Sect. 16.2. Also, clustering methods with user guidance are of interest in this regard, see Sect. 16.3.

With the obtained clustering result, the volume visualization parameters can be set to highlight the areas in object (or physical) space that correspond to the clusters in attribute space. The mapping from the clustering result to the volume visualization parameters can be obtained automatically, but due to occlusion one needs to restrict the volume visualization to a subset of clusters. It is desirable to have an interactive selection of these subsets. This selection mechanism replaces the interactive operation in a high-dimensional attribute space. However, it still requires some suitable visual encoding of the clustering result to allow for intuitive interactions, see Sect. 16.4. The requirements to this visual encoding are that it scales well in the number of dimensions and in the number of samples, as we may be dealing with a larger number of attributes and the underlying multivariate field is typically sampled at many positions of its volumetric domain.

The visual encodings of the clustering result lead to coordinated views of the object-space volume visualization and the attribute-space cluster visualization, see Sect. 16.5. Multiple visual encodings of the clustering results and its value distributions in attribute space may be coupled with the volume visualization to allow for an interactive exploration of the clustering result.

Finally, one may also need to consider that the automatic part of the pipeline, i.e., the clustering step, may not produce the optimal results. This may be due to some limitations of the clustering methods or due to the fact that the user may bring in some domain expertise that goes beyond what one can extract from the raw data. Consequently, an interactive modification of the clustering result is also of interest, see Sect. 16.6.

In this chapter, we present and discuss different approaches for the individual steps described above. We conclude the paper with open problems and future directions, see Sect. 16.7.

16.2 Automatic Clustering of Attribute Space

Cluster analysis divides data into meaningful or useful groups (clusters). Clustering algorithms can be categorized with respect to their properties of being based on partitioning, hierarchical, based on density, or based on grids. In partitioning methods, data sets are divided into an number of clusters and each object must belong to exactly one cluster. In hierarchical methods, data sets are represented using similarity trees and clusters are extracted from this hierarchical tree. In density-based methods, clusters are a dense region of points separated by low-density regions. In grid-based methods, the data space is divided into a finite number of cells that form a grid structure and all of the clustering operations are performed on the cells. A complete review of existing methods is beyond the scope of this chapter, but we refer to respective survey papers [13, 18].

In the context of multivariate volume data visualization, Maciejewski et al. [29] proposed to apply a clustering of the attribute space to the 2D histogram obtained by the range of a scalar function and the magnitude of its gradients. The clustering is being visualized as a segmented image of the 2D histogram using color coding of the segments. This segmented image serves as a user interface to select clusters and display them in a volume visualization. This approach is limited to two-dimensional attribute spaces.

When dealing with higher-dimensional attribute spaces, it is favorable to use a hierarchical clustering method, as the cluster hierarchy can be used for visual encoding of the clustering result. Moreover, a density-based approach is desirable, as the computed density values can be assigned to the respective sample points in object space leading to a volumetric density field. This density field can be exploited for volume visualization methods. Density-based clustering methods are either kernel-based or grid-based. Kernel-based methods convolve each point with some kernel function and sum up the contributions of all kernels to compute the density. Grid-based methods subdivide the space into cells, count the number of points that fall into each cell, and compute the density as the number of points within a cell divided by its area. Typically, the grid is a uniform, rectilinear one, i.e., it represents a multi-dimensional histogram. In multivariate volume data, attribute values may be given at different scales. Hence, when using kernel-based methods one would need to apply anisotropic kernels. How to choose the scaling in the individual dimensions becomes an issue. Thus, grid-based methods are favorable, as they split each dimension into a number of cells independent of the scales the attributes are given in.

Linsen et al. [25–28] presented an approach for multivariate volume data visualization that uses a hierarchical density-based approach where densities are computed over a grid. The main advantage of that approach over other techniques with similar properties is the direct identification of clusters without any threshold parameter of density level sets. This property is achieved by the observation that density is proportional to the number of points per cell when assuming cells of equal size. Hence, density becomes an integer value and one can iterate over the density values. To estimate all non-empty cells, a partitioning algorithm is used that iterates through all dimensions. Given the multi-dimensional histogram, clusters are defined as largest sets of neighboring non-empty cells, where neighboring refers to sharing a common vertex. To detect higher-density clusters within each cluster, all cells containing the minimum number of points in this cluster are removed and one detects among the remaining cells, again, largest sets of neighboring cells. This step may lead to splitting of a cluster into multiple higher-density clusters. This process is iterated until no more clusters split. Recording the splitting information, one obtains the cluster hierarchy. Those clusters that do not split anymore represent local maxima and are referred to as mode clusters. The approach scales well with the number of dimensions, since only non-empty cells are being stored. Moreover, it is quite efficient, since the histogram is computed by iterating over the dimensions and subdividing the non-empty cells only. For a detailed analysis and comparison of this clustering approach to other clustering approaches, we refer to the thesis by Long [40]. In particular, it is shown that this clustering approach is capable of extracting clusters of any shape including clusters with concave areas and even holes.

Another suitable approach for defining clusters in multivariate volume data is to use statistical methods to evaluate correlations between the different attributes and identify regions of the object space with matching correlations. Jänicke et al. [19] propose to use mutual information to detect regions of a certain behavior in multivariate volume data. This approach is not anymore build upon the idea of clustering in attribute space but rather on segmenting the object space directly. Similarly, one can also apply standard segmentation algorithms and generalize them such that they can deal with multivariate volume data. Ivanovska [16] presented a segmentation approach based on the concept of minimum description length of the encoding of the segments. It can be derived from information theory that the segmentation with minimum description length for the segments boundaries represents an optimum. The original idea was extended to a multivariate setting, where the minimum description length takes into account all attributes such that the optimization strives for getting the segments boundary to coincide. Figure 16.1 shows a result of the approach.

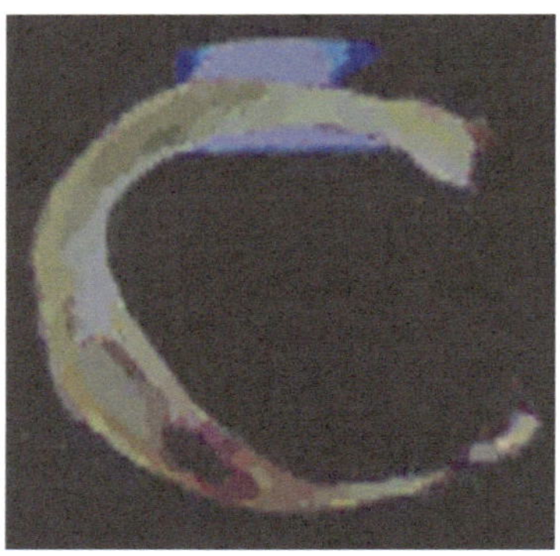

Fig. 16.1 Automatic segmentation of multivariate medical imaging data: segments are derived from three attributes obtained using T1-, T2-, and T2*-weighted MRI imaging of plaque using a multivariate version of the minimum description length approach. Segments are shown as regions of constant color. (Data courtesy of Andreas Harloff and Michael Markl, University Hospital Freiburg, Germany.)

16.3 User-Guided Clustering of Attribute Space

Instead of using completely automatic clustering approaches, it may be favorable to keep the user in the loop. A simple way of doing this is to start with a coarse clustering result and iteratively refine it in an adaptive manner. Ivanovska and Linsen [17] presented an approach, where simple and efficient clustering methods such as median cut [14], *k*-means [30], and *c*-means [7] are used to cluster multivariate volume data, the results are visualized using a 2D slice viewer, and clusters are selected to adaptively refine them by further splitting (i.e., clustering) operations. Because of the simple clustering methods used, the process is highly interactive. The approach was only applied to RGB color data, where the clusters can be represented by their average color, which makes it easy to identify clusters for adaptive refinement.

A different approach for user-guided clustering is to interact in object space. The user brushes in the volume visualization to select some samples and assigns to them a cluster ID. Based on this sparse information about the desired clustering result, a clustering of the full data set is obtained using machine learning techniques. Tzeng et al. [38] presented an approach, where the user brushes on 2D slices using different colors, where the colors function as cluster identifiers. The selected voxels of the data set are used as training set for a neural network. With the trained network, the entire data set in classified. The neural network is implemented on the GPU to assure that computations can be done in an interactive setting. Thus, the selection can be modified anytime by further brushing operations. The updated training set is then fed to the neural network classifier, again. Tzeng et al. applied their approach to scalar fields and RGB color data. El-Moasry et al. [11] presented an approach that builds on this idea, but uses a rough set classifier instead of a neural network. The rough set classifier reports back probabilities that a certain voxel belongs to a certain cluster. The probability information can be used for uncertainty visualization. Figure 16.2 shows a result of the approach.

Dobrev et al. [10] present an approach that, in principle, embeds an automatic clustering approach. However, it is shown that the clustering can also be generated completely interactively by brushing on a parallel coordinates plot representing the attribute space and seeing the selection in a linked volume rendering of the object space. Although it is possible to create some meaningful clusters, the approach would be very cumbersome for extracting higher-dimensional clusters with non-axis aligned shapes.

16.4 Visual Encoding of Clustering Result

16.4.1 Object-Space Representation

The simplest way of displaying the result of clustering a multi-dimensional attribute space is to use volume visualizations of the object-space representation of each cluster and show the visualizations all next to each other. Tzeng and Ma [39]

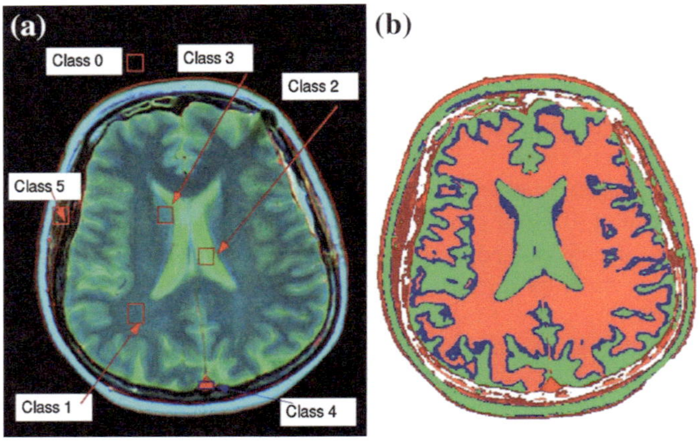

Fig. 16.2 User-guided clustering based on brushing in object space and using machine-learning techniques to classify all data from the brushed selection: **a** Assigning the training set by interactively selecting image regions and respective classes (or clusters). **b** Medical color imaging data segmentation based on rough classification, where *blue color* is used to visualize an area of uncertainty between the two adjacent clusters. (Data set courtesy of Art Toga, UCLA, USA.)

follow this idea. They applied ISODATA clustering to the attribute space and show object-space visualizations of all clusters by using one display per cluster. They allow the users to directly interact with the extracted clusters. User interactions allow for assigning material properties to clusters, i.e., one can show selected clusters in one display with different appearance properties. User interactions also include merging, splitting, growing, and shrinking of clusters.

16.4.2 Cluster Hierarchy

When using a hierarchical clustering approach, the cluster structure can be represented using a tree data structure. The cluster tree can be visualized using conventional tree visualization methods such as nodelink diagrams, where nodes represent clusters and links represent parent-child relationships. Different layouts of the nodelink diagram are applicable. Linsen et al. [25–27] visually encode the cluster tree using a nodelink diagram with radial layout. The size of the nodes encode the size of the clusters and the color of the nodes represent the respective colors that are used in the volume visualization of the clusters.

16.4.3 Projection

The clusters can also be visualized using an attribute space visualization. The attribute space structure is given in form of a set of points in a multi-dimensional space. As the number of dimensions is typically larger than three, the space must be projected into a 2D or 3D visual space. In the visual space, clusters can be indicated using enclosing curves or surfaces or by color coding.

Many algorithms for projections exist. The simplest projection of the multi-dimensional attribute space is to project all data points to a visual space that is spanned by two of the attribute dimension. This projection leads to a standard 2D scatter plot. In the context of multivariate volume data visualizations, standard 2D scatter plots can be extended to continuous scatter plots [5, 6, 15, 23, 24]. Continuous scatter plots generalize the concept of scatter plots to the visualization of spatially continuous input data by a continuous and dense plot. It uses the spatial neighborhood in the object space to allow for applying an interpolation between the points in the attribute space.

Blaas et al. [8] use scatter plots in attribute space, where the multi-dimensional data is projected into arbitrary planes. In general, one refers to multi-dimensional scaling for techniques that project high-dimensional data into a low-dimensional visual space. Given a set of d-dimensional data points $\{p_1, \ldots, p_n\}$, multidimensional projection techniques apply some criterion to generate a representation of the points in an m-dimensional space with $m \leq d$. A possible criterion is to preserve as much as possible the neighborhood relationships amongst the original points. The projected points are the input to visual representations that convey information about groups of elements with similar or dissimilar behavior. Several classical techniques like Sammon Mapping [34], FastMap [12] or more recent techniques like Nearest-Neighbor Projection [37] or ProjClus [31] are described in the literature to handle different high-dimensional data in different ways. A somewhat outdated survey paper is given by König [4]. Least Square Projection [32] is a multidimensional projection technique that effectively handles large data sets characterized by a sparse data distribution in the high-dimensional space.

Takanashi et al. [36] applied Independent Component Analysis (ICA) on a multi-dimensional histogram to classify the volume domain. Classification becomes equivalent to interactive clipping in the ICA space. Paulovich et al. [33] presented an efficient two-phase mapping approach that allows for fast projection of large data sets with good projection properties. They applied their approach to multivariate volume data and were even able to handle time-varying data.

The described projection techniques are based on a set of high-dimensional data points as input. When pre-clustering the data, one can optimize the projection such that clusters stay as much separated as possible. Linsen et al. [25] are using a respective projection technique for projecting clusters into a visual space whose layout is given in form of optimized star coordinates. The main idea of the approach is to represent the mode clusters of the cluster hierarchy by their barycenter and project the centers using a linear contracting projection that maximizes the distance between the clusters' barycenters. Since the method is linear, distances in the projected space still allow for some interpretations. The contraction property assures that clusters stay together and do not fall apart. The maximization of the distance between the barycenters when projecting assures that separated clusters stay separated as much as possible. Finally, the projection uses a star coordinate approach, where the directions,

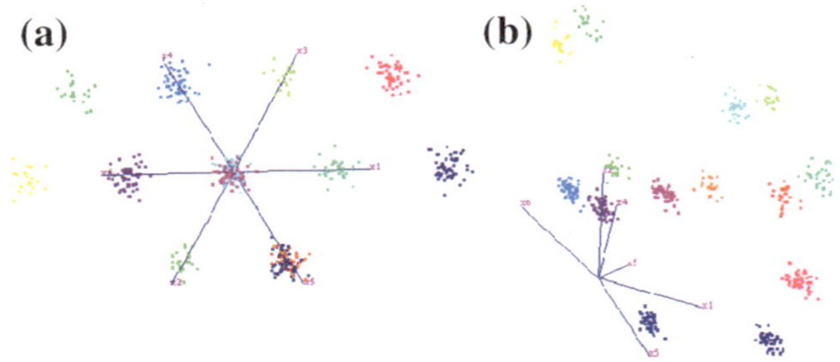

Fig. 16.3 Visual encoding of clustering result using color after projection into a 2D visual space. Comparison of projections into standard star coordinates (*left*) and optimized star coordinates (*right*). The optimized star coordinates minimize the overlap of projected clusters

scaling, and order of the axes are optimized by the derived projection. Displaying the projected clusters in this optimized star coordinates plot allows for relating the clusters to the original attribute dimensions. Figure 16.3 shows the projection of a six-dimensional data set into optimized star coordinates and compares the projection to standard star coordinates (with optimal permutation of the coordinate axes) [20]. The clusters are visually encoded in the projection by using color. The optimized star coordinates achieve to keep all clusters separated, while the standard star coordinates fail to do so. This can also be quantified using the distance consistency measure introduced by Sips et al. [35]: The standard star coordinates achieved a value of 82.71, while the optimized star coordinates has the optimal value of 100.

16.4.4 Parallel Coordinates

A common approach to visualize multi-dimensional data visualization is the use of parallel coordinates plots. When compared to projected views, parallel coordinates plots have the advantage that the attribute values can be visually retrieved in a non-ambiguous way. Moreover, correlations between attributes can be observed. However, the latter is subject to an appropriate ordering of the attribute dimensions in the parallel coordinates plot. While parallel coordinates plots scale well in the number of attribute dimensions (when compared to other techniques), one of the main drawbacks of parallel coordinates plots is that they do not scale so well in the number of samples. While scatter plots still work for large number of samples, parallel coordinates plots tend to suffer from visual clutter. Since we are typically dealing with many samples in the context of multivariate volume data, it is advisable to not display all the individual samples of the clusters, but rather have a more abstract representation of the clusters. For example, Dobrev et al. [10] display a band (in form of a sequence of quadrilaterals) for each cluster instead of set of polylines. The opacity of the band reflects the line density of the parallel coordinates plot for that cluster. Using parallel coordinates plots for the clustering result does not only exhibit the individual attribute values for the clusters, but also visually encodes how homogeneous or heterogeneous a cluster is. It can even reveal whether a cluster splits into subclusters in one of the attributes.

16.5 Coordinated Views for Visual Exploration of Clustering Result

The visual encodings presented in the previous section all represent the clustering result, but they do reveal different properties of the clusters: The object-space representation exhibits the distribution of the clusters within the volumetric domain, the cluster hierarchy encodes the structure of the clusters, a projection shows the distribution of the clusters within the multi-dimensional attribute space, and the parallel coordinates allow for retrieving individual attribute values. Consequently, it is desirable to have all those visual encodings embedded into a visual exploration system. The system would provide different views on the data, where the views shall be coordinated, i.e., any interactions like selection or filtering that are performed in one view shall simultaneously also be applied to all the other views. Then, all views provide one coherent snapshot of the data.

Akiba et al. [2, 3] presented a system that operates with coordinated views on multivariate volume data. In particular, they used a representation of the multi-variate data in parallel coordinates to allow the user to generate a transfer function by brushing regions of interest. They even add another aspect to it, as they are dealing with time-varying multivariate data, where another coordinated view shows a visual encoding of changes over time. However, their approach is not based on the clustering idea such that the amount of necessary user interaction increases with increasing dimensionality and may at some point get cumbersome.

Dobrev et al. [10] follows the idea of using clustering to provide intuitive operations in cluster space. Like for Akiba et al., the system is based on an intuitive user interface, but the combination with hierarchical density-based clustering has the benefit that it scales better to high dimensionality. Object-space representation of the clusters is achieved using a GPU-based volume rendering approach, to which only the cluster IDs and the density values from the clustering approach are handed. Each cluster is assigned a unique color and opacity value according to the user selections. The gradients of the density values can be used to obtain appropriate surface normals for the clusters, which are needed for illumination. The cluster tree is visually encoded as a radial nodelink diagram, which serves as the main interaction widget for selecting clusters and assigning material properties. Parallel coordinates plots are used to show the attribute values of the selected clusters. Figure 16.4 shows the system with the three coordinated views.

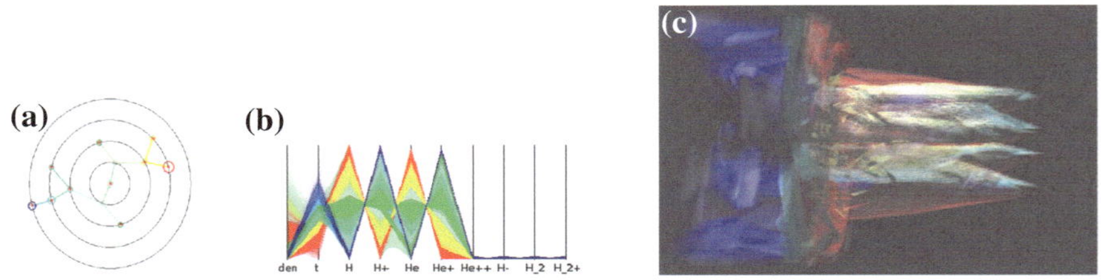

Fig. 16.4 Coordinated views for visual exploration of clustering result: **a** Cluster tree visualization of a hierarchical density-based clustering approach. **b** Parallel coordinates plots of selected clusters. **c** 3D texture-based volume rendering of selected clusters with selected material properties. System is applied to a physical ionization front instability simulation dataset with a ten-dimensional feature space. (Data set provided in the 2008 IEEE Visualization Design Contest [41])

16.6 Interactive Modification of Clustering Result

User-guided clustering approaches like the ones by Ivanovska and Linsen [17] or Tzeng and Ma [39] allow for a modification of the clustering result within their user-guided clustering framework as described above. Since automatic clustering approaches may also produce results that are subject to manual improvements and since the user may bring in some domain expertise that may go beyond what can be achieved with fully automatic approaches, it is also desirable to allow modifications of the automatic clustering results. In particular, it is desirable to couple the visual exploration described in the previous section with means to interactively adjust clusters. As the visual exploration process may lead to new insights about the clusters under investigation, those new insights shall be documented, e.g., by splitting a cluster into two smaller clusters.

The system of coordinated views by Dobrev et al. [10] as shown in Fig. 16.4 also allows for an interactive modification of the clusters. First, since density values are given, clusters can be shrunk (and grown again) by adjusting the density level that corresponds to the density cluster. Second, the parallel coordinates plot also serves as an interactive widget. Brushing on the individual axes of the parallel coordinates plot induces a selection that is directly reflected in the linked volume renderer. If the interactions on the parallel coordinates plot lead to the detection of a certain subcluster within the selected cluster, the cluster can be split appropriately. Third, clusters can be merged by selecting them in the cluster tree widget and applying a merge operation.

16.7 Conclusions and Future Directions

We have presented approaches for multivariate volume visualization that are based on clustering methods. Clustering is applied to capture the information in the multidimensional attribute space of the multivariate volume data into regions of similar behavior. This aggregation step allows for interactive visual analyses of the data. While purely interactive methods have limits when trying to operate in a multi-dimensional attribute space, aggregation via clustering allows for new visual encodings of the attribute space. Different visual encodings of the clustering result provide different information such that a system with different coordinated views on the clustering result proves to be most efficient. Those coordinated views shall also allow for interactive modification of the clustering result during the exploration process, especially if the user can bring in some additional domain expertise.

The presented approaches mainly deal with multivariate data in the sense of multiple scalar fields, where the scalar fields may be given dimensions or may be derived attributes. In particular, the attributes may be derived from vector or tensor fields. The most obvious future direction is to extend the approaches to time-varying data. Akiba and Ma [1] did incorporate the time dimension into their interactive system, however, it was not based on the clustering idea. Still, the ideas may be applicable to cluster-based approaches. Next, it is of interest to also incorporate spatial information into the clustering approach such that one cluster always represents a connected region in object space. Local spatial relationships have been used, e.g., to draw statistics [11], which allows for further differentiation between regions of similar values but different textures. However, these local measures do not guarantee any global properties. In terms of the clustering itself, scaling and normalization issues of the individual dimensions may need further investigation, especially when simultaneously considering original attributes, derived attributes, spatial dimensions, time dimension, etc. Non-linear scalings may be appropriate to use. Finally, it has not yet been investigated how these approaches generalize to multi-run or ensemble data, although a cluster-based visualization of the parameter space is also feasible.

References

1. Akiba, H., Ma, K.L.: A tri-space visualization interface for analyzing time-varying multivariate volume data. In: Proceedings of Eurographics/IEEE VGTC Symposium on Visualization, pp. 115–122, May 2007
2. Akiba, H., Ma, K.L: A tri-space visualization interface for analyzing time-varying multivariate volume data. In: EuroVis07—Eurographics/IEEE VGTC Symposium on Visualization, pp. 115–122, May 2007
3. Akiba, H., Ma, K.L., Chen, J.H., Hawkes, E.R.: Visualizing multivariate volume data from turbulent combustion simulations. Comput. Sci. Engg. **9**(2), 76–83 (2007)
4. Andreas K.ö.: A survey of methods for multivariate data projection, visualisation and interactive analysis. In: Yamakawa, T. (ed.) 5th International Conference on Soft Computing and Information/Intelligent Systems, pp. 55–59, Iizuka, Japan, Oct 16–20 (1998)
5. Bachthaler, S., Weiskopf, D.: Continuous scatterplots. IEEE Trans. Vis. Comput. Graph. (Proceedings Visualization/Information Visualization 2008), **14**(6), 1428–1435, Nov–Dec 2008
6. Bachthaler, S., Weiskopf, D.: Efficient and adaptive rendering of 2-d continuous scatterplots. vol. 28, pp. 743–750 (2009)
7. Bezdek, J.C.: Pattern recognition with fuzzy objective function algorithms. Plenum Press, New York (1981)

8. Blaas, J., Botha, C.P., Post, F.H.: Interactive visualization of multi-field medical data using linked physical and feature-space views. In: EuroVis, pp. 123–130 (2007)
9. Daniels, J. II., Anderson, E.W., Nonato, L.G., Silva, C.T.: Interactive vector field feature identification. IEEE Trans. Vis. Comput. Graph. **16**, 1560–1568 (2010)
10. Dobrev, P., Van Long, T., Linsen, L. : A cluster hierarchy-based volume rendering approach for interactive visual exploration of multi-variate volume data. In: Proceedings of 16th International Workshop on Vision, Modeling and Visualization (VMV 2011), pp. 137–144. Eurographics Association (2011)
11. Elmoasry, A., Maswadah, M.S., Linsen, L.: Semi-automatic rough classification of multichannel medical imaging data. In: Visualization in Medicine and Life Sciences II (to appear). Springer, Heidelberg (2011)
12. Faloutsos, C., Lin, K.I.: Fastmap: a fast algorithm for indexing, data-mining and visualization of traditional and multimedia datasets. In: SIGMOD '95: Proceedings of the 1995 ACM SIGMOD international Conference on Management of data, pp. 163–174. ACM Press, New York (1995)
13. Han, J., Kamber, M.: Data Mining: Concepts and Techniques. Morgan Kaufmann Publishers, Los Altos (2006)
14. Heckbert, P.: Color image quantization for frame buffer display. In: Computer Graphics (Proceedings of ACM SIGGRAPH 82), pp. 297–307 (1982)
15. Heinrich, J., Bachthaler, S., Weiskopf, D.: Progressive splatting of continuous scatterplots and parallel coordinates. Comput. Graph. Forum **30**(3), 653–662 (2011)
16. Ivanovska, T: Efficient multichannel image partitioning: theory and application. PhD thesis, School of Engineering and Science, Jacobs University, Bremen, Germany (2009)
17. Ivanovska, T., Linsen, L: A user-friendly tool for semi-automated segmentation and surface extraction from color volume data using geometric feature space operations. In: Visualization in Medicine and Life Sciences, pp. 145–162, 319. Springer, Heidelberg (2007)
18. Jain, A.K., Dubes, R.C.: Algorithms for Clustering Data. Prentice Hall, Englewood Cliffs (1988)
19. Jänicke, H., Wiebel, A., Scheuermann, G., Kollmann, W.: Multifield visualization using local statistical complexity. IEEE Trans. Vis. Comput. Graph. **13**(6), 1384–1391 (2007)
20. Kandogan, E.: Star coordinates: a multi-dimensional visualization technique with uniform treatment of dimensions. In: Proceedings of IEEE Information Visualization Symposium (Hot Topics), pp. 4–8 (2000)
21. Kindlmann, G., Durkin, J.W.: Semi-automatic generation of transfer functions for direct volume rendering. In: IEEE Symposium on Volume Visualization, pp. 79–86 (1998)
22. Kniss, J., Kindlmann, G., Hansen, C.: Interactive volume rendering using multi-dimensional transfer functions and direct manipulation widgets. In: VIS '01: Proceedings of the Conference on Visualization '01, pp. 255–262. IEEE Computer Society, Washington, DC, USA (2001)
23. Lehmann, D.J., Theisel, H.: Discontinuities in continuous scatter plots. IEEE Trans. Vis. Comput. Graph. **16**, 1291–1300 (2010)
24. Lehmann, D.J., Theisel, H.: Features in continuous parallel coordinates. IEEE Trans. Vis. Comput. Graph. **17**(12), 1912–1921 (2011)
25. Linsen, L., Van Long, T., Rosenthal, P., Rosswog, S.: Surface extraction from multi-field particle volume data using multi-dimensional cluster visualization. IEEE Tran. Vis. Comput. Graph. **14**(6), 1483–1490 (2008)
26. Linsen, L., Van Long, T., Rosenthal, P.: Linking multi-dimensional feature space cluster visualization to surface extraction from multi-field volume data. IEEE Comput. Graph. Appl. **29**(3), 85–89 (2009)
27. Van Long, T., Linsen, L.: MultiClusterTree: Interactive visual exploration of hierarchical clusters in multidimensional multivariate data. Comput. Graph. Forum **28**(3), 823–830 (2009)
28. Van Long, T., Linsen, L.: Visualizing high density clusters in multidimensional data using optimized star coordinates. Computat. Stat. **26**, 655–678 (2011)
29. Maciejewski, R., Woo, I., Chen, W., Ebert, D.: Structuring feature space: a non-parametric method for volumetric transfer function generation. IEEE Trans. Vis. Comput. Graph. **15**(6), 1473–1480 (2009)

30. MacQueen, J.B.: Some methods for classification and analysis of multivariate observations. In: Proceedings of 5-th Berkeley Symposium on Mathematical Statistics and Probability, vol. 1, pp. 281–297 (1967)
31. Paulovich, F.V., Minghim, R.: Text map explorer: a tool to create and explore document maps. In: Proocedings of the 10th International Conference on Information Visualisation—IV, pp. 245–251. IEEE CS Press. London (2006)
32. Paulovich, F.V., Nonato, L.G., Minghim, R., Levkowitz, H.: Least square projection: a fast high-precision multidimensional projection technique and its application to document mapping. IEEE Trans. Vis. Comput. Graph. **14**(3), 564–575 (2008)
33. Paulovich, F.V., Silva, C., Nonato, L.G.: Two-phase mapping for projecting massive data sets. IEEE Trans. Vis. Comput. Graph. **16**, 1281–1290 (2010)
34. Sammon, J.W.: A nonlinear mapping for data structure analysis. IEEE Trans. Comput. **18**(5), 401–409 (1969)
35. Sips, M., Neubert, B., Lewis, J.P., Hanrahan, P.: Selecting good views of high-dimensional data using class consistency. Comput. Graph. Forum **28**(3), 831–838 (2009)
36. Takanashi, I., Lum, E.B., Ma, K.L., Muraki, S.: Ispace: interactive volume data classification techniques using independent component analysis. In: Pacific Conference on Computer Graphics and Applications, p. 366 (2002)
37. Tejada, E., Minghim, R., Nonato, L.G.: On improved projection techniques to support visual exploration of multidimensional data sets. Inf. Vis. 2(4), 218–231 (2003)
38. Tzeng, F.Y., Lum, E.B., Ma, K.L.: A novel interface for higher-dimensional classification of volume data. In: VIS '03: Proceedings of the 14th IEEE Visualization 2003 (VIS'03), pp. 66. IEEE Computer Society, Washington, DC (2003)
39. Tzeng, F.Y., Ma, K.L.: A cluster-space visual interface for arbitrary dimensional classification of volume data. In: Proceedings of Joint Eurographics-IEEE TVCG Symposium on Visualization, pp 17–24 (2004)
40. Van Long, T.: Visualizing high-density clusters in multidimensional data. PhD thesis, School of Engineering and Science, Jacobs University, Bremen, Germany (2009)
41. Whalen, D. Norman, M.L.: Competition data set and description. In: IEEE Visualization Design Contest (2008). http://vis.computer.org/VisWeek2008/vis/contests.html

Chapter 17
Feature-Based Visualization of Multifields

Harald Obermaier and Ronald Peikert

Abstract Feature-based techniques are one of the main categories of methods used in scientific visualization. Features are structures in a dataset that are meaningful within the scientific or engineering context of the dataset. Extracted features can be visualized directly, or they can be used indirectly for modifying another type of visualization. In multifield data, each of the component fields can be searched for features, but in addition, there can be features of the multifield which rely on information form several of its components and which cannot be found by searching in a single field. In this chapter we give a survey of feature-based visualization of multifields, taking both of these feature types into account.

17.1 Feature Extraction in Scientific Visualization

Scientific visualization has adopted the concept of a *feature* from computer vision [6], where it describes a salient structure of an image. Some of the most common image features are edges, ridges, corners, and blobs. Features are important in various applications, such as object recognition and tracking.

In visualization, features are used to put a focus on those parts that are of interest in the context of a certain research or engineering problem. For example, in flow data sets, features can be shock waves, vortices, recirculation, boundary layers, and separation and attachment lines. By restricting the visualization of a dataset to its features, its visual complexity can be substantially reduced. A visualization can consist of only the extracted features, together with some context information, but

H. Obermaier (✉)
UC Davis, Davis, USA
e-mail: hobermaier@ucdavis.edu

R. Peikert
ETH Zurich, Zurich, Switzerland
e-mail: peikert@inf.ethz.ch

C.D. Hansen et al. (eds.), *Scientific Visualization*, Mathematics and Visualization,
DOI 10.1007/978-1-4471-6497-5_17

more often features, which are abstract objects, are visualized in combination with more traditional techniques. Furthermore, features can be used to guide the placement of visualization objects such as glyphs or streamline seeds [3, 36, 38]. Finally, in time-dependent datasets, features can be tracked over time [29], providing information on the dynamics of processes. Feature-based visualization methods have been developed for scalar, vector, and tensor fields in a wide range of application areas such as fluid flow [27] and medical visualization [4].

17.2 Multifield Feature Definitions

17.2.1 Single-Field Versus Multifield Features

Multifield features and their definitions can only be fully understood in the context of single-field feature definitions. While multifield and single-field features share common properties, they often present researchers and users with additional challenges with respect to data size, limitations of visual space, and conceptional differences as highlighted in the following.

Feature definitions and visualization techniques in a single-field setting benefit from a set of assumptions. Often features are free of spatial overlaps, have equal priority with respect to the visual space they occupy if no feature strength is measured, and, most importantly, the feature definition process is a one-to-many mapping from data to feature space. Contrarily, the feature extraction and visualization process for a multifield data set is inherently a many-to-many mapping, thus introducing an additional dimension of complexity. Where a single-field feature extraction technique may produce a set of features for different feature definitions on the same field that are combined into a final visualization, features in a multifield data set can be created by feature definitions based on a single-field, multiple heterogeneous fields, or, in the most complex case, multiple heterogeneous fields.

In situations, where feature based visualization consists of visual blending of single-field features, the feature extraction and visualization process is extremely similar to that in a single-field setting. True combined feature extraction techniques are unique to multifield data in such that the geometric representation of the final feature is dependent on a set of single fields.

17.2.2 Classes of Multifield Feature Definitions

Feature definitions in a multifield setting may be classified into one of two groups:

(a) **Isolated Feature Definitions**: Feature definition and extraction is performed on an independent per-field level. The multifield notion is obtained as the final visualization combines these isolated features into a common representation. The resulting multifield feature is essentially a collection of classic features.

(b) **Combined Feature Definitions**: Feature definition and extraction is performed on a combined set of fields, i.e., the feature definition depends on information encoded in multiple fields. These feature definitions extend the classic single-field feature definition and creates features that are not present in any single field.

There is generally no preferred way of multifield feature extraction and visualization, neither in theory nor in applications. Both feature definitions are widely used in the visualization community.

17.3 Classification of Visualization Techniques

17.3.1 Isolated and Modulated Features

The earliest and most straight-forward feature-based visualization techniques that rely on data from multiple fields make use of individual feature definitions for each field and combine these isolated features into a combined multifield visualization. Often, complex feature extraction is limited to one field only with other fields used for modulation of the extracted feature representation.

One of the earliest example is the enhanced display of streamlines in a flow field as streamtubes and streamribbons by modulating extracted line geometry with properties of the derived rotation and divergence fields [34]. Similarly, integral lines in time-varying flow fields have been enhanced with deformation information extracted from the derived velocity gradient field [24]. Both examples illustrate how the initial feature extraction process is limited to one field, whereas the final visualization conveys multifield data.

Love et al. [19] demonstrate the use of streamline extraction for multivalue data sets. While the feature extraction process is limited to an individual field, concurrent visualization of streamlines extracted from multiple related fields facilitates visual comparison of flow directions.

An approach that modulates the geometry of the shared computational mesh of multifield CFD data is presented by Henze [14]. Henze proposes to map the initially feature-less computational grid of the simulation domain into spaces of fields present in the data set. A single vector valued field as well as multiple scalar valued fields are used to displace vertices of the original mesh. Feature identification and extraction in this new coordinate system allows joint analysis of the affected fields.

17.3.2 Locally Defined Features

One of the classical sources of multifields is *Computational Fluid Dynamics (CFD)* simulations, the results of which contain physical variables such as velocity, pressure, and density. CFD solvers commonly output an extensive set of scalar-, vector-, or tensor-valued variables, up to several dozens [1]. Therefore, it is not surprising that multifield visualization is crucial for the understanding of flow data. This also holds for the class of feature-based visualization methods, as some of the physically most relevant features of flows rely on several physical quantities in their definition. For example, some of the classical vortex detectors are based on the combination of pressure and vorticity [2] or of velocity and vorticity [18]. Since vorticity can be derived from the velocity field using a numerical gradient estimation technique, the classification of such a method as a multifield technique might be questionable. However, the velocity gradient tensor is typically an original result of a CFD solver. In the case of vortex methods [7], it is even so that vorticity is an original variable, while velocity is a derived one.

Multifield feature detectors can be realized in a number of ways. An efficient class of methods requires only *local information*, i.e., the field values at a given point and, possibly, some low-order derivatives of them. One such approach is reduction to a scalar field, followed by an *isosurface* extraction. It was used by Levy et al. in their definition of a vortex as a region where *helicity* [18] is high. Helicity is defined as the scalar product of velocity and vorticity, and often normalized helicity is used where vectors are normalized.

Multifields lend themselves also to *predictor-corrector* approaches. The vortex detector developed by Banks and Singer [2] is based on the assumption that the vortex axis roughly follows the vorticity vector while passing through pressure minima in its cross section. Therefore, it is realized by taking steps along the vorticity vector alternating with steps along the pressure gradient in the orthogonal space. A variant of this method, proposed by Stegmaier et al. [32], replaced pressure by the λ_2 vortex indicator [17]. Both of these methods not only compute the vortex axis as a curve, but they also provide the vortex hull, which is obtained by a radial search in the normal plane of the vortex axis.

In many cases, a feature defined by a pair of variables can be expressed by the *parallel vectors operator* [25], which is a generic approach to extract line-like features from a pair of two- or three-dimensional vector fields. This way, the aforementioned methods [2, 18] can be reformulated. Further examples are the criteria by Sujudi and Haimes [33] for vortex axes in steady flow and its extension to unsteady flow by Fuchs et al. [11]. Here, the vortex axis is defined as the locus where velocity and acceleration are parallel vectors, subject to the additional condition that the velocity gradient has a pair of complex conjugate eigenvalues.

Related to the parallel vectors technique is the extraction of *creases* (ridges and valleys). Examples are the definition of vortex axes as valley lines of pressure by Miura and Kida [20] and the definition of *Lagrangian coherent structures* as ridge surfaces of the *finite-time Lyapunov exponent*, by Haller [13]. Extraction of creases requires first and second derivatives of a scalar field. Therefore, this is not a typical multifield method, but rather a method based on derived fields.

Fields of a multifield often represent different physical quantities. An interesting technique for comparing multiple scalar fields is to compare their gradients. For $k \leq n$ scalar fields in n-space, the comparison measure proposed by Edelsbrunner et al. [10] is the volume of the parallelotope spanned by the k gradient vectors. A feature can now be defined to consist of the places where all gradients are similar with respect to this measure. In the given application example, a pair of scalar quantities from a combustion simulation was used, and the resulting features represent the frame front. Nagaraj et al. [22] use as an alternative comparison measure the norm of the matrix composed of the gradient vectors.

Feature definitions can involve, besides physical quantities, also geometric information. An example are the spatial coordinates, which can serve to define features restricted to regions-of-interest. A less trivial information, that deserves to be called a "field" within the multifield, is the *wall distance*, the distance of each data point to the nearest point on a solid boundary. Wall distance is required for some turbulence models, therefore it is provided as output by most CFD solvers. It can be used to focus on features that are either close to walls or distant from walls. Furthermore, wall distance is needed for estimating limits, such as the *wall-shear stress*, a quantity which is of importance, e.g., for the study of aneurysms in medical data [23]. The wall-shear stress is amenable to 2D *vector field topology*, and the latter can be combined with the 3D topology of the velocity field, thus making vector field topology a two-field method [26].

17.3.3 Statistical Features

In multifield data sets with homogeneous field types, i.e., collections of scalar fields, the existence of a common dimensionality allows the application of statistical methods for visualization purposes. Especially in this context of the visualization of statistics of fields, data is often called *multi-variate data*, see [5]. While most features defined in this manner are representatives of the group of combined features, as is the case for multifield feature detectors or operators, some techniques compute statistics over time or space rather than different fields.

The most basic geometric feature in a scalar field is an isocontour or, in a three-dimensional setting, an isosurface. Their mathematical and computational simplicity makes them prime for feature analysis in sets of scalar fields. Gosink et al. [12] perform statistical multifield visualization on multiple scalar fields by extracting representative principle isosurfaces of one of the scalar fields and color it according to *correlation* of a pair of two other scalar fields present in the data set. These principle isosurfaces correspond to contours of the most frequently occurring scalar values. The required statistics for frequency determination are gathered by scalar-value histogram computation. The *variation* of a collection of scalar fields is another statistical quantity that can contribute to multifield visualization. In the work by Nagaraj et al. [21] relevant isovalues are identified by computing and plotting the variation density function of multiple scalar fields in a data set. Their method is able to compare $n \leq k$ scalar fields in a k-dimensional data set and helps in identifying characteristic isosurfaces in sets of scalar fields. Two scalar field comparison methods that extend to n dimensions are presented by Sauber et al. [30], namely *gradient similarity* and *local correlation coefficient*. The large number of correlation fields created by their method requires the use of a multifield graph for interactive correlation field selection. Regions with interesting field correlations can be easily selected manually.

A related problem that is not so much concerned with correlations between different fields, but with statistical and stochastic properties of different instances of the same field is known as *ensemble data* analysis. Ensembles are frequently used in meteorological data analysis, where a number of simulation outputs created by perturbed simulation input parameters are compared to investigate the likelihood of certain weather events. For small perturbations, the resulting ensemble data sets have high correlation and describe the same sets of variables. In the work by Smith et al. [31] outputs from multiple simulation runs are clustered before visualization to avoid the loss of distinct features when global averaging over all simulation runs is applied. Potter et al. [28] visualize standard deviation and mean of ensembles by color mapping and contouring. The simultaneous visualization of contours from multiple scalar fields called *spaghetti plots* facilitates analysis of relative distributions of scalar values.

Correlation and variation can not only be computed for sets of fields, but also for different instances of the same variable as common in time-varying data sets. In such a way Jänicke et al. [16] compute local statistical complexity for different quantities of CFD simulations. *Local statistical complexity* is a measure that describes the predictability of values of a local variable overt time. Extrema of this measure often correspond to interesting regions in the flow domain or feature regions as, for example, extracted by vortex core techniques. The mathematical framework allows application of the same visualization and feature extraction to different fields of the same simulation, while providing independent visualization output for each analyzed variable. Thus, it is a multifield visualization technique only in the sense that it can be applied to several fields in a multi field data set in a consistent way.

17.3.4 Interactive Feature Specification

Features in fields or multifields have in the ideal case a precise mathematical definition which does not depend on any "tuning" parameters. An example in hydrodynamics are cavitation zones, which occur where the local pressure falls below the vapor pressure at the given temperature. As another example, Hunt et al. [15] defined an eddy as the region with positive second invariant, Q, of $\nabla \mathbf{u}$, with the additional condition that the pressure be lower than the ambient value. While this is a precise and parameter-free definition, there are competing definitions of eddies, which in a visualization can be used as well, possibly even in combination. An example of a feature definition involving a parameter is the vortex definition of Jeong and Hussain [17]. In the original definition, the derived quantity λ_2 has to be negative, but practically, for better isolation of the vortices a negative number is used as a threshold for λ_2. Such feature definitions involving a parameter require a visualization system where parameters can be controlled by the user. For multifields the visualization system has to support several such parameters and the possibility to derive new fields from the given ones. A system for combining multiple volumetric scalar fields by applying a hierarchy of algebraic and logical operations has been proposed by Woodring et al. [37]. Not in all cases, the feature definition is known in advance. For an interactive feature specification, Doleisch et al. [8] presented a feature definition language, in which a hierarchy of subset operations is built up interactively by using "brushing" operations in a system offering multiple views, i.e., projections of the data space. The feature definition language has been applied to hurricane simulations [9] and chemical reactions in biological processes [35]. A similar system of linked views has been used by Blaas et al. [4] for medical applications, especially for studying tumors for surgery planning.

References

1. ANSYS CFX Reference Guide, Release 12. ANSYS Inc, Apr 2009
2. Banks, D.C., Singer, B.A.: A predictor-corrector technique for visualizing unsteady flow. IEEE Trans. Vis. Comp. Graph. **1**(2), 151–163 (1995)
3. Bauer, D.: Ronald Peikert, Mie Sato, and Mirjam Sick. A case study in selective visualization of unsteady 3d flow. In: IEEE Visualization '02 Proceedings, pp. 525–528. IEEE Computer Society, Oct 2002
4. Blaas, J., Botha, C.P., Post, F.H.: Interactive visualization of multi-field medical data using linked physical and feature-space views. In: Museth, K., Möller, T., Ynnerman, A. (eds.) Eurographics/IEEE-VGTC Symposium on Visualization, pp. 123–130. Eurographics Association, Norrköping (2007)
5. Bürger, R., Hauser, H.: Visualization of multi-variate scientific data. In: EuroGraphics State of the Art Reports (STARs), pp. 117–134 (2007)
6. Canny, J.: A computational approach to edge detection. Pattern Anal. Mach. Intell. IEEE Trans. **8**(6), 679–698 (1986)
7. Cottet, G.H., Koumoutsakos, P.: Vortex Methods—Theory and Practice. Cambridge University Press, Cambridge (2000)
8. Doleisch, H., Gasser, M., Hauser, H.: Interactive feature specification for focus+context visualization of complex simulation data. In: Proceedings of the Symposium on Data Visualisation, pp. 239–248. Eurographics Association (2003)
9. Doleisch, H., Muigg, P., Hauser, H.: Interactive visual analysis of hurricane isabel with SimVis. Technical Report TR-VRVis-2004-058, VRVis Research Center, Vienna (2004)
10. Edelsbrunner, H., Harer, J., Natarajan, V., Pascucci, V.: Local and Global Comparison of Continuous Functions. In: Proceedings of the Conference on Visualization '04, pp. 275–280. IEEE Computer Society (2004)
11. Fuchs, R., Peikert, R., Hauser, H., Sadlo, F., Muigg, P.: Parallel vectors criteria for unsteady flow vortices. IEEE Trans. Vis. Comput. Graph. **14**(3), 615–626 (2008)
12. Gosink, L., Anderson, J., Bethel, W., Joy, K.: Variable interactions in query-driven visualization. IEEE Trans. Vis. Comp. Graph. **13**, 1400–1407 (2007)
13. Haller, G.: Distinguished material surfaces and coherent structures in three-dimensional fluid flows. Physica D **149**, 248–277 (2001)
14. Henze, C.: Feature Detection in Linked Derived Spaces. In: Proceedings of the Conference on Visualization '98, VIS '98, pp. 87–94. Los Alamitos, IEEE Computer Society Press, CA (1998)
15. Hunt, J.C.R., Wray, A.A., Moin, P.: Eddies, Stream and Convergence Zones in Turbulent Flows. In: 2. Proceedings of the 1988 Summer Program, pp. 193–208 (1988)

16. Jänicke, H., Böttinger, M., Tricoche, X., Scheuermann, G.: Automatic detection and visualization of distinctive structures in 3d unsteady multi-fields. Comput. Graph. Forum **27**(3), 767–774 (2008)
17. Jeong, J., Hussain, F.: On the identification of a vortex. J. Fluid Mech. **285**, 69–84 (1995)
18. Levy, Y., Degani, D., Seginer, A.: Graphical visualization of vortical flows by means of helicity. AIAA J. **28**, 1347–1352 (1990)
19. Love, A.L., Pang, A., Kao, D.L.: Visualizing spatial multivalue data. IEEE Comput. Graph. Appl. **25**, 69–79 (2005)
20. Miura, H., Kida, S.: Identification of tubular vortices in turbulence. J. Phys. Soc. Japan **66**, 1331–1334 (1997)
21. Nagaraj, S., Natarajan, V.: Relation-aware isosurface extraction in multifield data. Vis. Comput. Graph. IEEE Trans. **17**(2), 182–191 (2011)
22. Nagaraj, S., Natarajan, V., Nanjundiah, R.S.: A gradient-based comparison measure for visual analysis of multifield data. Comput. Graph. Forum **30**(3), 1101–1110 (2011)
23. Neugebauer, M., Gasteiger, R., Beuing, O., Diehl, V., Skalej, M., Preim, B.: Combining map displays and 3D visualizations for the analysis of scalar data on cerebral aneurysm surfaces. Comput. Graph. Forum **28**(3), 895–902 (2009)
24. Obermaier, H., Hering-Bertram, M., Kuhnert, J., Hagen, H.: Volume deformations in grid-less flow simulations. Comput. Graph. Forum **28**(3), 879–886 (2009)
25. Peikert, R., Roth, M.: The "Parallel Vectors" Operator—A Vector Field Visualization Primitive. In: Proceedings of the 10th IEEE Visualization Conference (VIS '99), pp. 263–270. IEEE Computer Society, Washington (1999)
26. Peikert, R., Sadlo, F.: Topology-guided visualization of constrained vector fields. In: Hagen, H., Hauser, H., Theisel, H. (eds.) Topology-Based Methods in Visualization, pp. 21–34. Springer-Verlag, Berlin (2007)
27. Post, F., Vrolijk, B., Hauser, H., Laramee, R., Doleisch, H.: The state of the art in flow visualization: feature extraction and tracking. Comput. Graph. Forum **22**(4), 775–792 (2003)
28. Potter, K., Wilson, A., Bremer, P.T., Williams, D., Doutriaux, C., Pascucci, V., Johnson. C.R.: Ensemble-vis: A Framework for the Statistical Visualization of Ensemble Data. In: Proceedings of the 2009 IEEE International Conference on Data Mining Workshops, ICDMW '09, pp. 233–240. IEEE Computer Society, Washington (2009)
29. Reinders, F., Post, F.H., Spoelder, H.J.W.: Visualization of time-dependent data with feature tracking and event detection. Vis. Comput. **17**(1), 55–71 (2001)
30. Sauber, N., Theisel, H., Seidel, H.-P.: Multifield-graphs: an approach to visualizing correlations in multifield scalar data. Vis. Comput. Graph. IEEE Trans. **12**(5), 917–924 (2006)
31. Smith, K.M., Banks, D.C., Druckman, N., Beason, K., Hussaini, M.Y.: Clustered ensemble averaging: a technique for visualizing qualitative features of stochastic simulations. J. Comput. Theor. Nanosci. **3**(5), 752–760 (2006)
32. Stegmaier, S., Rist, U., Ertl, T.: Opening the Can of Worms: An Exploration Tool for Vortical Flows. In: Silva, C., Grer, E., Rushmeier, H. (eds.) Proceedings of IEEE Visualization '05, pp. 463–470 (2005)
33. Sujudi D., Haimes, R.: Identification of Swirling Flow in 3D Vector Fields. Technical Report, pp. 95–1715. AIAA (1995)
34. Ueng, S.K., Sikorski, C., Ma, K.L.: Efficient streamline, streamribbon, and streamtube constructions on unstructured grids. IEEE Trans. Vis. Comput. Graph. **2**, 100–110 (1996)
35. Unger, A., Muigg, P., Doleisch, H., Schumann, H.: Visualizing statistical properties of smoothly brushed data subsets. In: Information Visualisation. IV '08. 12th International Conference, pp. 233–239 (2008)
36. Weinkauf, T., Hege, H.-C., Noack, B.R., Schlegel, M., Dillmann, A.: Coherent structures in a transitional flow around a backward-facing step. Phys. Fluids **15**(9), S3 (2003)
37. Woodring, J., Shen, H.W.: Multi-variate, time varying, and comparative visualization with contextual cues. Vis. Comput. Graph. IEEE Trans. **12**(5), 909–916 (2006)
38. Ye, X., Kao, D., Pang, A.: Strategy for scalable seeding of 3d streamlines. In: Proceedings IEEE Visualization '05, pp. 471–478 (2005)

Chapter 18
Feature Analysis in Multifields

Hamish Carr

Abstract As with individual fields, one approach to visualizing multifields is to analyze the field and identify *features*. While some work has been carried out in detecting features in multifields, any discussion of multifield analysis must also identify techniques from single fields that can be extended appropriately.

18.1 Introduction

Analysis of features in multifields is a discipline in its infancy. As such, the relevant literature is somewhat scattered, but some broad categories can be seen in the existing work. Before examining these categories, however, it is useful to start with a working definition of a feature: a *feature* is a (usually) geometric feature of the underlying phenomenon that is of significance to the user.

Strictly speaking, feature analysis relies on the computer to detect objects for the human user to view or for further processing. In practice, it is more accurate to think of there being a spectrum of methods:

1. Visual Fusion
2. Interactive Definition
3. Derived Properties
4. Distributions
5. Abstract Structures

Of these, *visual fusion* maps individual properties of the multifield to different visual properties, then relies on the user's visual system to identify regions of interest. *Interactive definition* takes this one step further, with the user changing the visual mapping(s)—in essence, this is a manual search of a parameter space.

H. Carr (✉)
School of Computing, University of Leeds, Leeds, UK
e-mail: h.carr@leeds.ac.uk

C.D. Hansen et al. (eds.), *Scientific Visualization*, Mathematics and Visualization,
DOI 10.1007/978-1-4471-6497-5_18

Derived Properties such as gradient, local density, vorticity or information theoretic complexity are computed directly from the input data, then visualized separately. *Distributions* compute statistics of one or more properties of the data which are visualized separately, while *Abstract Structures* show summary information about relationships deduced from the data.

All these methods depend on implicit or explicit understanding of features, so that feature analysis in multifields depends strongly on interactive and human-centred methods. Moreover, it is characteristic in all methods that the result is confirmed by visualizing any features detected directly, and usually interactively.

In addition to this spectrum of methods, it is often the case that researchers in the application domain have an existing test for features which can be exploited directly. Moreover, visual fusion, interactive methods, derived properties and clustering are canvassed elsewhere in this volume, and are largely omitted, except where explicit feature extraction is used.

As a result, it is convenient to discuss multifield feature detection and analysis in the following categories:

Section 18.2 Scalar Features in Reduced Domains
Section 18.3 Scalar Features in the Range
Section 18.4 Manifold Features
Section 18.5 Overlapping Scalar Features
Section 18.6 Joint Feature Analysis

18.2 Scalar Features in Reduced Domains

A related approach is to choose a feature in one property of the multifield, then restrict another property to that feature, and analyse its restriction.

Bremer, Weber et al. [2] apply this to combustion simulations. One property (temperature) is restricted to define an isosurface. Features are then identified and tracked over time for a second property (combustion rate) restricted to the underlying isosurface. Subsequently, Bremer, Weber et al. [3] analysed the topology of one property (fuel consumption), but annotated the features so discovered with values derived from the other properties.

Ropinski et al. [34] extract aortic arches from mouse PET/CT scans, then use image-processing techniques to recognize standard features and register the scans. Once this is done, however, secondary properties such as vessel diameter are mapped onto the detected geometry for human visualization.

To date, however, there has been relatively little work performed under this heading. Instead, many researchers have concentrated on detecting features in the range of the multifield, and we consider these approaches next.

18.3 Scalar Features in the Range

Instead of reducing the multifield to a scalar field defined over the same domain, it is possible to reduce the multifield to a scalar field whose domain is the range of the multifield. Once this has been done, it becomes possible to perform scalar feature detection this new scalar field, provided that there is a solid understanding of the relationship between features in the domain and features in the range of the function.

Work in this area includes many papers on transfer function design, and in particular multi-dimensional transfer function design. This work started with the observation by Levoy [25] that aliasing (also called partial volume effects) mean that an isovalued boundary is rarely seen clearly in a simple isovalue-based transfer function. Levoy's solution was to define a tent function in which isovalues close to a selected isovalue were given opacity proportional to both the proximity to the isovalue and to the local gradient: much subsequent work on transfer functions has been based on variations of this.

Kindlmann and Durkin [21] then showed that the sampled isovalues can be plotted against gradient values in a scatterplot, and that, due to partial volume effects, edges show up in these plots as parabolic arcs. Sato et al. [35] elaborated this with several characteristic types of object showing up as different types of curves, then applied fuzzy logic to assign individual samples to different classes of curves (i.e. features).

Kniss et al. [22, 23] built an interface in which rectangular and triangular regions were used in the isovalue-gradient histogram to identify features in the domain, and related their triangular regions to Levoy's tent functions. In contrast, Tenginakai et al. [39] computed secondary moments such as skew and kurtosis, then displayed two-dimensional histograms of isovalue plus one other property and visually identified isovalues of interest: in some of these plots, the features of interest are ellipsoidal.

Independently, Kettner et al. [20] showed an interface where the number of components in the isosurfaces was mapped to isovalue and time to produce a two-dimensional plot of topological complexity, in which topological complexity was recognizable in the forms of peaks for individual time-slices, or ridges when considered over time. Again, these features were simply observed, rather than detected automatically.

Similarly, Rezk-Salama et al. [32] described using trapezoids, paraboloids and quadrilaterals for designing transfer functions in the 2D isovalue-gradient histogram. Hadwiger et al. [16] used rectangular blocks in 3D transfer functions: here, two of the three properties were derived properties from an underlying scalar field, but the same approach to transfer function design could be used to define features in multifields.

Šereda et al. [40] used a span space representation instead of gradient, then used region-growing to identify peaks as features in the span space. However, span space is now know to be equivalent to the isovalue-gradient histogram [8], so this can now be seen as a variation on the general theme of feature detection in the range. Similarly, Roettger et al. [33] performed region detection in a smoothed version of the isovalue-gradient histogram.

Rautek et al. [30] used fuzzy logic on multiple derived properties such as gradient and curvature. Because this implicitly uses ranges in each, the effect of this is to apply blocks or Gaussian ellipsoids in the histograms to define features, even if they are not shown directly. Johnson and Huang [19] compute histograms separately for individual properties, and compare them using boolean predicates to identify features of interest.

Song et al. [38] provided a volume-renderer for meteorological data that incorporate an equation parser so that scientists could compute derived properties at runtime in order to identify features. In a similar vein, Glatter et al. [14] assumed that the underlying numerical data could be represented in textual form, then used grep-style pattern matching to identify features. Gosink et al. [15] also took a similar approach, giving it the name of Query-Driven Visualization.

Maciejewski et al. [27] combined this with clustering to define features as arbitrary shapes in the isovalue-gradient histogram. It thus becomes clear that it is feasible, although not necessarily desirable, to discuss clustering in the context of n-D histograms.

More recently, Lindholm et al. [26] observed that small spatial neighborhoods in medical data normally intersect only a few distinct materials. Thus, by identifying ellipsoidal Gaussian peaks in local histograms, material types are identified and mapped to transfer functions. These peaks are mapped manually, except in simple cases, where they can be detected by iterative peak detection, or in established workflows, where templates from other data sets can be adapted. Similarly, Correa and Ma [7] used manual sums of Gaussians to define features in 2D histograms.

18.4 Manifold Features

We have just seen that features can be detected in either the domain of the function or in its range. Interestingly, this allows us to see in retrospect that multiple researchers have converged on similar solutions: defining features to be compact sets (preferably peaks) in n-D histograms or n-D feature space.

More recently, work that relates n-D feature space or n-D histograms back to the domain started with the observation by Carr et al. [5] that statistics of data sampled from a continuous function are directly related to geometric properties of contours in the domain. Subsequent work by Scheidegger et al. [36] refined the formulation to include a missing gradient factor. Simultaneously, Bachthaler and Weiskopf [1] showed that the observation could be extended to multifields, and that doing so resulted in the *continuous scatterplot*: i.e. that multi-dimensional histograms are discrete approximations of a projection of the function manifold to the range.

Subsequently, Lehman and Theisel [24] observed that a prominent feature of continuous scatterplots was the presence of visible edges, or *boundary curves*, and extracted these curves with Canny edge-detection. This ties in with recent work by Duffy et al. [9] which formalises the relationship of histogram statistics with the function manifold and Geometric Measure Theory [13], the underlying mathematical theory of integration. As part of the connection between integration, mensuration and statistics, these authors identified the role in integration of the *multiplicity function* which measures the number of distinct input values mapping to a single output value. From this viewpoint, it then becomes apparent that the boundary curves detected by Lehman and Theisel [24] are the folds in the manifold when projected into the range. It then follows that this form of feature detection is related to domain-based topological analysis.

Nagaraj and Natarajan [28] have also recently defined a variation density function for multifields: while the details have not been explored, it seems likely that this function is also connected to mensuration.

Heinrich and Weiskopf [18] have also extended continuous scatterplots to parallel coordinates: by implication feature-recognition techniques are likely to develop in this area as well.

In summary, then, three broad trends are visible in recent work in this area: that transfer function papers often rely in practice on human perception to detect features in multi-dimensional histograms, that theoretical work is providing strong linkages between the multi-dimensional histograms and the underlying multifield, and that feature recognition methods are increasingly exploiting these linkages to identify significant features.

18.5 Feature Overlap

Since feature detection algorithms now exist in scalar and vector fields, one simple approach is to detect features separately in each field, then overlap them spatially to see how well they match.

Woodring and Shen [41] allowed the user to perform set operations on (iso-)surfaces defined by individual properties. Similarly, Navrátil et al. [29] mapped isosurfaces of different properties to different colours. In each case, we can see that isovalued-features (i.e. isosurfaces) are in effect being overlapped either logically or visually.

Schneider et al. [37] took the next step by using contour trees to recognize features of each of two fields separately. For each pair of features, the spatial overlap was computed as a similarity measure. An interface then showed all pairings above a similarity threshold as a bipartite graph for user selection. Heine et al. [17] then extended this to arbitrary numbers of fields, showing strong correlations as cliques.

18.6 Joint Feature Analysis

We have noted above that scalar topological analysis can be applied to one or more properties of the multifield. The obvious next question is whether there exist forms of topological analysis that can be applied directly to the multifield. To date, several approaches have been reported, all based on the existing work on scalar data, in particular the contour tree and Reeb graph approaches.

Edelsbrunner and Harer [10] introduced *Jacobi Sets*, which produce a skeletal representation of the data by tracking the critical points of individual properties as other properties are varied. Edelsbrunner, Harer, Natarajan and Pascucci integrated gradients over Jacobi Sets to find κ, a measure of topological persistence, and used this to identify features. However, since the Jacobi Sets were $1 - D$ structures, the features identified could not be space-filling.

In subsequent work, Edelsbrunner et al. [11] computed time-varying Reeb graphs for continuous space-time data and related them to Jacobi Sets, while Nagaraj and Natarajan [28] showed how to simplify Jacobi Sets.

Carlsson, Singh and Zomorodian [4] also described a formal algorithm for multidimensional persistence, a generalization of persistence for scalar fields.

A more general approach is to take Reeb graphs and generalise them. For a functions $f : IR^m \rightarrow IR^n$, a contour can be defined as a connected component of the inverse image $f^{-1}(h)$ for some $h \in IR^m$. Following the definition of the Reeb graph [31], Edelsbrunner, Harer & Patel defined the *Reeb space* [12] to be the continuous contraction of each contour to a single point. Since the domain has n dimensions instead of the one originally assumed by Reeb, the result is not a graph but a manifold of dimension $min(m, n)$. More recently, Carr and Duke [6] have described the *Joint Contour Net*—a graph approximation of the Reeb space.

Since the Reeb graphs and contour trees of individual properties can be extracted directly from the Reeb space or Joint Contour Net, many of the approaches described above will turn out to be special cases of this more general approach. Similarly, both Jacobi Sets and multidimensional persistence are likely to be special cases.

While none of these methods is complete, multivariate analysis is beginning to become possible, and visualizations based on these abstractions can be expected to follow in due course.

References

1. Bachthaler, S., Weiskopf, D.: Continuous scatterplots. IEEE Trans. Vis. Comput. Graph. **14**(6), 1428–1435 (2008)
2. Bremer, P.T., Weber, G., Pascucci, V., Day, M.S., Bell, J.: Analyzing and tracking burning structures in lean premixed hydrogen flames. IEEE Trans. Vis. Comput. Graph. **16**(2), 248–260 (2009)
3. Bremer, P.T., Weber, G., Tierny, J., Pascucci, V., Day, M.S., Bell, J.: Interactive exploration and analysis of large-scale simulations using topology-based data segmentation. IEEE Trans. Vis. Comput. Graph. **17**(9), 1307–1324 (2011)
4. Carlsson, G., Singh, G., Zomorodian, A.: Computing multidimensional persistence. J. Comp. Geom. **1**(1), 72–100 (2010)
5. Carr, H., Duffy, B., Denby, B.: On histograms and Isosurface statistics. IEEE Trans. Vis. Comput. Graph. **12**(5), 1259–1266 (2006)
6. Carr, H., Duke, D.J.: Joint contour nets: topological analysis of multi-variate data. IEEE VisWeek 2011 Poster (2011)

7. Correa, C.D., Ma, K.L.: Visibility histograms and visibility-driven transfer functions. IEEE Trans. Vis. Comput. Graph. **17**(2), 192–204 (2011)
8. Duffy, B., Carr, H.: Interval based data structure optimisation. In: Proceedings of Theory & Practice of Computer Graphics, pp. 151–158 (2010)
9. Duffy, B., Carr, H., Möller, T.: Integrating isosurface statistics and histograms. IEEE Trans. Visual Comput. Graphics **19**(2), 263–277 (2013)
10. Edelsbrunner, H., Harer, J.: Jacobi sets of multiple morse functions. In: Foundations in Computational Mathematics, pp. 37–57. Cambridge University Press, Cambridge (2002)
11. Edelsbrunner, H., Harer, J., Mascarenhas, A., Pascucci, V., Snoeyink, J.: Time-varying Reeb graphs for continuous space-time data. Comput. Geom.: Theor. Appl. **41**, 149–166 (2008)
12. Edelsbrunner, H., Harer, J., Patel, A.K.: Reeb spaces of piecewise linear mappings. In: Proceedings of ACM Symposium on Computational Geometry, pp. 242–250 (2008)
13. Federer, H.: Geometric Measure Theory. Springer-Verlag, Berlin (1965)
14. Glatter, M., Huang, J., Ahern, S., Daniel, J., Lu, A.: Visualizing temporal patterns in large multivariate data using textual pattern matching. IEEE Trans. Vis. Comput. Graph. **14**(6), 1467–1474 (2008)
15. Gosink, L.J., Anderson, J.C., Bethel, E.W., Joy, K.I.: Variable interactions in query-driven visualization. IEEE Trans. Vis. Comput. Graph. **13**(6), 1400–1407 (2007)
16. Hadwiger, M., Fritz, L., Rezk-Salama, C., Höllt, T., Geier, G., Pabel, T.: Interactive volume exploration for feature detection and quantification in industrial CT data. IEEE Trans. Vis. Comput. Graph. **14**(6), 1507–1514 (2008)
17. Heine, C., Schneider, D., Carr, H., Scheuermann, G.: Interactive Comparison of Multi-field Scalar Data Based on Mutual Information of Largest Contours. In: Foundations of Topological Analysis Workshop (Salt Lake City, Utah, Oct. 24, 2010) (2010)
18. Heinrich, J., Weiskopf, D.: Continuous parallel coordinates. IEEE Trans. Vis. Comput. Graph. **15**(6), 1531–1538 (2009)
19. Johnson, C.R., Huang, J.: Distribution-driven visualization of volume data. IEEE Trans. Vis. Comput. Graph. **15**(5), 734–746 (2009)
20. Kettner, L., Rossignac, J., Snoeyink, J.: The safari interface for visualizing time-dependent volume data using Iso-surfaces and contour spectra. Comput. Geom.: Theor. Appl. **25**(1–2), 97–116 (2003)
21. Kindlmann, G., Durkin, J.W.: Semi-Automatic Generation of Transfer Functions for Direct Volume Rendering. In: Proceedings of IEEE Visualization, pp. 79–86 (1998)
22. Kniss, J., Kindlmann, G., Hansen, C.D.: Interactive Volume Rendering Using Multi-Dimensional Transfer Functions and Direct Manipulation Widgets. In: Proceedings of Visualization 2001, pp. 255–262, 562 (2001)
23. Kniss, J., Kindlmann, G., Hansen, C.D.: Multidimensional transfer functions for interactive volume rendering. IEEE Trans. Vis. Comput. Graph. **8**(3), 270–285 (2002)
24. Lehmann, D.J., Theisel, H.: Discontinuities in continuous scatterplots. IEEE Trans. Vis. Comput. Graph. **16**(6), 1291–1300 (2010)
25. Levoy, M.: Volume rendering: display of surfaces from volume data. IEEE Comput. Graph. Appl. **8**(3), 29–37 (1988)
26. Lindholm, S., Ljung, P., Lundström, C., Persson, A., Ynnerman, A.: Spatial conditioning of transfer functions using local material distributions. IEEE Trans. Vis. Comput. Graph. **16**(6), 1301–1310 (2010)
27. Maciejewski, R., Woo, I., Chen, W., Ebert, D.S.: Structuring feature space: a non-parametric method for volumetric transfer function generation. IEEE Trans. Vis. Comput. Graph. **15**(6), 1473–1480 (2009)
28. Nagaraj, S., Natarajan, V.: Simplification of jacobi sets. In: Pascucci, V., Tricoche, X., Hagen, H., Tierny, J. (eds.) Topological Data Analysis and Visualization: Theory, Algorithms and Applications, Mathematics and Visualization, pp. 91–102. Springer, Berlin (2011)
29. Navrátil, P.A., Johnson, J.L., Bromm, V.: Visualization of cosmological particle-based datasets. IEEE Trans. Vis. Comput. Graph. **13**(6), 1712–1718 (2007)

30. Rautek, P., Bruckner, S., Gröller, E.: Semantic layers for illustrative volume rendering. IEEE Trans. Vis. Comput. Graph. **13**(6), 1336–1343 (2007)
31. Reeb, G.: Sur les points singuliers d'une forme de Pfaff complètement intégrable ou d'une fonction numérique. Comptes Rendus de l'Acadèmie des Sciences de Paris **222**, 847–849 (1946)
32. Rezk-Salama, C., Keller, M., Kohlmann, P.: High-level user interfaces for transfer function design with semantics. IEEE Trans. Vis. Comput. Graph. **12**(5), 1021–1028 (2006)
33. Roettger, S., Bauer, M., Stamminger, M.: Spatialized Transfer Functions. In: Eurographics/IEEE Symposium on Visualization (EuroVis), pp. 271–278 (2005)
34. Ropinski, T., Hermann, S., Reich, R., Schäfers, M., Hinrichs, K.: Multimodal vessel visualization of mouse aorta PET/CT scans. IEEE Trans. Vis. Comput. Graph. **15**(6), 1515–1522 (2009)
35. Sato, Y., Westin, C.F., Bhalerao, A., Nakajima, S., Shiraga, N., Tamura, S., Kikinis, R.: Tissue classification based on 3d local intensity structure for volume rendering. IEEE Trans. Vis. Comput. Graph. **6**(2), 160–180 (2000)
36. Scheidegger, C.E., Schreiner, J.M., Duffy, B., Carr, H., Silva, C.T.: Revisiting histograms and isosurface statistics. IEEE Trans. Vis. Comput. Graph. **14**(6), 1659–1666 (2008)
37. Schneider, D., Wiebel, A., Carr, H., Hlawitschka, M., Scheuermann, G.: Interactive comparison of scalar fields based on largest contours with applications to flow visualization. IEEE Trans. Vis. Comput. Graph. **14**(6), 1475–1482 (2008)
38. Song, Y., Ye, J., Svakhine, N., Lasher-Trapp, S., Baldwin, M., Ebert, D.S.: An atmospheric visual analysis and exploration system. IEEE Trans. Vis. Comput. Graph. **12**(5), 1157–1164 (2006)
39. Tenginakai, S., Lee, J., Machiraju, R.: Salient Iso-surface detection with model-independent statistical signatures. Proc. Vis. **2001**, 231–238 (2001)
40. Šereda, P., Bartrolí, A.V., Serlie, I.W., Gerritsen, F.A.: Visualization of boundaries in volumetric data sets using LH histograms. IEEE Trans. Vis. Comput. Graph. **12**(2), 208–218 (2006)
41. Woodring, J., Shen, H.W.: Multi-variate, time-varying and comparative visualization with contextual cues. IEEE Trans. Vis. Comput. Graph **12**(5), 909–916 (2006)

Chapter 19
Future Challenges and Unsolved Problems in Multi-field Visualization

Robert S. Laramee, Hamish Carr, Min Chen, Helwig Hauser, Lars Linsen, Klaus Mueller, Vijay Natarajan, Harald Obermaier, Ronald Peikert and Eugene Zhang

Abstract Evaluation, solved and unsolved problems, and future directions are popular themes pervading the visualization community over the last decade. The top unsolved problem in both scientific and information visualization was the subject of an IEEE Visualization Conference panel in 2004. The future of graphics hardware

---

R.S. Laramee (✉)
Computer Science Department, Swansea University, Swansea, UK
e-mail: r.s.aramee@swansea.ac.uk

H. Carr
University of Leeds, Leeds, UK
e-mail: h.carr@leeds.ac.uk

M. Chen
Oxford e-Research Centre, University of Oxford, Oxford OX1 3QG, UK
e-mail: min.chen@oerc.ox.ac.uk

H. Hauser
Department of Informatics, University of Bergen, Bergen, Norway
e-mail: helwig.hauser@UiB.no

L. Linsen
School of Engineering and Science, Jacobs University, Bremen, Germany
e-mail: l.linsen@jacobs-university.de

K. Mueller
Department of Computer Science, Stony Brook University, Stony Brook, NY, USA
e-mail: mueller@cs.sunysb.edu

V. Natarajan
Department of Computer Science and Automation, IIS, Bengaluru, India
e-mail: vijayn@csa.iisc.ernet.in

H. Obermaier
Department of Computer Science, UC Davis, Davis, CA, USA
e-mail: hobermaier@ucdavis.edu

R. Peikert
ETH Zurich, Scientific Visualization Group, Zurich, Switzerland
e-mail: peikert@inf.ethz.ch

E. Zhang
Oregon State University, Corvallis, USA
e-mail: zhange@eecs.oregonstate.edu

C.D. Hansen et al. (eds.), *Scientific Visualization*, Mathematics and Visualization,
DOI 10.1007/978-1-4471-6497-5_19

was another important topic of discussion the same year. The subject of how to evaluate visualization returned a few years later. Chris Johnson published a list of 10 top problems in scientific visualization research. This was followed up by report of both past achievements and future challenges in visualization research as well as financial support recommendations to the National Science Foundation (NSF) and National Institute of Health (NIH). Chen recently published the first list of top unsolved information visualization problems. Future research directions of topology-based visualization was also a major theme of a workshop on topology-based methods. Laramee and Kosara published a list of top future challenges in human-centered visualization.

19.1 Introduction

Robert S Laramee:

Evaluation, solved and unsolved problems, and future directions are popular themes pervading the visualization community over the last decade. The top unsolved problems in both scientific and information visualization was the subject of an IEEE Visualization Conference panel in 2004 [10]. The future of graphics hardware was another important topic of discussion the same year [6]. The subject of how to evaluate visualization returned a few years later [3, 12]. Chris Johnson published a list of top problems in scientific visualization research [4]. This was followed up by report of both past achievements and future challenges in visualization research as well as financial support recommendations to the National Science Foundation (NSF) and National Institute of Health (NIH) [5]. Chen recently published the first list of top unsolved information visualization problems [1]. Future research directions of topology-based visualization was also a major theme of a workshop on topology-based methods [2, 11]. Laramee and Kosara published a list of top future challenges in human-centered visualization [7].

These pervasive themes coincide roughly with the 20th anniversary of what is often recognized as the start of visualization in computing as a distinct field of research [8]. Consensus is growing that some fundamental problems have been solved and a re-alignment including new directions is sought. In accordance to this redirection, we present a list of top unsolved problems and future challenges in multi-field visualization. Our list draws upon discussions at the Dagstuhl Workshop in Scientific Visualization 2011 as well as our own first hand experiences.

19.2 Challenges

Hamish Carr on Topology:

While scalar and vector topology have received a lot of attention, multifield topology and visualization techniques based on it have not. Moreover, where a large body of literature existed on topological analysis of scalar or vector data, the same is not true for multi-field topology. For example, Morse-Smale complexes are based on gradient lines, but in multifield data, the gradient is replaced by the Jacobian, a tensor quantity, and it is far from clear what the equivalent of a gradient line might be. Even were there to be an equivalent, the mapping to features in the underlying phenomena is not clear—where the Morse-Smale complex can be understood in terms of drainage patterns, such metaphors are not immediately obvious for s. As a result, the challenges related to multifield topology are manifold, including developing the underlying mathematics, insight and metaphors, as well as the usual topological feature descriptions, algorithms, data structures, visualization methods, and interfaces.

Min Chen on Standard Protocols:

One of the most fundamental challenges in multi-field visualization is to establish a set of intuitive and effective protocols for using visual channels. Given a multi-field data set, a "brute-force" visual design would be to juxtapose the visualizations of individual fields. However, such a visual design cannot support many comparative or combinational tasks effectively because of the difficulties in visual search for spatially corresponding positions across many images. An alternative approach is to depict information in the multi-fields in a comparative or combinational manner. However, as existing visual representations have largely been developed for single field visualization, combining such visual representations into a single visualization will inevitable cause conflicts in using visual channels. For instance, if the color channels are being used for one field, the other fields may have to make use of less desirable channels. Furthermore, there is no commonly agreeable means to depict the effect of constructive operations on different fields. For example, if one has used the texture channel to depict the similarity and difference between two scalar fields, perhaps one should not use such a channel for depicting the addition or union of these two fields in the same application. Hence, we may challenge ourselves with the following questions. Should there be some standard (or de facto standard) visual designs or visual metaphors for depicting different constructive operators (e.g., addition, subtraction, mean, OR, AND, etc.)? Should there be some standard (or de facto standard) protocols for visualizing some common configurations of multi-fields, such as two or a few scalar fields, on scalar field and one vector field, and so on? Can we evolve such protocols from some ad hoc visual effects, to commonly adopted visual metaphors, and eventually to standardized visual languages?

Helwig Hauser on Multi-dimensional, Scientific Visualization:

One common notion of scientific data is to consider it as a mapping of independent variables—usually space and/or time in scientific visualization—to a set of dependent values, very often resembling some measurements or computational simulation results that represent different aspects of a natural or man made phenomenon. Traditionally, neither the spatio-temporal domain nor the dependent variables are of higher dimensionality. A larger number of dependent values, however, leading to multi-variate data (as a special case of multi-field data), however, has recently lead to interesting visualization research. Highly interesting and very challenging, also, the emergence of higher-dimensional scientific data (in the sense of a higher-dimensional domain) leads to new visualization questions. Multi-run/ensemble simulation data, for example, includes parameters as additional independent variables. New approaches are needed to deal with this situation, especially in the context of scientific visualization, where generally a stronger and more immediate relation is present between the domain of the data and the visualization space (and to establish this relation in an effective way becomes more challenging, obviously, the more dimensions the data domain has). The integration of descriptive statistics, for example, is one opportunity that allows to perform a linked interactive visual analysis both on aggregation level as well as on the original multi-run data. It seems clear, however, that more research is needed to more thoroughly discuss, what the best possible approaches are.

Robert S Laramee on Spatial Integration:

Another major challenge of multi-field visualization is the integration (or coupling) of two or more data fields into the same spatial domain from which they originate. A common example is from computational fluid dynamics (CFD) [9]. CFD simulation data generally contains many attributes, e.g., flow velocity, pressure, temperature, kinetic energy, etc. And each multi-attribute data sample is associated with the same spatial domain. It is tempting to separate each attribute into its own visualization space, either abstract or scientific. However, integration of the data attributes into the same spatial domain from which they stem offers distinct advantages. However, how can such an integration be done in a meaningful and helpful way without overcrowding the visualization space?

Lars Linsen on Intuitive Visual Exploration of Multi-variate Features:

Features may have a complicated geometrical structure in the multi-dimensional attribute space. Extracting those features interactively is often tedious, if not impossible. Automatic components can help to compute such features. However, an intuitive visual exploration of such features is crucial to the user's understanding. What is the object space representation and, more importantly, what attribute values correspond to such a feature? Are their other features that are related, which possibly should have been merged by the automatic component? How homogeneous is a feature? Are their sub-regions within a feature that allow for further splitting of the feature? Such questions shall a user be able to answer when exploring the multi-field data. Intuitive visual encodings in object- and attribute-space as well as intuitive interaction mechanisms need to be provided.

Klaus Mueller on Channel Fusion:

The term "channel" is often used in the context of color images, comprised of a regular array of RGB color pixels. By mapping these 3D vector data to the three display primaries, channel fusion can occur directly in the viewer's visual system, engaging the tristimulus processes of color perception. However, once the number of channels exceeds three, the fusion must be externalized via some analysis and subsequent transformation to RGB color for display. In essence, one may regard this fusion as a mapping from H to L where H is the original and L the reduced number of channels, with the latter being three in this case. These types of reductive mappings are often encountered in low-dimensional embeddings of high-dimensional data. Such embeddings are ill-defined once the number of significant principal components in H is greater than L, which is most often the case. Hence, when applying such techniques for channel fusion, one must make certain trade-offs which are also determined by the type of dimension reduction technique used. There are a great many of these, some linear (PCA, LDA, and others) and some non-linear (MDS, LLE, and others). The former require some kind of component thresholding for channel reduction, while the latter suffer from distortion problems. Since in our specific case, both thresholding and distortion will affect the color composition of the display—as opposed to the spatial layout—the effects are possibly more noticeable. This leaves much room for further study. For example, it will be interesting to examine to what extent feature analysis and user-defined or learned constraints can be used to alleviate or control the adverse effects of dimension reduction in color display. A targeted and intuitive user interface might be needed to determine the appropriate fusion mapping. Finally, since gradients and higher-order derivatives are often employed in the graphics rendering of the data, it will be beneficial to study how the tensor resulting from high-dimensional derivative calculus can be interpreted for shading and other gradient-enhancements in 3D.

Vijay Natarajan on Categorizing Relationships between Fields:

Scientists try to understand physical phenomena by studying the relationship between multiple quantities measured over a region of interest. A characterization of the relationship between the measured/computed quantities will greatly enable the design of effective techniques for multifield visualization. For example, the dependence between fields could be linear or non-linear, the fields could be statistically correlated, or the relationship can be inferred using information theoretic measures. A challenging problem in this context is the categorization of different types of relationships and the design of measures that quantify the relationship in each case.

Harald Obermaier on Field Prioritization:

Modern simulation and measurement techniques can generate large numbers of fields spanning a wide range of types. While some of these fields may be crucial for the understanding and analysis of the behavior of the system, others may be used to enhance or extend the insights gained by multi-field visualization, while further others are largely irrelevant from an application or visualization point-of-view. Such a static prioritization of fields in a multi-field setting limits the potential of in-depth visual analysis especially in the area of application-driven data analysis, where the focus of interest can change during exploration. Future research in (interactive) multi-field visualization has to develop and integrate techniques that allow for a dynamically changing focus or field prioritization. Especially for inhomogeneous field types the question remains, how and whether multi-field visualization can incorporate such dynamic changes in an intuitive and expressive way.

Ronald Peikert on Feature-based Visualization:

The challenges of multifield visualization also extend to the area of feature-based visualization. Many useful techniques have been developed for finding inherent features in scientific data. They typically operate on one or at most two scalar, vector or tensor fields. In most cases, such feature detectors are not based on concepts that easily generalize to larger multi-fields containing additional variables. A feature can in the simplest case be represented by scalar field indicating the presence or absence of the feature or, alternatively, a probability for the feature to be present at a given location. But even with this simple notion of a feature, it is not clear how to combine a large number of them in a single visualization. To visualize their statistics, e.g., using uncertainty visualization techniques, can be a solution, but only if the features are based on the same physical quantities and can therefore be directly compared. New approaches are needed if the underlying multi-field represents a multitude of physical quantities, in which case features having different meanings are to be combined in one visualization. Extending other feature concepts, such as geometric or topological ones, to multi-fields will be an additional challenge.

Eugene Zhang on Tensor Fields and their Derived Fields:

Given a tensor field of some order, it is possible to derive a number of tensor fields from it. Examples of this includes the spatial gradient, the Laplacian, and the divergence. The derived fields contain rich information and provide great insight to the original field. However, the derived fields often are of a different order. This leads to the need of simultaneous analysis and visualization of multiple tensor fields of different types. Most existing work on multi-field analysis focuses on fields of the same type, and there has not been much research on higher-order tensor fields due to the mathematical and physics background it often requires.

References

1. Chen, C.: Top 10 unsolved information visualization problems. IEEE Comput. Graph. Appl. **25**(4), 12–16 (2005)
2. Hauser, H., Bremer, P.T., Theisel, H., Trener, M., Tricoche, X.: Panel: What are the most demanding and critical problems, and what are the most promising research directions in Topology-Based Flow Visualization? In Topology-Based Methods in Visualization Workshop. Budmerice, Slovakia (2005)
3. House, D., Interrante, V., Laidlaw, D., Taylor, R., Ware, C.: Panel: Design and evaluation in visualization research. Proc. IEEE Vis. **2005**, 705–708 (2005)
4. Johnson, C.R.: Top scientific visualization research problems. IEEE Comput. Graph. Appl. **24**(4), 13–17 (2004)
5. Johnson, C.R., Moorehead, R., Munzner, T., Pfister, H., Rheingans, P., Yoo, T. S.: NIH/NSF Visualization Research Challenges (Final Draft, Jan 2006). Technical report (2006)
6. Johnson, G., Ebert, D., Hansen, C., Kirk, D., Mark, B., Pfister, H.: Panel: the future visualization platform. Proc. IEEE Vis. **2004**, 569–571 (2004)

7. Laramee, R.S., Kosara, R.: Human-Centered Visualization Environments, Chapter Future Challenges and Unsolved Problems Springer Lecture Notes in Computer Science (LNCS) 4417, pp. 231–254. Springer, Berlin (2007)
8. McCormick, B.H., DeFanti, T.A., Brown, M.D.: Visualization in Scientific Computing. Technical report, The National Science Foundation (NSF) (1987)
9. Peng, Z., Grundy, E., Laramee, R.S., Chen, G., Croft, N.: Mesh Driven Vector Field Clustering and Visualization: An Image-Based Approach. IEEE Transactions on Visualization and Computer Graphics. **24**(2), 283–298 (2012)
10. Rhyne, T.-M., Hibbard, B., Johnson, C., Chen, C., Eick, S.: Panel: can we determine the top unresolved problems of visualization? Proc. IEEE Vis. **2004**, 563–565 (2004)
11. Scheuermann, G., Garth, C., Peikert, R.: Panel: Even More Theory, or More Practical Applications to Particular Problems: In Which Direction will Topology-Based Flow Visualization go? In: Topology-Based Methods in Visualization Workshop. Budmerice, Slovakia (2005)
12. Van Wijk, J. J.: The Value of Visualization. In: Proceedings IEEE Visualization '05, pp. 79–86. IEEE Computer Society (2005)

Part III
Biomedical Visualization

Chapter 20
Overview of Visualization in Biology and Medicine

Arie E. Kaufman, Gerik Scheuermann and Jos B.T.M. Roerdink

Similar to all other areas of visualization, visualization in biology and medicine is driven to a large extent by developments in the application domain itself. In recent years, new experimental techniques have increased measured data by orders of magnitude. These technological improvements continue to change biomedical research, and consequently, biomedical visualization. Thus far, there is no end in sight to this continuous challenge. One special aspect in this development is that biomedical research uses more and more heterogeneous data which need to be integrated to gain insight. This requires visualization systems that deal with these different data types and provide a unified view of this variety of aspects of the same problem. Especially, we continue to see measured images (but with an enormous increase in resolution), networks (with rapid updates and increasing size), electromagnetic signals, simulations of partial differential equations, and rapidly increasing genomics data that are partially or all present in typical application scenarios. Publicly available databases enhance this trend substantially, since the data in an application scenario typically comes only partially from measurements and simulations of the researcher or medical doctor—the rest of the picture comes from databases. In addition, most modern research in life sciences requires consideration of different scales spanning many orders of magnitude in space and time. Furthermore, we see the ongoing trend of data analysis and data cleaning incorporated into the visualization process. Overall, we observe that visualization has become a key technology for larger parts

A.E. Kaufman (✉)
Department of Computer Science, Stony Brook University, New York, USA
e-mail: ari@cs.stonybrook.edu

G. Scheuermann
Institute of Computer Science, University of Leipzig, P.O. Box 100920, 04009 Leipzig, Germany
e-mail: scheuermann@informatik.uni-leipzig.de

J.B.T.M. Roerdink
Johann Bernoulli Institute for Mathematics and Computer Science, University of Groningen, P.O. Box 407, 9700 AK Groningen, The Netherlands
e-mail: j.b.t.m.roerdink@rug.nl

C.D. Hansen et al. (eds.), *Scientific Visualization*, Mathematics and Visualization,
DOI 10.1007/978-1-4471-6497-5_20

of biological and medical research. This underlines the high practical relevance of the on-going quest in visualization to find the optimal tool for human insight. Even as various issues become solved problems, further developments in technology and new questions posed by researchers require an unceasing and continuous spectrum of research in the fields of data acquisition, management, analysis, and visualization in order to achieve further insight.

Chapter 21 provides a general overview of the emerging field of *connectomics*. Connectomics is a branch of neuroscience that attempts to create a *connectome*, that is, a complete map of the neuronal system and all connections between neuronal structures. Such a holistic representation can then be used to understand how functional brain states emerge from their underlying anatomical structures and how dysfunction and neuronal diseases arise.

The notion of brain connectivity by itself is not straightforward. In fact, different types of connectivity can be distinguished at different spatial scales. *Structural* or anatomical connectivity usually refers to the physical connections between neural elements. *Functional* connectivity refers to the temporal correlation between spatially remote neurophysiological events; it does not necessarily imply an anatomical connection. Finally, *effective* connectivity concerns causal interactions between distinct units within a nervous system. One can also differentiate between macro-, meso-, and microscale connectomes. At the *macroscale*, a whole brain can be divided into anatomically distinct areas with specific patterns of connectivity. One order of magnitude smaller is the *mesoscale* connectome that describes local neuronal circuits, such as, cortical columns. At the finest *microscale*, the connectome involves mapping single neuronal cells and their connectivity patterns. Ultimately, connectomes from all scales should be merged into one hierarchical representation.

Since the field of connectomics is to a large extent based on image data, visualization is an important task for the analysis of brain structures and their functional connections. Therefore, this chapter reviews the current state-of-the-art of visualization and image processing techniques in the field of connectomics and associated challenges. This chapter first presents some biological background into the concepts of neural systems and model systems. Relevant imaging modalities are also introduced, including electroencephalography, magnetoencephalography, magnetic resonance imaging, positron emission tomography, and diffusion-weighted imaging. Then, current techniques to extract connectivity information from the image data at the macro-, meso-, and microscale are reviewed. Based on this extraction, integration of the data for the important topics of brain mapping and neural network modeling by reverse engineering are discussed. Lastly, techniques for visual analysis, measurements, and comparative visualization are discussed.

Chapter 22 concerns visualization in biology. Similar to Chap. 21, the notion of scale is very important. Basic biological research spans a huge range of scales, from the genome level up to the cellular and population level. Advances in high-throughput measuring devices such as genome sequencers and the public availability of large amounts of data have fundamentally changed the way that biologists conduct research. Access to this data has made visualization a key component in almost every biological discovery workflow.

Several common themes can be identified in this field. First, to keep up with the accelerated experimental process in biology, visualization tools should be developed rapidly to be relevant. A second challenge is the integration of many different types of data. Visualization should support the discovery of complex patterns in such heterogeneous data. Thirdly, close collaboration between biologists and visualization researchers is essential for requirement elicitation and prototype design. The chapter discusses visualization in the context of comparative genomics and functional genomics, as well as evolutionary and developmental biology. The common types of biological data, questions and methods in each of these fields are covered, along with visualization challenges and case studies that highlight the biological impact of visualization tools.

In the field of comparative genomics, scientists compare the genomes of organisms to answer questions about evolution and how the genome encodes cellular functions. Here, several challenges arise. First, the number of features to be compared can easily run into the thousands. Second, the size of the features is often orders of magnitude smaller than the size of the chromosomes. Third, it is often important to understand the location and size of paired features in the context of their similarity scores. Many current visualization tools do not support this.

Functional genomics studies how genes work together in a cell to perform different cellular functions, such as metabolism or reproduction, and how these are controlled by many interrelated chemical reactions which form complicated networks. Finding differences and similarities in networks from different experimental conditions, in different cell types, and in different species is an important component of functional genomics. Again, scale is a major challenge. The number of nodes and links in the network can become very large. Major questions are how to visualize such networks on different scales, how to support interactive exploration and pattern discovery, how to understand changes over time, and, maybe most difficult of all, how to integrate the various data.

Finally, in evolutionary and developmental biology, scientists can nowadays capture data about living organisms at an unprecedented level of detail in time and space. For example, it is possible to identify each single cell of a complex organism, follow its development over time, and connect this with genetic information. This allows the scientist to study how a single cell evolves into a complex organism, how internal regulatory processes cause differentiation, or how genomic differences relate to differences in physiological structure. Due to the wealth of data, robust automatic preprocessing and a sophisticated visualization framework are central requirements to allow for future advances in this field.

Chapter 23 is about medical visualization. Given the ubiquitous nature of medical volume data, medical visualization is now an established branch of visualization, with applications in medical diagnosis, treatment, research and education. During the past decades, medical image acquisition technology has undergone continuous and rapid development. It is now possible to acquire larger and more complex data than ever before by techniques such as computed tomography, (functional) magnetic resonance imaging, electroencephalography, diffusion tensor imaging, etc. The questions that users attempt to answer using medical visualization have also become significantly more complex.

This chapter first gives a brief overview of developments in medical visualization over the past three decades. Basic techniques such as isosurface and direct volume rendering are discussed, as well as more advanced methods for multi-modal, multi-field, multi-subject, and time-dependent data visualization. These techniques are useful for therapy planning, predictive simulation, and diagnosis. In addition, illustrative medical visualization is discussed, which is useful for presentation and exploration purposes. Then, major medical visualization research challenges for the coming decade are discussed. These arise first because of advances in hardware and data acquisition, such as combined CT-PET scanners, simultaneous EEG-fMRI acquisition, high angular resolution diffusion imaging (HARDI), molecular imaging, high-resolution microscopy imaging, etc., leading to ever larger and more complex data sets. Mobile display and computing devices also may have a great impact on medical practice, leading to demands for new interaction and visualization for such devices, and introducing tele-medicine. Other challenges are the interactive segmentation of medical data, the integration of predictive simulation models and uncertainty visualization in surgical planning, intelligent data mapping and reformatting, and evaluation of illustrative versus (hyper)realistic visualization for diagnostic and treatment planning purposes. Furthermore, visual analysis in healthcare as well as visualization of population data are expected to grow in importance.

Chapter 24 is devoted to ultrasound imaging. Ultrasound is one of the most frequently used imaging modalities in medicine due to its high spatial resolution, interactive nature, and patient-friendliness. The main challenge of ultrasound is image interpretation for diagnostic purposes, which requires extensive training. Special problems arise because of the low dynamic range, noise and speckle occurring in ultrasound images.

Ultrasound imaging presents several challenges for visualization. For example, in functional ultrasonography, that is, ultrasound imaging of physiology and/or organ function, information on motility, biomechanics and perfusion can be obtained non-invasively. A set of 2D images can be aligned to form 3D data sets for which volume visualization provides added value in terms of a more holistic understanding of the data. Typical examples are demonstration of complex anatomy and pathology, pre-operative surgical planning or virtual training of medical students. In addition, matrix 3D probes are now on the market that allow real-time 3D acquisition. To benefit from the high temporal resolution, advanced graphics techniques are required to develop fast and efficient algorithms for rendering on-the-fly. Co-registration techniques enable the use of multi-modal data sets. Fusion imaging, where ultrasound is combined with either CT, MRI, or PET images, allows for more precise navigation in ultrasound-guided interventions. This demands advanced visualization research to enlighten how different data types can be combined and presented in novel ways.

This chapter presents the process-pipeline for ultrasound visualization, with an overview of the specific tasks performed. A technique-based taxonomy is presented based on a set of significant publications. In pre-processing, the ultrasound data is reconstructed and oftentimes enhanced to improve quality. Segmentation techniques are used to extract important features. Methods for registration, both rigid and non-rigid, are discussed to align the ultrasound data with other modalities. The rendering task presents a visual presentation of the data to the user, and important techniques in transfer function design, multi-modal rendering, shading, and illumination are discussed. Lastly, augmented reality projects are discussed which, though currently not popularly available in clinical systems, show great potential for the future. The differences between ultrasound-based techniques and techniques for other imaging modalities are also discussed.

Chapter 25, finally, concerns visual exploration of simulated and measured blood flow. This is of high importance in diagnosis and treatment planning for severe cardiovascular diseases. Assessment of cardiovascular disease is facilitated by various imaging modalities. Vascular diseases occur primarily at regions with complex or unstable flow, which significantly influences the morphology of cardiovascular tissue. The flow behavior is therefore of vital importance to the cardiovascular system and potentially harbors a considerable value for both diagnosis and risk assessment. The analysis of haemodynamic characteristics involves qualitative and quantitative inspection of the blood-flow field. Visualization plays an important role in the qualitative exploration, as well as the definition of relevant quantitative measures and its validation.

There are two main approaches to obtain information about the blood flow: simulation by computational fluid dynamics, and real measurements. Although research on blood flow simulation has been conducted for decades, many open problems remain concerning accuracy and patient-specific solutions. Possibilities for real measurement of blood flow have recently increased considerably through new developments in magnetic resonance imaging which enable the acquisition of 3D quantitative measurements of blood-flow velocity fields. MRI scanners with higher magnetic field strengths (7–9 Tesla) provide the required resolution and signal-to-noise ratios to analyze blood flow in smaller vessels than the main arteries around the heart.

This chapter presents the visualization challenges for both simulation and real measurements of unsteady blood-flow fields. For simulation, challenges arise because of the many assumptions made, the difficulty to make it patient specific, and the validation. Measured flow data, on the other hand, although being patient specific, has many limitations regarding resolution, artifacts, and noise in the data. An interesting direction is to combine both methods for higher performance. Recent blood-flow visualization techniques involve ad-hoc decisions with respect to seeding, segmentation, or the use of illustration techniques, which need to be better linked to the user needs. A major challenge is the novelty of this type of data for the domain experts. Many existing methods involve rather complex visual representations that might overwhelm a considerable portion of the target user group. Future research should address simplifications of the blood flow and aim at a better understanding of specific tasks, decisions and relevant information necessary to support blood flow exploration with a guided workflow-based interaction.

Chapter 21
Visualization in Connectomics

Hanspeter Pfister, Verena Kaynig, Charl P. Botha, Stefan Bruckner, Vincent J. Dercksen, Hans-Christian Hege and Jos B.T.M. Roerdink

Abstract Connectomics is a branch of neuroscience that attempts to create a *connectome*, i.e., a complete map of the neuronal system and all connections between neuronal structures. This representation can be used to understand how functional brain states emerge from their underlying anatomical structures and how dysfunction and neuronal diseases arise. We review the current state-of-the-art of visualization and image processing techniques in the field of connectomics and describe a number of challenges. After a brief summary of the biological background and an overview of relevant imaging modalities, we review current techniques to extract connectivity

H. Pfister (✉) · V. Kaynig
School of Engineering and Applied Sciences, Harvard University,
33 Oxford St., Cambridge, MA 02138, USA
e-mail: pfister@seas.harvard.edu

V. Kaynig
e-mail: vkaynig@seas.harvard.edu

C.P. Botha
vxlabs, Somerset West, SA
e-mail: cpbotha@vxlabs.com

S. Bruckner
Institute of Computer Graphics and Algorithms, Vienna University of Technology,
Favoritenstraße 9-11/E186, 1040 Wien, Austria
e-mail: bruckner@cg.tuwien.ac.at

V.J. Dercksen · H.-C. Hege
Department of Visualization and Data Analysis,
Zuse Institute Berlin, Takustraße 7, 14195 Berlin, Germany
e-mail: dercksen@zib.de

H.-C. Hege
e-mail: hege@zib.de

J.B.T.M. Roerdink
Johann Bernoulli Institute for Mathematics and Computer Science,
University of Groningen, P.O. Box 407, 9700, AK Groningen, The Netherlands
e-mail: j.b.t.m.roerdink@rug.nl

C.D. Hansen et al. (eds.), *Scientific Visualization*, Mathematics and Visualization,
DOI 10.1007/978-1-4471-6497-5_21

information from image data at macro-, meso- and microscales. We also discuss data integration and neural network modeling, as well as the visualization, analysis and comparison of brain networks.

21.1 Introduction

Connectomics is a field of neuroscience that analyzes neuronal connections. A connectome is a complete map of a neuronal system, comprising all neuronal connections between its structures. The term ‘connectome’ is close to the word ‘genome’ and implies completeness of all neuronal connections, in the same way as a genome is a complete listing of all nucleotide sequences. The goal of connectomics is to create a complete representation of the brain’s wiring. Such a representation is believed to increase our understanding of how functional brain states emerge from their underlying anatomical structure [89]. Furthermore, it can provide important information for the cure of neuronal dysfunctions like schizophrenia or autism [83].

Different types of connectivity can be distinguished. *Structural* or anatomical connectivity usually refers to the “wiring diagram” of physical connections between neural elements. These anatomical connections range in scale from those of local circuits of single cells to large-scale networks of interregional pathways [87]. *Functional* connectivity is defined as “the temporal correlation between spatially remote neurophysiological events” [32]. This can be seen as a statistical property; it does not necessarily imply direct anatomical connections. Finally, *effective* connectivity concerns causal interactions between distinct units within a nervous system [32].

Sporns et al. [89] differentiate between macro-, meso- and microscale connectomes. At the *macroscale*, a whole brain can be imaged and divided into anatomically distinct areas that maintain specific patterns of interconnectivity. Spatial resolution at the macroscale is typically in the range of millimeters. One order of magnitude smaller is the *mesoscale* connectome that describes connectivity in the range of micrometers. At this scale, local neuronal circuits, e.g., cortical columns, can be distinguished. At the finest *microscale*, the connectome involves mapping single neuronal cells and their connectivity patterns. Ultimately, connectomes from all scales should be merged into one hierarchical representation [89].

Independently of the scale, the connectivity can be represented as a *brain graph* $G(N;E)$ with nodes N and weighted edges E representing anatomical entities and the degree of structural or functional interactions, respectively. Associated to each abstract graph is a graph in real space that connects real anatomical entities. Neural systems can be investigated by analyzing topological and geometrical properties of these graphs and by comparing them. An equivalent way of representing an undirected or directed brain graph is a *connectivity* or *association matrix* C, whose entries c_{ij} represent the degrees of interactions. Thresholding and sometimes also binarizing them reveals the essential interactions. A spatial connectivity graph can be depicted in real space, showing the actual physical structure of the neural system. A connection matrix is usually visualized using a color-coded matrix view. For more details and examples see, e.g., the recent reviews [12, 30].

In contrast to genomics, the field of connectomics is to a large extent based on image data. Therefore, visualization of image data can directly support the analysis of brain structures and their structural or functional connections.

In this chapter, we review the current state-of-the-art in visualization and image processing techniques in the field of connectomics and describe some remaining challenges. After presenting some biological background in Sect. 21.2 and an overview of relevant imaging modalities in Sect. 21.3, we review current techniques to extract connectivity information from image data at macro-, meso- and microscale in Sects. 21.4–21.6. Section 21.7 focuses on integration of anatomical connectivity data. The last section discusses visually supported analysis of brain networks.

21.2 Biological Background

Neural systems. Functionally, neurons (or nerve cells) are the elementary signaling units of the nervous system, including the brain. Each neuron is composed of a cell body (soma), multiple dendritic branches and one axonal tree, which receive input from and transfer output towards other neurons, respectively. This transfer is either chemical (synapses) or electrical (gap junctions). Generally, during synaptic transmission, vesicles containing neurotransmitter molecules are released from terminals (boutons) on the axon of the presynaptic neuron, diffuse across the synaptic cleft, and are bound by receptors on dendritic spines of the postsynaptic neuron, inducing a voltage change, i.e., a signal.

These basic building blocks can mediate complex behavior, as potentially large numbers of them are interconnected to form local and long-range neural microcircuits. At the meso-level, local neuron populations, e.g., *cortical minicolumns*, can be identified that act as elementary processing units. At the macroscale, neurons in the human cortex are arranged in a number of anatomically distinct areas, connected by interregional pathways called *tracts* [89].

Model systems. An important neuroscientific goal is to understand how the human brain works. However, due to its complexity (with an estimated 10^{11} neurons with 10^{15} connections [89]), brain function at the circuit or cellular level is often studied in other organisms that are more amenable in complexity and size.

Conserved genes and pathways between different species offer the potential of elucidating the mechanisms that affect complex human traits based on similar processes in other organisms. This problem is particularly tractable in the roundworm *Caenorhabditis elegans*, whose brain with 302 neurons has been completely mapped [106], or in insects. In these organisms brain structure and function can be studied at the level of single identifiable neurons. Classical insect model organisms that are well understood and allow easy genetic manipulations are the fruit fly *Drosophila melanogaster* and the honeybee. Drosophila, for example, has been shown to be an experimentally amenable model system even for the study of such quintessential human physiological traits as alcoholism, drug abuse, and sleep [63].

Rodents, being mammals, have a brain structure that is similar but much smaller than the human brain, and that therefore can be used to study cortical networks. The mouse brain is an attractive model system to study, for example, the visual system, due to the abundant availability of genetic tools allowing monitoring and manipulating certain cell types or circuits [38]. The whisker-barrel pathway of the rat is a relatively small and segregated circuit that is amenable to studying sensory information processing at the molecular/synaptic, cell, and circuit/region levels.

21.3 Imaging Modalities Employed in Connectomics

We now provide an overview of imaging modalities that are used in obtaining connectivity information. They differ in the spatial and temporal resolution at which connectivity is captured. At the *macroscale* there is a wide range of structural and functional imaging modalities, with applications in medical settings and anatomical research. Functional imaging modalities include electroencephalography (EEG), magnetoencephalography (MEG), functional magnetic resonance imaging (fMRI), and positron emission tomography (PET). Modalities such as single-photon emission computed tomography (SPECT) and magnetic resonance imaging (MRI) provide structural information on the macroscale. Section 21.4 gives a detailed introduction to the relevant modalities in the context of connectomics. At the *mesoscale*, light microscopy (LM) techniques provide sufficient resolution to image single neurons. Most light microscopy techniques focus on structural imaging. Techniques such as wide-field fluorescence microscopy allow for the imaging of living cells, and computational optical sectioning microscopy techniques [17] enable non-destructive acquisition of 3D data sets. Section 21.5 provides further details about light microscopy techniques. At the *microscale*, the sufficient resolution is offered by electron microscopy techniques (EM) such as Transmission Electron Microscopy (TEM) and Scanning Electron Microscopy (SEM). These methods require technically complex specimen preparation and are not applicable to live cell imaging. Imaging of 3D volumes requires ultra-thin sectioning of the brain tissue followed by computational realignment of the acquired images into one image volume [46]. More information about electron microscopy in the connectomics setting can be found in Sect. 21.6. Figure 21.1 provides an overview of the different imaging modalities and their spatial and temporal resolution.

21.4 Macroscale Connectivity

First, we discuss the main acquisition techniques for revealing macroscopic functional and structural connectivity. We start with MEG and EEG, as these were used for functional connectivity before fMRI, then diffusion-weighted MRI for structural

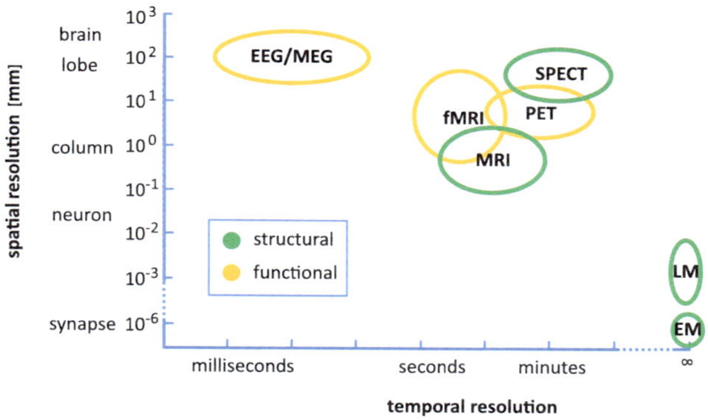

Fig. 21.1 Different brain imaging modalities and their spatial and temporal resolutions. For connectomics, light-(*LM*) and electron microscopy (*EM*) are mostly performed in vitro. The color indicates functional versus structural information in the acquired data

connectivity, and finally fMRI for functional connectivity. Besides the visualization approaches discussed here, the reader is also referred to Sect. 21.8 for more detail on network analysis and comparative visualization techniques.

21.4.1 EEG and MEG

Developed in the 1920s, electroencephalography (EEG) is the oldest noninvasive functional neuroimaging technique, which records electrical brain activity from electrodes on the scalp. Nowadays, the number of electrodes can be as large as 128 or even 512; in that case one speaks of *multichannel* or *high-density* EEG [81, 93]. By contrast, magnetoencephalography (MEG) measures magnetic fields outside the head induced by electrical brain activity [35]. The temporal frequency of these signals ranges from less than 1 Hz to over 100 Hz. The spatial resolution is lower than that of fMRI. Sometimes, MEG is preferred over EEG because the electrical signals measured by EEG depend on the conduction through different tissues (e.g., skull and skin). However, EEG has much lower costs and higher equipment transportability than MEG (and fMRI). Moreover, EEG allows participants more freedom to move than MEG and fMRI. In Sect. 21.8 we will discuss the use of EEG to discover functional brain networks. Therefore, we will focus on EEG for the remainder of this section.

Electrical potentials generated within the brain can be measured with electrodes at the scalp during an EEG recording. The measured EEG signals reflect rhythmical activity varying with brain state. Specific brain responses can be elicited by the presentation of external stimuli. For EEG analysis, one often studies activity in various frequency bands, such as alpha, beta, theta or delta bands. As a result of *volume conduction*, an electrical current flows from the generator in the brain through different tissues (e.g., brain, skull, skin) to a recording electrode on the scalp. The measured EEG is mainly generated by neuronal (inhibitory and excitatory) postsynaptic potentials and burst firing in the cerebral cortex. Measured potentials depend on the source intensity, its distance from the electrodes, and on the conductive properties of the tissues between the source and the recording electrode.

Several visualization methods are applied to assist in the interpretation of the EEG [93]. In a conventional EEG visualization, the time-varying EEG data are represented by one time series per electrode, displaying the measured potential as a function of time. Synchronous activity between brain regions is associated with a functional relationship between those regions. EEG coherence, calculated between pairs of electrode signals as a function of frequency, is a measure for this synchrony. A common visualization of EEG coherence is a graph layout. In the case of EEG, graph vertices (drawn as dots) represent electrodes and graph edges (drawn as lines between dots) represent similarities between pairs of electrode signals. Traditional visual representations are, however, not tailored for multichannel EEG, leading to cluttered representations. Solutions to this problem are discussed in Sect. 21.8.

21.4.2 MRI

In magnetic resonance imaging, or MRI, unpaired protons, mostly in hydrogen atoms, precess at a frequency related to the strength of the magnetic field applied by the scanner. When a radio-frequency pulse with that specific frequency is applied, the protons resonate, temporarily changing their precession angle. They eventually regain their default precession angle, an occurrence that is measured by the scanner as an electromagnetic signal. By applying magnetic field gradients throughout three-dimensional space, protons at different positions will precess and hence resonate at different frequencies, enabling MRI to generate volume data describing the subject being scanned.

21.4.2.1 Diffusion-Weighted Imaging

Water molecules at any temperature above absolute zero undergo Brownian motion or molecular diffusion [23]. In free water, this motion is completely random, and water molecules move with equal probability in all directions. In the presence of constraining structures such as the axons connecting neurons together, water molecules move more often in the same direction than they do across these structures. When such a molecule moves, the two precessing protons its hydrogen nucleus contains move as well. When this motion occurs in the same direction as the diffusion gradient q (an extra magnetic field gradient that is applied during scanning) of a diffusion-weighted MRI scan, the detected signal from that position is weakened. By applying diffusion gradients in a number of different directions, a dataset can be built up showing the 3D water diffusion at all points in the volume, which in turn is related to the directed structures running through those points.

Diffusion tensor imaging. When at least six directions are acquired, a 3×3 symmetric diffusion tensor can be derived, in which case the modality is described as Diffusion Tensor Imaging (DTI). Per voxel DTI, often visualized with an ellipsoid, is not able to represent more than one major diffusion direction through a voxel. If two or more neural fibers were to cross, normal single tensor DTI would show either planar or more spherical diffusion at that point. The left image of Fig. 21.2 shows a 3-D subset of such a dataset, where each tensor has been represented with a superquadric glyph [50].

DTI visualization techniques can be grouped into the following three classes [102]: *Scalar metrics* reduce the multi-valued tensor data to one or more scalar values such as fractional anisotropy (FA), a measure of anisotropy based on the eigenvalues of the tensor, and then display the reduced data using traditional techniques, for example multi-planar reformation (slicing) or volume rendering. An often-used technique is to map the FA to intensity and the direction of the principal tensor eigenvector to color and then display these on a slice. Multiple anisotropy indices can also be used to define a transfer function for volume rendering, which is then able to represent the anisotropy and shape of the diffusion tensors [49].

Glyphs can be used to represent diffusion tensors without reducing the dimensionality of the tensor. In its simplest form, the eigensystem of the tensor is mapped directly to an ellipsoid. More information can be visually represented by mapping diffusion tensors to superquadrics [50] (see Fig. 21.2).

Vector- and tensor-field visualization techniques visualize global information of the field. The best known is probably fiber tractography, where lines are reconstructed that follow the tensor data in some way and hence are related to the major directions of neural fibers. In its simplest form, streamlines, tangent to the principal eigenvec-

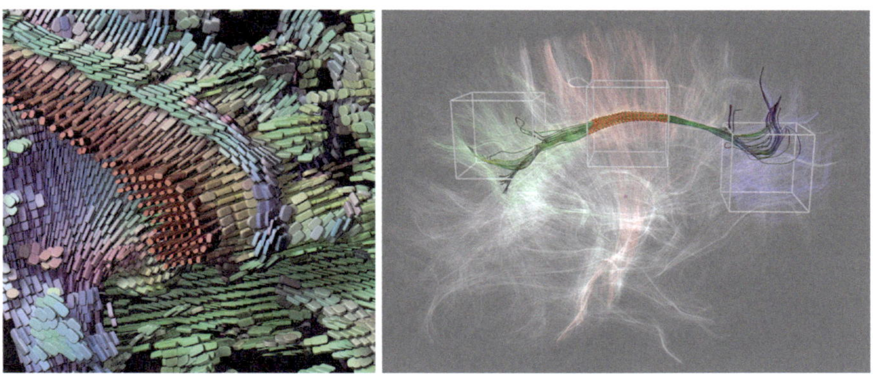

Fig. 21.2 *On the left*, superquadric glyphs have been used to represent the diffusion tensors in a 3-D region of a brain dataset (image courtesy of Gordon Kindlmann, University of Chicago). *On the right*, the cingulum neural fiber bundle has been highlighted in a full-brain tractography [8]

tors of the diffusion tensors, are extracted and displayed [2]. Care must to be taken to terminate the streamlines in areas of isotropic or planar diffusion. Hyperstreamlines take into account more of the tensor information [109]. Many tractography approaches require one or more regions of interest to be selected *before* tracts can be seeded starting only from those regions, while more recent efforts allow for full-brain fiber tracking followed by more intuitive interactive selection within the brain's tracked fiber bundles [8, 85] (see the right image in Fig. 21.2 for an example). For a simplified visual representation, the envelopes of clustered streamline bundles can be shown [25], or illustrative techniques such as depth-dependent halos can be used [26]. With probabilistic tractography, local probability density functions of diffusion or connectivity are estimated and can in turn be used to estimate the global connectivity, that is, the probability that two points in the brain are structurally connected [4]. This type of data is arguably a higher fidelity representation of structural connectivity. Connectivity between two points can be visualized with, e.g., constant-probability isosurfaces, with direct volume rendering of the probability field, or using topological methods from flow visualization [82]. Calculating and effectively visualizing a full-brain probabilistic tractography would be challenging.

DSI and HARDI. As explained above, DTI is not able to capture more than one principal direction per sample point. In order to reconstruct the full diffusion probability density function (PDF), that is, the function describing the probability of water diffusion from each voxel to all possible displacements in the volume, about 500 or more diffusion-weighted MRI volumes need to be acquired successively. This is called diffusion spectrum imaging or DSI [34] and is the canonical way of acquiring the complete 3-D water diffusion behavior. However, the time and processing required to perform full DSI complicate its use in research and practice.

In High Angular Resolution Diffusion Imaging, or HARDI, 40 or more directions are typically acquired in order to sample the 3-D diffusion profile around every point [95]. Based on such data, multiple diffusion tensors can be fit to the data [95], higher order tensors can be used [69], or a model-free method such as Q-Ball imaging [96] can be applied. Q-Ball yields as output an orientation distribution function, or ODF. The ODF is related to the diffusion PDF in that it describes for each direction the sum of the PDF values in that direction. It can be visualized as a deformed sphere whose radii represent the amount of diffusion in the respective direction.

HARDI visualization follows much the same lines as DTI visualization, except that the data are more complex. Analogous to DTI, HARDI scalar metrics, such as generalized (fractional) anisotropy and fractional multifiber index, can be used to reduce the data to one or more scalar values that can be visualized with traditional techniques. Multiple diffusion tensors can be represented as glyphs, or the diffusion ODF can be directly represented using a tessellated icosahedron or by raycasting the spherical harmonics describing the ODF [70]. This results in a field of complex glyphs representing at each point the diffusion profile at that position. In contrast to DTI glyph techniques, regions of crossing fibers can in general be identified.

Although there are fewer examples, especially in the visualization literature, (probabilistic) fiber tracking can be performed based on HARDI data [72]. More recently, HARDI glyphs have been combined dynamically with DTI glyphs and fiber tracts based on local data characteristics [73].

21.4.3 Functional MRI

Blood-oxygen-level dependence, or BOLD, is a special type of MRI that is able to measure increased levels of blood oxygenation [67]. Due to requiring more glucose from the bloodstream, active neurons cause higher blood oxygenation in nearby veins. Based on this principle, functional MRI, or fMRI, uses BOLD to image time-dependent 3-D neural activity in the brain [68].

fMRI can also be used to derive functional or *effective connectivity* in the brain. Functional connectivity is determined by calculating the temporal correlations between the fMRI signals originating from different parts of the brain [32]. This is done either whilst the subject performs a specific task, in order to assess how the brain network is applied during that task, or during resting state, in order to derive the baseline functional brain network. Connectivity data can be determined between a specific seed region or voxel and one or more other regions or voxels, or exhaustively between all regions or voxels in the brain.

Effective connectivity, defined as the causal influence one neuronal system exerts over another, is dependent on a model of the connectivity between the participating regions. For example, the signal at one position could be expressed as the weighted sum of the signals elsewhere [32]. If the model is invalid, the effective connectivity derived from fMRI is also invalid.

Visualization of fMRI-derived connectivity information is quite varied, often combining techniques from scientific and information visualization. Scatter plots have been used to plot correlation strength over distance, dendrograms and multi-dimensional scaling to represent correlations between regions in 2D [80], matrix bitmaps to represent region-wise correlation matrices [28], 2-D and 3-D (pseudo-) anatomical node-link diagrams to show the derived brain networks [107], and coupled-view visual analysis techniques to explore resting state fMRI data [99]. When connectivity is determined between all pairs of voxels in the cortex, visualization and knowledge extraction pose perceptual and computational challenges that have not yet been fully explored.

21.5 Mesoscale Connectivity

Light microscopy was the first modality that allowed for imaging of single neuronal cells. While the resolution of a light microscope is not sufficient to resolve synapses, it allows for the identification of major cell parts, such as dendrites, somas, axons, and also boutons as possible locations for synaptic connections. Imaging whole neuronal cells and analyzing their geometry enables neuroanatomists to identify different types of cells and to come to conclusions about their function. Following the motto "the gain in the brain lies mainly in the stain" [1], the three following main techniques are employed to map neuronal circuits with light microscopy [60].

Single-cell staining by dye impregnation. This is the oldest staining method and it laid the foundation for modern neuroscience. As neuronal tissue is densely packed with cells, a complete staining of the whole sample would not allow one to discriminate single cells in light microscopy images. Instead, the so-called *Golgi stain* enables stochastic marking of just a few individual nerve cells. The stained cells appear dark in the light microscopy images, discriminating them from a bright background formed by the unstained tissue. This staining method, combined with the ability of the light microscope to focus on different depth of the sample, allows for 3D imaging of the cell geometry. The famous neuroscientist Cajal (1852–1934) was able to identify different types of neurons and also describe connectivity patterns and principles of neuronal circuit organization using Golgi's method [60].

Diffusion or transport staining. Diffusion staining techniques enable biologists to analyze the projective trajectory of brain regions. For this technique, different staining markers are injected into different regions of the brain *in vivo*. The staining is then diffused along the connected neurons. Finally, a sample of brain tissue is extracted from a different region, in which no marker has been injected. The color code in the staining of different neurons in this area then reveals the projection of these neurons back to the initial staining areas, providing information about long-distance connectivity [33]. The range of possible colors for this method is limited to three or four different stainings.

Multicolor or brainbow. This staining technique does not involve application or injection of staining to brain tissue. Instead, transgenic mice are bred to produce photophysical fluorescent proteins. A confocal laser-scanning microscope activates the fluorescent proteins with a laser beam and records an image with the expressed light. Brainbow mice are bred to express three fluorescent proteins of different colors. By different stochastic expression of these three colors, the single neurons of the mice are colored with one out of >100 labels. The main advantage of this method is that it allows one to uniquely identify dendrites and axons belonging to the same neuron in densely colored tissue [60], see also Fig. 21.3.

All of these three staining methods allow imaging the geometry of neurons at the micrometer scale. The different staining protocols all aim at visually separating single neurons out of the complex and dense neuronal tissue. Visualization techniques for connectomics need to enhance the visual separation further, e.g., by providing contrast enhancement and enabling flexible mappings of image data to varying amounts of transparency in the transfer function [51]. Especially for the brainbow staining it is useful to have visual enhancement of color differences in regions of interest where two neurons with a similar staining combination need to be distinguished. For diffusion staining this problem is less pronounced than for brainbow data, as typically only three to four easily distinguishable colors are used. But this also leads to the challenge of distinguishing two neighboring cells that are

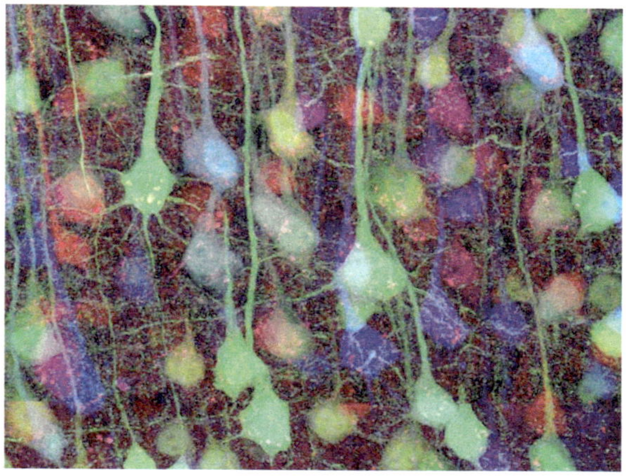

Fig. 21.3 Brainbow image of mouse cerebral cortex tissue. The different color stainings facilitate the differentiation of neuronal cells. Image courtesy of Jean Livet and Jeff Lichtman

stained with the same color. This problem also arises in the Golgi stain, as only one color is applicable for this staining. Thus, visualization needs to focus on providing a good impression of the neurons' geometry. The user needs to be able to access the three-dimensional structure on different scale levels to infer the connectivity of dendritic parts and axons. In order to analyze the neuron geometry further, dendritic and axonal trees have to be identified and segmented. This task is typically performed either semi-automatically or fully automatically with a final proof-reading step [97].

An additional major challenge for the visualization of microscopy data sets in the field of connectomics is the large data volume required to analyze the geometry of full neurons. Microscopes typically only record regions of interest at the required resolution. Afterwards the acquired images or image stacks need to be stitched into one large data volume. While this problem is well known and automatic methods for image stitching and alignment exist [24, 74], these tools typically work offline, assembling all images into one large image file for later visualization. But with image volumes in the gigapixel range this method is no longer applicable. Instead, visualization tools are required to perform operations like image stitching, alignment, contrast enhancement, and denoising *on-demand* in the region of interest shown to the user. To allow for interactive visualization, these operations do not only need to be executed fast, but also on multiple scales, allowing the user to zoom in and out of the displayed data volume. Recent work by Jeong et al. [41] provides this demand-driven visualization approach and combines it with a client server architecture. The client can visualize the data with user interaction and annotation while computations are performed on a high-performance server transparently to the user. Multiple client instances can connect to the same server to allow multiple users to access the data at the same time and cooperatively work on the same data set.

21.6 Microscale Connectivity

In contrast to light microscopy, which is limited in its resolution by the wavelength of light, electron microscopy enables imaging of neuronal tissue at the nanometer scale. Hence, electron microscopy is the only imaging modality so far that can resolve single synapses. However, the sample preparation and image acquisition in electron microscopy is labor-intensive and time-consuming. As a consequence, the analysis of the connectivity between single neurons has been limited to sparse analysis of statistical properties such as average synapse densities in different brain regions [20]. Little is known about the complete connectivity between single neurons. Information about the individual strength of synapses or the number of connections between two cells can have important implications for computational neuroanatomy and theoretical analysis of neuronal networks [98].

Recently, significant progress has been made in the automation of ultra-thin serial sectioning [36] and automatic image acquisition [21, 52]. These techniques allow neuroanatomists to acquire large datasets of multiple terabytes (TB) in size. With a resolution of 5 nm per pixel, and a section thickness of 50 nm, one cubic millimeter of brain tissue requires imaging of 20,000 sections with 40 gigapixels per image, leading to an image volume of 800 TB. With data sets of this size new challenges emerge for automatic computed analysis and visualization techniques. Important processing tasks include demand-driven image stitching and alignment, cell segmentation and 3D reconstruction, as well as multi-scale visualization and multi-user interaction via client server architectures.

Electron microscopy samples are typically densely stained. While in light microscopy sparse staining is necessary to visually separate a cell of interest from unstained background tissue (see Sect. 21.5), the fine resolution of electron microscopy allows one to discriminate structures according to shape, size, and texture. Electron microscopy images are limited to gray scale and typically do not have a uniform background. Instead, the background is noisy and highly variable, which imposes an important challenge for the visualization of electron microscopy image stacks. The image data cannot be visualized according to gray values alone, as the densely stained tissue forms a nearly solid block. Instead, higher order features that discriminate texture and shape, e.g., gradient histograms, are necessary to enhance the visibility of different structures of interest in the visualization [42]. Ultimately, full segmentation of the image data is necessary to allow the user visual inspection of different biological structures, from small structures such as vesicles or mitochondria to entire neuronal cells. Figure 21.4 shows example reconstructions of different neuronal structures from electron microscopy images. A number of software packages have been developed to aid the user in manual segmentation of cell structures in the images [14, 29, 37]. More recent semi-automatic methods greatly facilitate this time-intensive process [16, 76, 77, 91].

Progress has also been made on fully automatic segmentation of EM brain images [39, 44, 47, 48, 101, 103]. However, all methods developed so far require manual interaction and inspection by users. Thus, visualization tools should not only provide the ability to inspect the original EM data and the computed segmentations, but also

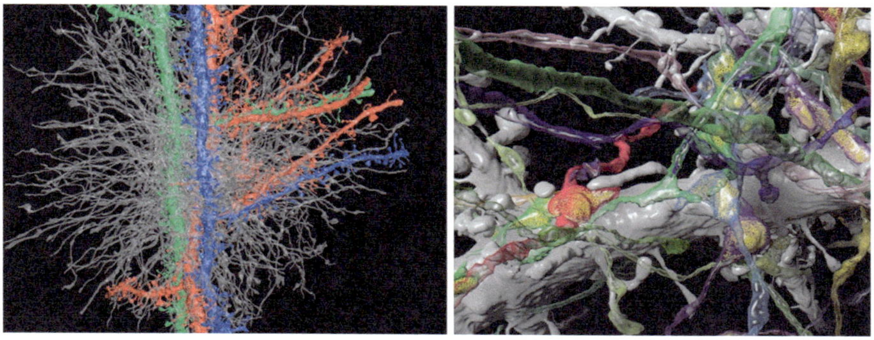

Fig. 21.4 Three dimensional reconstructions of neuronal structures from electron microscopy data. *Left* three dendrites (colored) and all intervening axons (transparent), *right* different axons (colored) with vesicle filled boutons (*yellow*)

provide a user interface to detect and correct segmentation errors, a process called *proofreading*.

Another interesting challenge for the visualization of neuronal microscopy images is the concurrent display of light and electron microscopy data acquired from the same sample. Correlative microscopy is a newly developing field, which allows for inspection of the same neuronal tissue using both light and electron microscopes. Thus the fine resolution of the electron microscopy images can be combined with the advantage of color staining and information about long-range connectivity in, e.g., diffusion stained light microscopy images. Visualization of this data requires multi-modal registration of both data sets, which has not yet been addressed for correlative microscopy.

Currently, most research efforts in connectomics at the microscale concentrate on the image acquisition and segmentation of electron microscopy images. Little research has been done in the visualization of entire connectomes, i.e. the wiring diagram of neurons, their types and the connectivity for detailed analysis of neuronal circuits. Connectomes, like the manually reconstructed circuit of C. elegans, are visualized by connectivity matrices or connection graphs [100].

21.7 Data Integration and Neural Network Modeling

As described in the previous sections, neurobiological data can be acquired from many different sources. Relating these different kinds of data by integrating them in a common reference frame offers interesting opportunities to infer new knowledge about the relation between structure and function. In this section, we describe two approaches and their visualization aspects for such data integration with the purpose of inferring functional properties: brain mapping and network modeling by reverse engineering.

21.7.1 Brain Mapping

A major goal in neuroscience is to define the cellular architecture of the brain. Mapping the fine anatomy of complex neuronal circuits is an essential first step in investigating the neural mechanisms of information processing. The term *brain mapping* describes a set of neuroscience techniques predicated on the mapping of biological quantities or properties onto spatial representations of the brain resulting in maps. While all of neuroimaging can be considered part of brain mapping, the term more specifically refers to the generation of atlases, i.e., databases that combine imaging data with additional information in order to infer functional information. Such an undertaking relies on research and development in image acquisition, representation, analysis, visualization, and interaction. Intuitive and efficient visualization is important at all intermediate steps in such projects. Proper visualization tools are indispensable for quality control (e.g., identification of acquisition artifacts and misclassifications), the sharing of generated resources among a network of collaborators, or the setup and validation of an automated analysis pipeline. Data acquired to study brain structure captures information on the brain at different scales (e.g., molecular, cellular, circuitry, system, behavior), with different focus (e.g., anatomy, metabolism, function), and is multi-modal (text, graphics, 2D and 3D images, audio, video) [15, 53]. The establishment of spatial relationships between initially unrelated images and information is a fundamental step towards the exploitation of available data [7]. These relationships provide the basis for the visual representation of a data collection and the generation of further knowledge.

Databases and atlases. A neuroanatomical atlas serves as a reference frame for comparing and integrating data from different biological experiments. Maye et al. [64] give an introduction and survey on the integration and visualization of neural structures in brain atlases. Such atlases are an invaluable reference in efforts to compile a comprehensive set of anatomical and functional data, and in formulating hypotheses on the operation of specific neuronal circuits.

A classical image-based neuroanatomical atlas of Drosophila is the FlyBrain atlas,[1] spatially relating a collection of 2D drawings, microscopic images, and text. One approach in generating a digital atlas of this kind is by acquiring confocal microscope images of a large number of individual brains. In each specimen, one or more distinct neuronal types are highlighted using appropriate molecular genetic techniques. Additionally, a general staining is applied to reveal the overall structure of the brain, providing a reference for non-rigid registration to a standard template. After registration, the specific neuronal types in each specimen are segmented, annotated, and compiled into a database linked to the physical structure of the brain. Jenett et al. [40] describe techniques for quantitative assessment, comparison, and presentation of 3D confocal microscopy images of Drosophila brains and gene expression patterns within these brains. Pereanu and Hartenstein [71] and Rybak et al. [79] described 3D atlases of the developing Drosophila brain and the honeybee brain.

[1] http://flybrain.neurobio.arizona.edu

The Neuroterrain 3D mouse brain atlas [5] consists of segmented 3D structures represented as geometry and references a large collection of normalized 3D confocal images.

Visual exploration and analysis. 3D microscopy data is often visualized using Maximum Intensity Projection (MIP), which displays the maximum values along viewing rays. Direct Volume Rendering (DVR) enables better perception of spatial relationships, but has the disadvantage of added complexity, as an additional transfer function is required. It can lead to problems with occlusions, particularly when multiple channels need to be visualized simultaneously. Maximum Intensity Difference Accumulation (MIDA) [9] improves this situation by combining the simplicity of MIP with additional spatial cues provided by DVR. Wan et al.[105] presented a tool for the visualization of multi-channel data tailored to the needs of neurobiologists. As acquired volumetric data is typically visualized together with segmented structures, it is important to avoid occlusions as well as visual clutter. Kuß et al. [56] proposed and evaluated several techniques to make spatial relationships more apparent.

However, to enable the exploration of large-scale collections of neuroanatomical data, massive sets of data must be presented in a way that enables them to be browsed, analyzed, queried and compared. An overview of a processing and visualization pipeline for large collections of 3D microscopy images is provided in a study by de Leeuw et al. [59]. NeuARt II [13] provides a general 2D visual interface to 3D neuroanatomical atlases including interactive visual browsing by stereotactic coordinate navigation. Brain Explorer [58], an interface to the Allen Brain Atlas, allows the visualization of mouse brain gene expression data in 3D. The CoCoMac-3D Viewer developed by Bezgin et al. [6] implements a visual interface to two databases containing morphology and connectivity data of the macaque brain for analysis and quantification of connectivity data. An example of an interface to neuroanatomical image collections and databases that features basic visual query functionalities is the European Computerized Human Brain Database (ECHBD) [31]. It connects a conventional database with an infrastructure for direct queries on raster data. Visual queries on image contents can be performed by interactive definition of a volume of interest in a 3D reference image. Press et al. [75] focused on the graphical search within neuroanatomical atlases. Their system, called XANAT, allows for the study, analysis, and storage of neuroanatomical connections. Users perform searches by graphically defining a region of interest to display the connectivity information for this region. Furthermore, their system also supports textual search using keywords describing a particular region. Kuß et al. [55] proposed ontology-based high-level queries in a database of bee brain images based on pre-generated 3D representations of atlas information. In the BrainGazer system [9] anatomical structures can be visually mined based on their spatial location, neighborhood, and overlap with other structures. By delineating staining patterns in a volume rendered image, for example, the database can be searched for known anatomical objects in nearby locations (see Fig. 21.5). Lin et al. [61] presented an approach to explore neuronal structures forming pathways and circuits using connectivity queries. In order to explore the similarity and differences of a large population of anatomical variations, Joshi et

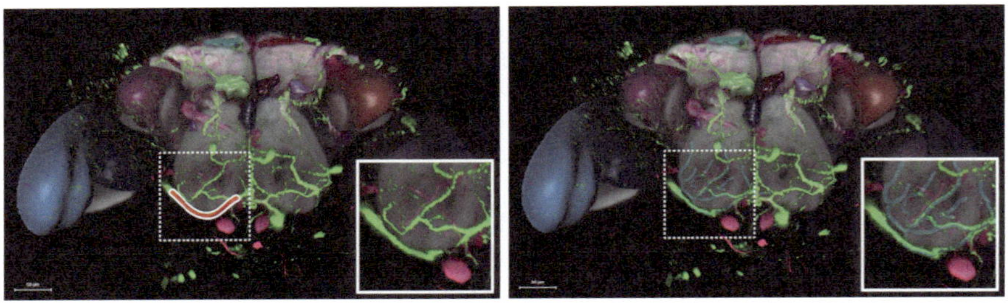

Fig. 21.5 Visual query for neural projection in the Drosophila brain using the BrainGazer system [9]. *Left* The query is specified by sketching a path on *top* of a Gal4 expression pattern. *Right* An existing segmented neural projection that matches the query is displayed

al. [43] proposed a similarity-space approach that embeds individual shapes in a meta-space for content-driven navigation.

While these efforts represent promising directions, many challenges remain. As noted by Walter et al. [104], a major goal is the integration of brain mapping data with other resources such as molecular sequences, structures, pathways and regulatory networks, tissue physiology and micromorphology. The ever-growing amount of data means that distributed solutions are required. The integration of computational and human resources gives significant benefits: each involved partner may bring computational resources (in terms of hardware and tools), human resources (in terms of expertise), and data to analyze. Advances in web technology, such as HTML5 and WebGL, provide new opportunities for visualization researchers to make their work accessible to the neuroscience community.

21.7.2 Neural Network Modeling

A complete reconstruction of the connectivity at the synapse level is currently possible for small brain volumes using electron microscopy techniques, but not yet feasible for volumes the size of a cortical column. Oberlaender et al. [65] therefore pursue a reverse engineering approach: A computational model of a cortical column in the rat somatosensory cortex, consisting of ~18,000 neurons, is created by integration of anatomical data acquired by different imaging and reconstruction techniques into a common reference system. As the data is acquired from different animals in a population, the network represents an "average" cortical column: some model parameters are given as probabilistic densities. By generating realizations of these stochastic parameters, concrete network models are created.

The number of neurons and their distribution in a cortical column is obtained by automatic counting of neural soma (cell bodies) in confocal images [66]. The 3D dendritic morphologies of ~100 neurons of different cell types in the column as well as axons are reconstructed from transmitted light bright field images [22]. The column model is created by generating soma positions satisfying the given neuron

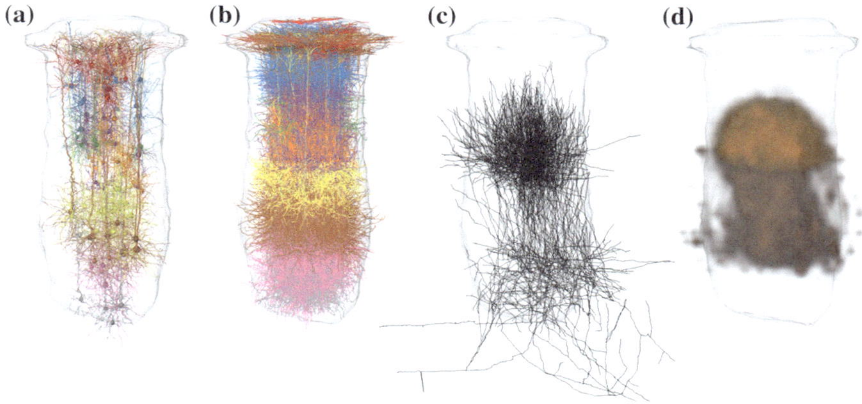

Fig. 21.6 Reverse engineering of a cortical column. Reconstructed dendrites (**a**) are replicated and inserted into the column reference frame according to a given neuron density (**b**). By determining the local structural overlap with axons projecting into the column (**c**), the number of synapses for different post-synaptic cell types can be estimated. (**d**) Shown are synapse densities for two cell types. Figure created from data published in [65]

density and replicating and inserting the dendrite morphologies into the reference frame according to the given cell type frequency (see Fig. 21.6). Differences in synaptic densities between cell types can be quantified and visualized [65]. Based on the estimated number of synapses per cell, a complete network wiring is established to study network function using numerical simulation [57].

Extracting relevant neurobiological knowledge from such network models is a challenging task. Whereas computation of specific quantities for comparison with literature results in order to validate the model is straightforward, exploratory knowledge discovery within such large, complex networks is not. Easy-to-use tools are needed to let the neurobiologist query and visualize the structural and functional properties of such networks or ensembles of network realizations. As network models are increasing in size, large data handling will be a challenging issue as well.

21.8 Network Analysis and Comparative Visualization

A recent innovation in neuroimaging is connectivity analysis, in which the anatomical or functional relation between different (underlying) brain areas is calculated from data obtained by various modalities, allowing researchers to study the resulting *networks* of interrelated brain regions. Of particular interest are *comparisons* of functional brain networks under different experimental conditions and between groups of subjects.

21.8.1 Network Measures

For each of the brain connectivity types (anatomical, functional, effective), one can extract networks from data obtained by an appropriate brain imaging modality [10, 54]. The next step is to characterize such networks. In the last decade, a multitude of *topological* network measures have been developed in an attempt to characterize and compare brain networks [11, 45, 78, 90]. Such measures characterize aspects of global, regional, and local brain connectivity.[2] Examples of global measures are characteristic path length, clustering coefficient, modularity, centrality, degree distribution, etc. Some of them, such as clustering coefficient or modularity, refer to *functional segregation* in the brain, i.e., the ability for specialized processing to occur in densely interconnected groups of brain regions. Others characterize *functional integration*, i.e., the ability to rapidly combine specialized information from distributed brain regions [78, 90]. Typical measures in this class are based on the concept of paths in the network, e.g., characteristic path length or global efficiency (average inverse shortest path length). It is believed that both anatomical and functional brain connectivity exhibit *small-world properties*, i.e., they combine functionally segregated modules with a robust number of intermodular links [3, 88]. The degree distribution can be used as a measure of network resilience, i.e., the capacity of the network to withstand network deterioration due to lesions or strokes.

For characterizing networks on a local scale one uses single node features such as in-degree and out-degree, or the local clustering coefficient. Typical regional network measures are *network motifs*, which are defined as patterns of local connectivity. A typical motif in a directed network is a triangle, consisting of feedforward and/or feedback loops. Both anatomical and functional motifs are distinguished. The significance of a certain motif in a network is determined by its frequency of occurrence, and the frequency of occurrence of different motifs around a node is known as the motif fingerprint of that node.

21.8.2 Brain Network Comparison and Visualization

The comparison of different brain networks presents challenging problems. Usually the networks differ in number and position of nodes and links, and a direct comparison is therefore difficult. One possible approach is to compute a network measure for each of the networks, and then compare the network measures. However, this loses spatial information. For interpretation and diagnosis it may be essential that local differences can be visualized in the original network representation [27, 86]. This asks for the development of mathematical methods, algorithms and visualization tools for the *local comparison* of complex networks—not necessarily of the same size—obtained under different conditions (time, frequency, scale) or pertaining to different (groups of) subjects.

---

[2] Similar approaches have been used in genomics [62, 84] and other areas.

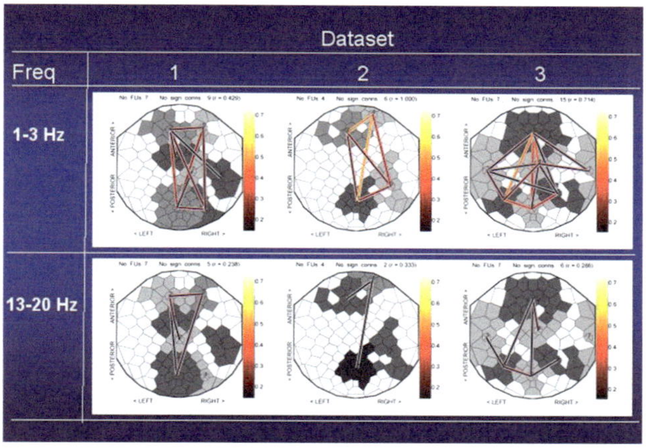

Fig. 21.7 FU maps for multichannel EEG coherence visualization. Brain responses were collected from three subjects using an EEG cap with 119 scalp electrodes. During a so-called P300 experiment, each participant was instructed to count target tones of 2,000 Hz (probability 0.15), alternated with standard tones of 1,000 Hz (probability 0.85) which were to be ignored. After the experiment, the participant had to report the number of perceived target tones. Shown are FU maps for target stimuli data, with FUs larger than 5 cells, for the 1-3Hz EEG frequency band (*top row*) and for 13–20 Hz (*bottom row*), for three datasets (Figure adapted from [92], Fig. 3)

Several methods exist for spatial comparison of brain networks, which assume that the position and number of network nodes is the same in the networks to be compared. For example, Salvador et al. [80] use a brain parcellation based on a prior standard anatomical template, dividing each cerebral hemisphere into 45 anatomical regions that correspond to the nodes of the brain network. Another possibility is to consider each voxel a network node, but in this way the networks become very large. Links between the nodes can then be defined by several measures of node-node association, such as correlation or mutual information of temporal signals. Using the same construction for two or more data sets enables a direct network comparison [108].

A method to perform network comparison in the original network representation was recently proposed for the case of multichannel EEG by Crippa et al. [19]. This approach is based on representation of an EEG coherence network by a so-called *functional unit* (FU), which is defined as a spatially connected clique in the EEG graph, i.e., a set of electrodes used in the EEG experiment that are spatially close and record pairwise significantly coherent signals [92, 94]. To each electrode a Voronoi cell is associated and all cells belonging to an FU are given a corresponding color. Lines connect FU centers if the inter-FU coherence exceeds a significance threshold. The color of a line depends on the inter-FU coherence. Such a representation of the FUs in an EEG recording is called a FU map. FU maps can be constructed for different frequency bands or for different subjects (see Fig. 21.7).

Comparison of multiple FU maps can be done visually when displayed next to each other, but this method is limited as humans are notoriously weak in spotting visual differences in images. An alternative, which is more quantitative although it still involves visual assessment to a certain degree, is to compute a *mean* FU map, based upon the concept of *graph averaging* [19]. The mean of a set of input FU maps is defined in such a way that it not only represents the mean group coherence during a certain task or condition, but also to some extent displays individual variations in brain activity. The definition of a mean FU map relies on a graph dissimilarity measure that takes into account both node positions and node or edge attributes. A visualization of the mean FU map is used with a visual representation of the frequency of occurrence of nodes and edges in the input FUs. This makes it possible to investigate which brain regions are more commonly involved in a certain task, by analyzing the occurrence of an FU of the mean graph in the input FUs.

In [19] the graph averaging method was applied to the analysis of EEG coherence networks in two case studies, one on mental fatigue and one on patients with corticobasal ganglionic degeneration. An extension of the method to resting state fMRI data was presented in [18].

21.9 Conclusions

There is currently great scientific interest in connectomics, as it is believed to be an important prerequisite for understanding brain function. As much of the data for obtaining neural connectivity is image-based, visualization techniques are indispensable. Great effort has been put recently into extraction of connectivity information from images, integration of multimodal information into reference systems, and visual analysis of such data and systems at different scales. These efforts will need to be intensified in the future, as data is being produced at a much larger scale, also by new imaging modalities. New methods to integrate this data across modalities and scales to attain the ultimate goal, a description of the human connectome, will be the main challenge for visualization in connectomics.

References

1. Appel, N.M.: Classical and contemporary histochemical approaches for evaluating central nervous system microanatomy. Ann. NY Acad. Sci. **820**(1), 14–28 (1997)
2. Basser, P.J., Pajevic, S., Pierpaoli, C., Duda, J., Aldroubi, A.: In vivo fiber tractography using DT-MRI data. Magn. Reson. Med. **44**(4), 625–632 (2000)
3. Bassett, D.S., Bullmore, E.: Small-world brain networks. Neuroscientist **12**(6), 512–523 (2006)
4. Behrens, T.E., Woolrich, M.W., Jenkinson, M., JohansenBerg, H., Nunes, R.G., Clare, S., Matthews, P.M., Brady, J.M., Smith, S.M.: Characterization and propagation of uncertainty in diffusion-weighted MR imaging. Magn. Reson. Med. **50**(5), 1077–1088 (2003)

5. Bertrand, L., Nissanov, J.: The neuroterrain 3D mouse brain atlas. Front. Neuroinform. **2**, 3 (2008)
6. Bezgin, G., Reid, A., Schubert, D., Kötter, R.: Matching spatial with ontological brain regions using java tools for visualization, database access, and integrated data analysis. Neuroinformatics **7**(1), 7–22 (2009)
7. Bjaalie, J.G.: Localization in the brain: new solutions emerging. Nat. Rev. Neurosci. **3**, 322–325 (2002)
8. Blaas, J., Botha, C.P., Peters, B., Vos, F.M., Post, F.H.: Fast and reproducible fiber bundle selection in DTI visualization. In: Silva, C., Gröller, E., Rushmeier, H. (eds.) Proceedings of IEEE Visualization 2005, pp. 59–64 (2005)
9. Bruckner, S., Gröller, M.: Instant volume visualization using maximum intensity difference accumulation. Comput. Graph. Forum **28**(3), 775–782 (2009)
10. Bullmore, E., Harrison, L., Lee, L., Mechelli, A., Friston, K.(eds.): Special issue on functional connectivity. Neuroinformatics **2**(2) (2004)
11. Bullmore, E., Sporns, O.: Complex brain networks: graph theoretical analysis of structural and functional systems. Nat. Rev. Neurosci. **10**, 186–198 (2009). see also Corrigendum (March 3, 2009)
12. Bullmore, E., Bassett, D.: Brain graphs: graphical models of the human brain connectome. Ann. Rev. Clin. Psychol. **7**, 113–140 (2011)
13. Burns, G.a.P.C., Cheng, W.C., Thompson, R.H., Swanson, L.W.: The NeuARt II system: A viewing tool for neuroanatomical data based on published neuroanatomical atlases. BMC Bioinform. **7**, 531–549 (2006)
14. Cardona, A., Saalfeld, S., Preibisch, S., Schmid, B., Cheng, A., Pulokas, J., Tomancak, P., Hartenstein, V.: An integrated micro- and macroarchitectural analysis of the drosophila brain by computer-assisted serial section electron microscopy. PLoS Biol **8**(10), e1000,502 (2010)
15. Chicurel, M.: Databasing the brain. Nature **406**(6798), 822–825 (2000)
16. Chklovskii, D.B., Vitaladevuni, S., Scheffer, L.K.: Semi-automated reconstruction of neural circuits using electron microscopy. Curr. Opin. Neurobiol. **20**(5), 667–675 (2010)
17. Conchello, J.A., Lichtman, J.: Optical sectioning microscopy. Nat. Methods **2**(12), 920–931 (2005)
18. Crippa, A., Roerdink, J.B.T.M.: Data-driven visualization of functional brain regions from resting state fMRI data. In: Eisert, P., Polthier, K., Hornegger, J. (eds.) Proceedings Vision, Modeling and Visualization Workshop (VMV), 4–6 Oct, Berlin, pp. 247–254 (2011)
19. Crippa, A., Maurits, N.M., Roerdink, J.B.T.M.: Graph averaging as a means to compare multichannel EEG coherence networks and its application to the study of mental fatigue and neurodegenerative disease. Comput. Graph. **35**(2), 265–274 (2011)
20. Da Costa, N.M., Fürsinger, D., Martin, K.A.C.: The synaptic organization of the claustral projection to the cat's visual cortex. J. Neurosci. **30**(39), 13,166–13,170 (2010)
21. Denk, W., Horstmann, H.: Serial block-face scanning electron microscopy to reconstruct three-dimensional tissue nanostructure. PLoS Biol. **2**(11), e329 (2004)
22. Dercksen, V.J., Oberlaender, M., Sakmann, B., Hege, H.C.: Interactive Visualization—a Key Prerequisite for Reconstruction of Anatomically Realistic Neural Networks. In: Linsen, L. (ed.) Visualization in Medicine and Life Sciences II. Springer, Berlin (2012)
23. Einstein, A.: Investigations on the Theory of the Brownian Movement. Dover (1956)
24. Emmenlauer, M., Ronneberger, O., Ponti, A., Schwarb, P., Griffa, A., Filippi, A., Nitschke, R., Driever, W., Burkhardt, H.: XuvTools: free, fast and reliable stitching of large 3D datasets. J. Microsc. **233**(1), 42–60 (2009)
25. Enders, F., Sauber, N., Merhof, D., Hastreiter, P., Nimsky, C., Stamminger, M.: Visualization of white matter tracts with wrapped streamlines. In: IEEE Visualization, 2005. VIS 05, pp. 51–58. IEEE (2005)
26. Everts, M.H., Bekker, H., Roerdink, J.B.T.M., Isenberg, T.: Depth-dependent halos: illustrative rendering of dense line data. IEEE Trans. Vis. Comput. Graph. **15**(6), 1299–1306 (2009)
27. Fair, D.A., Cohen, A.L., Power, J.D., Dosenbach, N.U.F., Church, J.A., Miezin, F.M., Schlaggar, B.L., Petersen, S.E.: Functional brain networks develop from a "local to distributed" organization. PLoS Comput. Biol. **5**(5), e1000,381 (2009)

28. Fair, D.A., Cohen, A.L., Dosenbach, N.U.F., Church, J.A., Miezin, F.M., Barch, D.M., Raichle, M.E., Petersen, S.E., Schlaggar, B.L.: The maturing architecture of the brain's default network. Proc. Natl. Acad. Sci. **105**(10), 4028–4032 (2008)
29. Fiala, J.C.: Reconstruct: a free editor for serial section microscopy. J. Microsc. **218**(1), 52–61 (2005)
30. Fornito, A., Zalesky, A., Pantelis, C., Bullmore, E.T.: Schizophrenia, neuroimaging and connectomics. NeuroImage **62**, 2296–2314 (2012)
31. Fredriksson, J.: Design of an Internet accessible visual human brain database system. In: Proceedings of IEEE International Conference on Multimedia Computing and Systems, vol. 1, pp. 469–474 (1999)
32. Friston, K.J.: Functional and effective connectivity in neuroimaging: a synthesis. Hum. Brain Mapp. **2**, 56–78 (1994)
33. Gan, W.B., Grutzendler, J., Wong, W.T., Wong, R.O.L., Lichtman, J.W.: Multicolor "DiOlistic" Labeling Neurotechnique of the Nervous System Using. Neuron **27**, 219–225 (2000)
34. Hagmann, P., Jonasson, L., Maeder, P., Thiran, J., Wedeen, V.J., Meuli, R.: Understanding diffusion MR imaging techniques: from scalar diffusion-weighted imaging to diffusion tensor imaging and beyond. Radiographics **26**(Suppl 1), S205–S223 (2006)
35. Hämäläinen, M., Hari, R., Ilmoniemi, R.J., Knuutila, J., Lounasmaa, O.V.: Magnetoencephalography—theory, instrumentation, and applications to noninvasive studies of the working human brain. Rev. Mod. Phys. **65**, 413–497 (1993)
36. Hayworth, K.J., Kasthuri, N., Schalek, R., Lichtman, J.W.: Automating the collection of ultrathin serial sections for large volume TEM reconstructions. Microsc. Microanal. **12**(S02), 86–87 (2006)
37. Helmstaedter, M., Briggman, K.L., Denk, W.: High-accuracy neurite reconstruction for high-throughput neuroanatomy. Nat. Neurosci. **14**, 1081–1088 (2011)
38. Huberman, A.D., Niell, C.M.: What can mice tell us about how vision works? Trends Neurosci. **34**(9), 464–473 (2011)
39. Jain, V., Bollmann, B., Richardson, M., Berger, D.R., Helmstaedter, M.N., Briggman, K.L., Denk, W., Bowden, J.B., Mendenhall, J.M., Abraham, W.C., Harris, K.M., Kasthuri, N., Hayworth, K.J., Schalek, R., Tapia, J.C., Lichtman, J.W., Seung, H.S.: Boundary learning by optimization with topological constraints. IEEE Conference on Computer Vision and Pattern Recognition pp. 2488–2495 (2010)
40. Jenett, A., Schindelin, J.E., Heisenberg, M.: The virtual insect brain protocol: creating and comparing standardized neuroanatomy. BMC Bioinform. **7**(1), 544–555 (2006)
41. Jeong, W.K., Johnson, M.K.: Display-aware Image Editing. IEEE International Conference on Computational Photography (2011)
42. Jeong, W.K., Beyer, J., Hadwiger, M., Blue, R., Law, C., Vazquez-Reina, A., Reid, R.C., Lichtman, J., Pfister, H.: Ssecrett and neurotrace: interactive visualization and analysis tools for large-scale neuroscience data sets. IEEE Comput. Graph. Appl. **30**(3), 58–70 (2010)
43. Joshi, S.H., Horn, J.D.V., Toga, A.W.: Interactive exploration of neuroanatomical meta-spaces. Front. Neuroinform. **3**, 38 (2009)
44. Jurrus, E., Paiva, A.R.C., Watanabe, S., Anderson, J.R., Jones, B.W., Whitaker, R.T., Jorgensen, E.M., Marc, R.E., Tasdizen, T.: Detection of neuron membranes in electron microscopy images using a serial neural network architecture. Med. Image Anal. **14**(6), 770–783 (2010)
45. Kaiser, M.: A tutorial in connectome analysis: Topological and spatial features of brain networks. NeuroImage **57**(3), 892–907 (2011) (Special Issue: Educational Neuroscience)
46. Kaynig, V., Fischer, B., Buhmann, J.M.: Probabilistic image registration and anomaly detection by nonlinear warping. In: IEEE Conference on Computer Vision and Pattern Recognition, pp. 1–8 (2008)
47. Kaynig, V., Fuchs, T., Buhmann, J.M.: Geometrical consistent 3D tracing of neuronal processes in ssTEM data. In: International Conference on Medical Image Computing and Computer Assisted Intervention (2010)

48. Kaynig, V., Fuchs, T., Buhmann, J.M.: Neuron geometry extraction by perceptual grouping in ssTEM images. In: T. Fuchs, J.M. Buhmann (eds.) IEEE Conference on Computer Vision and Pattern Recognition, pp. 2902–2909. IEEE (2010)
49. Kindlmann, G., Weinstein, D.: Hue-balls and lit-tensors for direct volume rendering of diffusion tensor fields. In: Proceedings of Visualization '99, pp. 183–524. IEEE (1999)
50. Kindlmann, G.: Superquadric tensor glyphs. In: Proceedings of IEEE TVCG/EG Symposium on Visualization, pp. 147–154 (2004)
51. Klein, J., Friman, O., Hadwiger, M., Preim, B., Ritter, F., Vilanova, A., Zachmann, G., Bartz, D.: Visual computing for medical diagnosis and treatment. Comput. Graph. **33**(4), 554–565 (2009)
52. Knott, G., Marchman, H., Wall, D., Lich, B.: Serial section scanning electron microscopy of adult brain tissue using focused ion beam milling. J. Neurosci. **28**(12), 2959–2964 (2008)
53. Koslow, S.H., Subramaniam, S.: Databasing the Brain: From Data to Knowledge. Wiley (2005)
54. Kötter, R.: Online retrieval, processing, and visualization of primate connectivity data from the cocomac database. Neuroinformatics **2**(2), 127–44 (2004)
55. Kuß, A., Prohaska, S., Meyer, B., Rybak, J., Hege, H.C.: Ontology-based visualization of hierarchical neuroanatomical structures. In: Proc. Eurographics Workshop on Visual Computing for Biomedicine, VCBM, Delft, The Netherlands, pp. 177–184 (2008)
56. Kuß, A., Gensel, M., Meyer, B., Dercksen, V., Prohaska, S.: Effective techniques to visualize filament-surface relationships. Comput. Graph. Forum **29**(3), 1003–1012 (2010)
57. Lang, S., Dercksen, V.J., Sakmann, B., Oberlaender, M.: Simulation of signal flow in three-dimensional reconstructions of an anatomically realistic neuronal network in rat vibrissal cortex. Neural Netw. **24**(9), 998–1011 (2011)
58. Lau, C., Ng, L., Thompson, C., Pathak, S., Kuan, L., Jones, A., Hawrylycz, M.: Exploration and visualization of gene expression with neuroanatomy in the adult mouse brain. BMC Bioinform. **9**, 153–163 (2008)
59. de Leeuw, W., Verschure, P.J., van Liere, R.: Visualization and analysis of large data collections: a case study applied to confocal microscopy data. IEEE Trans. Vis. Comput. Graph. **12**(5), 1251–1258 (2006)
60. Lichtman, J., Livet, J., Sanes, J.: A technicolour approach to the connectome. Nat. Rev. Neurosci. **9**(6), 417–422 (2008)
61. Lin, C.Y., Tsai, K.L., Wang, S.C., Hsieh, C.H., Chang, H.M., Chiang, A.S.: The Neuron Navigator: Exploring the information pathway through the neural maze. In: Proceedings of IEEE Pacific Visualization 2011, pp. 35–42. IEEE (2011)
62. Luscombe, N.M., Babu, M.M., Yu, H., Snyder, M., Teichmann, S.A., Gerstein, M.: Genomic analysis of regulatory network dynamics reveals large topological changes. Nature **431**, 308–312 (2004)
63. Mackay, T.F., Anholt, R.R.: Of flies and man: drosophila as a model for human complex traits. Ann. Rev. Genomics Hum. Genet. **7**, 339–367 (2006)
64. Maye, A., Wenckebach, T.H., Hege, H.C.: Visualization, reconstruction, and integration of neuronal structures in digital brain atlases. Int. J. Neurosci. **116**(4), 431–459 (2006)
65. Oberlaender, M., de Kock, C.P.J., Bruno, R.M., Ramirez, A., Meyer, H.S., Dercksen, V.J., Helmstaedter, M., Sakmann, B.: Cell Type-Specific Three-Dimensional Structure of Thalamocortical Circuits in a Column of Rat Vibrissal Cortex. Cerebral Cortex. **22**(10), 2375–2395 (2012) doi:10.1093/cercor/bhr317
66. Oberlaender, M., Dercksen, V.J., Egger, R., Gensel, M., Sakmann, B., Hege, H.C.: Automated three-dimensional detection and counting of neuron somata. J. Neurosci. Methods **180**(1), 147–160 (2009)
67. Ogawa, S., Lee, T.M., Kay, A.R., Tank, D.W.: Brain magnetic resonance imaging with contrast dependent on blood oxygenation. Proc. Natl. Acad. Sci. **87**(24), 9868–9872 (1990)
68. Ogawa, S., Tank, D.W., Menon, R., Ellermann, J.M., Kim, S.G., Merkle, H., Ugurbil, K.: Intrinsic signal changes accompanying sensory stimulation: functional brain mapping with magnetic resonance imaging. Proc. Natl. Acad. Sci. **89**(13), 5951–5955 (1992)

69. Özarslan, E., Mareci, T.H.: Generalized diffusion tensor imaging and analytical relationships between diffusion tensor imaging and high angular resolution diffusion imaging. Magn. Reson. Med. **50**(5), 955–965 (2003)
70. Peeters, T.H., Prčkovska, V., van Almsick, M., Vilanova, A., ter Haar Romeny, B.M.: Fast and sleek glyph rendering for interactive HARDI data exploration. In: Visualization Symposium, 2009. PacificVis '09. IEEE Pacific, pp. 153–160. IEEE (2009)
71. Pereanu, W., Hartenstein, V.: Neural lineages of the drosophila brain: a three-dimensional digital atlas of the pattern of lineage location and projection at the late larval stage. J. Neurosci. **26**(20), 5534–5553 (2006)
72. Perrin, M., Poupon, C., Cointepas, Y., Rieul, B., Golestani, N., Pallier, C., Rivire, D., Constantinesco, A., Bihan, D., Mangin, J.F.: Fiber tracking in q-Ball fields using regularized particle trajectories. In: Christensen, G.E., Sonka, M. (eds.) Inf. Process. Med. Imaging, vol. 3565, pp. 52–63. Springer, Berlin (2005)
73. Prčkovska, V., Peeters, T.H., van Almsick, M., ter, : Haar Romeny, B., Vilanova i Bartroli, A.: Fused DTI/HARDI visualization. IEEE Trans. Vis. Comput. Graph. **17**(10), 1407–1419 (2011)
74. Preibisch, S., Saalfeld, S., Tomancak, P.: Globally optimal stitching of tiled 3D microscopic image acquisitions. Bioinformatics **25**(11), 1463–1465 (2009)
75. Press, W.A., Olshausen, B.A., Essen, D.C.V.: A graphical anatomical database of neural connectivity. Philos. Trans. R. Soc. **356**(1412), 1147–1157 (2001)
76. Reina, A.V., Miller, E., Pfister, H.: Multiphase geometric couplings for the segmentation of neural processes. IEEE Conference on Computer Vision and Pattern Recognition, pp. 1–8 (2009)
77. Roberts, M., Jeong, W.K., V, A., Unger, M.: Neural Process Reconstruction from Sparse User Scribbles. In: Medical Image Computing and Computer Assisted Intervention, pp. 1–8 (2011)
78. Rubinov, M., Sporns, O.: Complex network measures of brain connectivity: Uses and interpretations. Neuroimage **52**, 1059–1069 (2010)
79. Rybak, J., Kuss, A., Lamecker, H., Zachow, S., Hege, H., Lienhard, M., Singer, J., Neubert, K., Menzel, R.: The digital bee brain: integrating and managing neurons in a common 3d reference system. Frontiers in systems neuroscience 4 (2010)
80. Salvador, R., Suckling, J., Coleman, M.R.: Pickard John, D., Menon, D., Bullmore, E.: Neurophysiological architecture of functional magnetic resonance images of human brain. Cereb Cortex **15**, 1332–1342 (2005)
81. Schomer, D.L., Lopes da Silva, F.: Niedermeyer's Electroencephalography: Basic Principles, Clinical Applications, and Related Fields. Wolters Kluwer/Lippincott Williams & Wilkins (2010)
82. Schultz, T., Theisel, H., Seidel, H.P.: Topological visualization of brain diffusion MRI data. IEEE Trans. Vis. Comput. Graph. **13**(6), 1496–1503 (2007)
83. Seung, S.: Connectome. Houghton Mifflin Harcourt (2011). In press
84. Sharan, R., Ideker, T.: Modeling cellular machinery through biological network comparison. Nat. Biotechnol. **24**(4), 427–433 (2006)
85. Sherbondy, A., Akers, D., Mackenzie, R., Dougherty, R., Wandell, B.: Exploring connectivity of the brain's white matter with dynamic queries. IEEE Trans. Vis. Comput. Graph. **11**(4), 419–430 (2005)
86. Shu, N., Liu, Y., Li, J., Li, Y., Yu, C., Jiang, T.: Altered anatomical network in early blindness revealed by diffusion tensor tractography. PLoS One **4**(9), e7228 (2009)
87. Sporns, O.: Networks of the brain. MIT Press, Cambridge (2010)
88. Sporns, O., Zwi, J.: The small world of the cerebral cortex. Neuroinformatics **2**, 145–162 (2004)
89. Sporns, O., Tononi, G., Kötter, R.: The human connectome: a structural description of the human brain. PLoS Comput. Biol. **1**(4), e42 (2005)
90. Stam, C.J., Reijneveld, J.C.: Graph theoretical analysis of complex networks in the brain. Nonlinear Biomed. Phys. **1** (2007)

91. Straehle, C., Köthe, U., Knott, G., Hamprecht, F.: Carving: Scalable Interactive Segmentation of Neural Volume Electron Microscopy Images. In: MICCAI, pp. 657–664 (2011)
92. Ten Caat, M., Maurits, N.M., Roerdink, J.B.T.M.: Functional unit maps for data-driven visualization of high-density EEG coherence. In: Proceedings of Eurographics/IEEE VGTC Symposium on Visualization (EuroVis), pp. 259–266 (2007)
93. Ten Caat, M.: Multichannel EEG visualization. Ph.D. thesis, Institute of Mathematics and Computing Science, University of Groningen, The Netherlands (2008)
94. Ten Caat, M., Maurits, N.M., Roerdink, J.B.T.M.: Data-driven visualization and group analysis of multichannel EEG coherence with functional units. IEEE Trans. Vis. Comput. Graph. **14**(4), 756–771 (2008)
95. Tuch, D.S., Reese, T.G., Wiegell, M.R., Makris, N., Belliveau, J.W., Wedeen, V.J.: High angular resolution diffusion imaging reveals intravoxel white matter fiber heterogeneity. Magn. Reson. Med. **48**(4), 577–582 (2002)
96. Tuch, D.S.: Qball imaging. Magn. Reson. Med. **52**(6), 1358–1372 (2004)
97. Türetken, E., González, G., Blum, C., Fua, P.: Automated reconstruction of dendritic and axonal trees by global optimization with geometric priors. Neuroinformatics **9**(2), 279–302 (2011)
98. Valiant, L.G.: A quantitative theory of neural computation. Biol. Cybern. **95**(3), 205–211 (2006)
99. Van Dixhoorn, A., Vissers, B., Ferrarini, L., Milles, J., Botha, C.P.: Visual analysis of integrated resting state functional brain connectivity and anatomy. In: Proc. Eurographics Workshop on Visual Computing for Biomedicine, VCBM, Leipzig, Germany, pp. 57–64 (2010)
100. Varshney, L.R., Chen, B.L., Paniagua, E., Hall, D.H., Chklovskii, D.B.: Structural properties of the caenorhabditis elegans neuronal network. PLoS Comput. Biol. **7**(2), 21 (2011)
101. Vazquez-Reina, A., Pfister, H., Miller, E.L.: Segmentation Fusion for Connectomics. International Conference on Computer Vision pp. 1–8 (2011)
102. Vilanova, A., Zhang, S., Kindlmann, G., Laidlaw, D.H.: An introduction to visualization of diffusion tensor imaging and its applications. In: Visualization and Image Processing of Tensor Fields. Springer (2005)
103. Vitaladevuni, S.N.: Co-clustering of image segments using convex optimization applied to em neuronal reconstruction. IEEE Conference on Computer Vision and Pattern Recognition pp. 2203–2210 (2010)
104. Walter, T., Shattuck, D.W., Baldock, R., Bastin, M.E., Carpenter, A.E., Duce, S., Ellenberg, J., Fraser, A., Hamilton, N., Pieper, S.: Ragan, M.a., Schneider, J.E., Tomancak, P., Hériché, J.K.: Visualization of image data from cells to organisms. Nat. Methods **7**(3s), S26–S41 (2010)
105. Wan, Y., Otsuna, H., Chien, C.B., Hansen, C.: An interactive visualization tool for multi-channel confocal microscopy data in neurobiology research. IEEE Trans. Vis. Comput. Graph. **15**(6), 1489–1496 (2009)
106. White, J.G., Southgate, E., Thomson, J.N., Brenner, S.: The structure of the nervous system of the nematode caenorhabditis elegans. Philos. Trans. R. Soc. B Biol. Sci. **314**(1165), 1–340 (1986)
107. Worsley, K.J., Chen, J., Lerch, J., Evans, A.C.: Comparing functional connectivity via thresholding correlations and singular value decomposition. Philos. Trans. R. Soc. B Biol. Sci. **360**(1457), 913–920 (2005)
108. Zalesky, A., Fornito, A., Bullmore, E.T.: Network-based statistic: identifying differences in brain networks. Neuroimage **53**(4), 1197–1207 (2010)
109. Zhang, S., Demiralp, C., Laidlaw, D.H.: Visualizing diffusion tensor MR images using streamtubes and streamsurfaces. IEEE Trans. Vis. Comput. Graph. **9**(4), 454–462 (2003)

Chapter 22
Visualization in Biology and Medicine

Heike Leitte and Miriah Meyer

Abstract Basic biological research spans a huge range of scales, from studying the genome up to studying populations of people. Biological data is just as expansive. Advances in measuring devices and the public dissemination of large amounts of data has fundamentally changed the way that biologists conduct research and make scientific discoveries. Access to this data has made visualization a key component in almost every biological discovery workflow. In this chapter we focus on visualization in just a few areas of biology and highlight the challenges inherent within each. We present case studies from our own work to illustrate the impact that thoughtfully designed visualization systems can have on complex biological problems and highlight challenges for visualization research in these areas.

22.1 From Genomes to Cells: Visualization in Biology

Basic biological research spans a huge range of scales, from studying the genome up to studying populations of people. Biological data is just as expansive. Advances in measuring devices and the public dissemination of large amounts of data has fundamentally changed the way that biologists conduct research and make scientific discoveries. Access to this data has made visualization a key component in almost every biological discovery workflow.

In this section we focus on visualization in just a few areas of biology and highlight the challenges inherent within each. We present case studies from our own work to illustrate the impact that thoughtfully designed visualization systems can have on

H. Leitte (✉)
Interdisciplinary Center for Scientific Computing (IWR),
Heidelberg University, Im Neuenheimer Feld 368,
69120 Heidelberg, Germany
e-mail: heike.leitte@iwr.uni-heidelberg.de

M. Meyer
School of Computing, University of Utah, 72 S. Central Campus Dr,
Salt Lake City, UT 84112, USA
e-mail: miriah@cs.utah.edu

C.D. Hansen et al. (eds.), *Scientific Visualization*, Mathematics and Visualization,
DOI 10.1007/978-1-4471-6497-5_22

complex biological problems. Despite the variety of scientific goals and data types in these areas, several common themes have emerged from our combined work. First, the accelerated experimental process in biology due to high-throughput technologies is requiring that visualization tools be developed rapidly and nimbly to be relevant to the discovery process. Second, many interesting visualization challenges now exist around the idea of integrating very different types of data and finding complex patterns within. And third, close collaboration with biologists is essential for developing visualization design requirements—we have found that these requirements often require many interviews and multiple prototypes to articulate clearly.

The rest of this chapter discusses visualization in the context of comparative genomics (Sect. 22.2), functional genomics (Sect. 22.3), and evolutionary and developmental biology (Sect. 22.4). In each section we will cover the common types of biological questions and data in each of these fields, along with typical methods and techniques for visualizing the data. We also discuss visualization challenges as well as present case studies that highlight the biological impact of visualization tools.

22.2 Comparative Genomics

In the field of comparative genomics, scientists compare the genomes of organisms to answer questions about evolution and how the genome encodes cellular functions. These comparisons look for regions of similar DNA sequences which provide evidence of common ancestry as well as potential shared function, giving insight into the Tree of Life, the discovery of new genes, and the understanding of how our DNA makes us who we are.

When studying the similarities and differences between genomes, these scientists are most often looking for similar *features* of interests, such as genes. Finding similar features implies a *conservation* relationship between the features, meaning these features were conserved through evolution as the individual species diverged from a common ancestor. Finding these similar features within large genomic data sets relies on sophisticated computational algorithms. These algorithms characterize conservation across a range of scales, from the genome down to the gene. Finding patterns of conservation, across multiple scales, allows scientists to answer questions like: Is there evidence of larger segments of conservation that could indicate a whole genome duplication? What changes to a genome can account for species variation? What segments of the genome account for the ability of a species to adapt to different environments?

22.2.1 Data in Comparative Genomics

A genome is physically composed of multiple, distinct chromosomes. Each chromosome is made up of a string of nucleotides which come in four common types represented by the letters A, T, C and G. Over evolutionary time, mutations arise in these strings as they are passed on from one organism to another, sometimes resulting in an increased survival rate and a possible divergence into a new species. Understanding how and when these mutations occurred is a major topic in comparative genomics and informs scientists about the relatedness of species, both genomically and functionally.

The sequence of a genome is determined from the output of a sequencing pipeline. In this pipeline, the genomes from many cells of an organism are extracted and chopped up into very small segments. These segments are *read* using a variety of techniques in which the individual nucleotides of each segment are determined. Using computational alignment algorithms, the sequences from the segments are then pieced together to produce the string of letters representing each chromosome in the sequenced genome.

Using the genomic sequences from different species, or sometime from just a single species, algorithms look for regions that have similar genomic sequences while taking into account models of how genomes evolve over time. These algorithms produce pairs of conserved features, giving each pair a strength based on the amount of similarity. Some algorithms will further group the paired features into larger regions based on characteristics of the features like proximity and orientation (features such as genes have an orientation along the genome). The result of these algorithms is a multiscale list of features and regions for each chromosome that are paired with features and regions on a different chromosome. Each of these pairs has a score that represents the strength of the similarity.

22.2.2 Challenges for Visualization

The challenge of visualizing comparative genomics data arises on several fronts. First, these data sets can easily contain thousands of paired features, scattered over dozens of chromosomes. Second, the size of the features is often orders of magnitude smaller than the size of the chromosomes. And third, it is often important to understand the location and size of paired features in the context of their similarity scores.

22.2.3 Visualization for Comparative Genomics

Visualization tools for exploring comparative genomics data represent chromosomes as distinct, 1D coordinate systems, with a set of chromosomes representing a complete genome. These chromosomes are most often represented linearly and in a series. Conserved features and regions are shown as subregions along the chromosomes. These tools generally compare two genomes at a time, where the genome of interest is considered the *source* and the comparison genome is considered the *destination*.

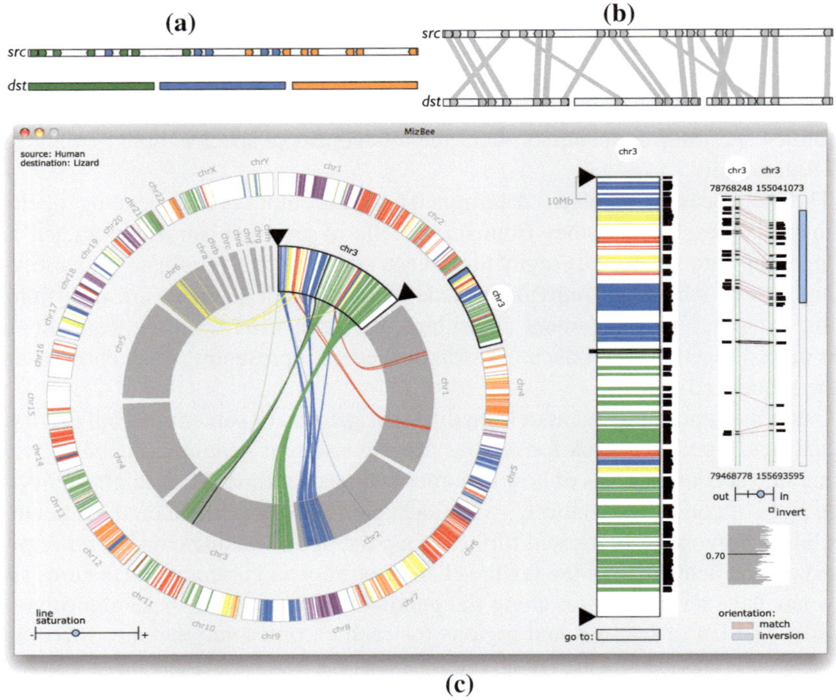

Fig. 22.1 Visualization methods for comparative genomics data. Most comparative genomics data is shown using two kinds of encoding: **a** color encoding, where each chromosome in the destination genome is assigned a unique color, and paired features in the source genome are colormapped according to their paired feature's chromosome; **b** connection encoding, where lines are drawn between the source and destination genomes connecting paired features. **c** MizBee [28] uses both color and connection for showing conserved genomics data across multiple scales and in conjunction with similarity scores

There are two predominate ways for visually representing the conserved pairs: using color and using connection. To use color, as shown in Fig. 22.1a, each chromosome in the destination genome is assigned a unique color. Each conserved feature in the source genome is colormapped according to the chromosome its paired feature is associated with. Examples of visualization tools that use this encoding are SyntenyVista [14], Sybil [46], and Cinteny [42]. Using color to encode conserved features benefits from minimal visual clutter as the number of pairs increases, but does so at the expense of only showing the location of the features on the source genome. This encoding, however, suffers from color indistinguishability. We are able to distinguish less than a dozen colors when showing categorical data [50], but most genomes of interest will have on the order of dozens of chromosomes.

The second method for encoding conserved pairs is connection, shown in Fig. 22.1b. In this encoding, the source and destination genomes are aligned, usually in parallel, and connecting lines are drawn between paired features. Visualization tools that use a connection encoding include Ensemble [6], SynBrowse [33], Mauve [8], and Apollo [24]. A connection encoding shows both the size and location of paired features on the source and destination genomes. The encoding, however, suffers from visual clutter when more than a few dozen paired features are shown—many comparative genomics datasets contain thousands of pairs.

A limitation of many of the visualization tools for comparative genomics data is that they do not encode similarity scores in conjunction with the size and location of paired features. A common method for looking at these scores is using a scatterplot [21, 35]. In this method, each axis is the coordinate system for a chromosome, and paired features are indicated with a dot at the appropriate coordinates. These dots are colormapped based on their associated similarity scores.

22.2.4 Case Study: MizBee

In this case study, we worked with two biologists to design a visualization tool to look at comparative genomics data across multiple scales and in the context of similarity scores. We conducted a series of interviews with the biologists from which we compiled a list of fourteen low-level, data-centric questions they were asking in their research. For each of these questions, we characterized the scale at which it operated. From an extensive review of existing tools for visualizing comparative genomics data, we created a taxonomy of the design space for showing conserved features as well as amenable ways to layout the chromosomes from the source and destination genomes.

Our question characterization and design space taxonomy guided the design of MizBee [28], the first interactive tool to visualize comparative genomics data across multiple different scales in the context of similarity scores. MizBee, shown in Fig. 22.1c uses multiple linked views [36] to support guided navigation of the data by incorporating overviews with detailed views [41]. The tool includes three levels of views: the genome view (left), chromosome view (middle), and region view (right). The *genome view* uses a circular layout of the source and destination genomes, with the destination chromosomes on the inner ring along with a single selected source chromosome from the outer ring. Both color and connection are used as a redundant encoding of paired features. The connecting edges are bundled [13] to reduce visual clutter and to highlight spurious pairs. In the *chromosome view*, the selected source chromosome is shown in more detail to see the size of conserved regions along with a bar chart that indicates each region's similarity score. The lowest level *region view* shows individual paired features within a conserved region. This view gives information about the size, location, and orientation of the features, along with a bar chart of similarity scores. All three views are linked together through selection and highlighting.

To validate MizBee, we conducted a series of case studies with our biology collaborators. These case studies highlight how MizBee improved communication of comparative genomics research findings, as well as supported the validation of data and computational algorithms that determine paired features. More information about MizBee, as well as source code, executables, and example data, can be found at http://mizbee.org.

22.3 Functional Genomics

The field of functional genomics is, at a high level, about answering the question: how do genes work together in a cell to perform different functions? Cellular functions, such as metabolism or reproduction, are controlled by many interrelated chemical reactions that are catalyzed by genes, or more precisely the products of genes called proteins. These chemical reactions are complex and highly connected, forming complicated networks that biologist need to discover, unravel, and understand. Finding differences and similarities in networks from different experimental conditions, in different cell types, and in different species is an important component of functional genomics.

22.3.1 Data in Functional Genomics

Scientists working in functional genomics predominately work with two kinds of data: gene expression and molecular networks. *Gene expression* is a continuous measurement of how much a gene is on or off in a cell. It is primarily derived using microarray technology where the expression levels of many genes are measured at once. The resulting data is stored as a table of values where most often the rows are genes and the columns are different samples such as time points, experimental conditions, tissue types, or species.

Molecular networks are large graphs representing chemical reactions that occur in a cell. Some of the very well-characterized networks, such as that for metabolism, are generated from years of careful experimentation. These networks are shared through curated libraries such as the BioCyc or KEGG databases [4, 17]. These large networks are often broken into smaller, more manageable subsets called *molecular pathways*, usually consisting of a dozen reactions or fewer. Other networks, such as protein-protein interaction networks, are generated using machine-learning algorithms that look for correlations in large sets of gene expression measurements.

22.3.2 Challenges for Visualization

Whether visualizing gene expression or molecular networks, scale is a major challenge. For gene expression, the table of measurements can contain thousands of data points, with current trends moving towards tables with more than two dimensions. For molecular networks, the graphs can contain many nodes with high connectivity.

How do you find patterns in these large data sets? How do you understand changes over time? An even bigger challenge is: how do you integrate these data types together?

22.3.3 Visualization for Comparative Genomics

The visual convention for looking at gene expression data from microarrays is nearly-universally the heatmap display, where the data is laid out in a matrix and each value is encoded with a color [10, 52, 53]. Heatmaps are often augmented with clustering algorithms to enhance the perception of trends in the data [9], as in Java TreeView [38] and the Hierarchical Clustering Explorer [39]. These visualizations have a very high data density, allowing many data points to be viewed at a single time. The use of color, however, makes fine scale analysis difficult due to perceptual limitations [7] as well as the relativity of color [54].

Tools designed to visualize molecular networks usually focus on showing the topological structure of the graph. In these systems, networks are most often visualized using a node-link graph, such as is Cytoscape [40], MicrobesOnline [1], and iPath [23]. These systems support the visualization of an additional dimension of data by colormapping values on the nodes and edges of the graph. Other recently developed tools support the visualization of an entire set of values, for example a full time series, for each node and edge using techniques like small multiples [48], animation, or glyphs—example tools are Cerebral [2], Pathway Tools [18], VANTED [16], PathwayExplorer [31], and GENeVis II [3].

22.3.4 Case Study: Pathline

We collaborated with a group of biologists who are pioneering the new field of comparative functional genomics. This field extends the questions of functional genomics to understand how gene interactions vary across species. Our collaborators are interested in understanding how evolutionary mechanisms affect gene regulation for metabolism in yeast. To probe these questions, the biologists collected data for multiple genes, at multiple time points, and in multiple related species. They need to integrate this data in order to find patterns in gene expression levels belonging to multiple pathways over time and across multiple species. When we started working with this group, the problem they faced was that existing visualization tools only look at subsets of this data at a time.

To address the challenge of data integration we designed a tool called Pathline [29], shown in Fig. 22.2. Pathline was designed in a user-centered process using iterative refinement based on feedback from our biology collaborators. Each design decision was motivated by the specific needs of the biologists.

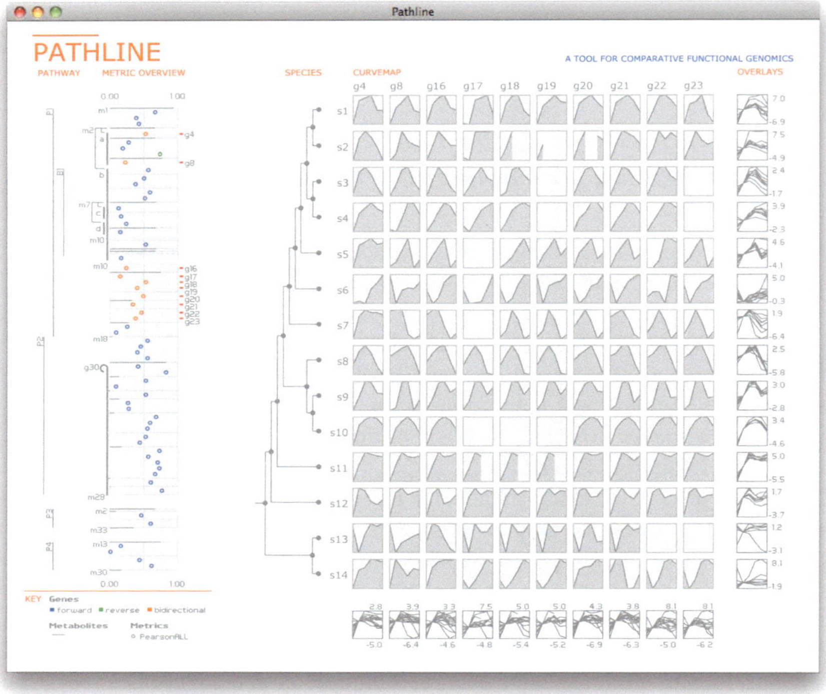

Fig. 22.2 Pathline [29], an interactive tool for the visualization of comparative functional genomics data. The *left side* shows a linearized pathway representation and the *right side* shows a detailed curvemap display. Pathline is the first tool to integrate data for multiple genes, time points, species, and pathways

Pathline consists of two novel visual representations, a linearized pathway representation and a detailed curvemap display, shown on the left and right sides of Fig. 22.2 respectively. The linearized pathway representation shows molecular pathways as an ordered list of nodes and edges for a visually concise overview that supports the comparison of quantitative data along pathways. The topology of the pathway is shown as secondary information using several types of stylized marks. The curvemap display shows time series data with small multiples of filled line charts and overlaid curves that support comparison of temporal gene expression across multiple species. The columns of the curvemap display are populated with genes selected by the user within the pathway representation view.

Through a series of case studies with our biology collaborators, we were able to validate that Pathline can show known information more clearly than could be seen with their previous tools. Furthermore, the biologists directly attribute new insights into their data to the use of Pathline. More information about Pathline, as well as source code, executables, and example data, can be found at http://pathline.org.

22.4 Evolutionary and Developmental Biology

A combination of novel microscopes with improved computer hardware and software has enabled scientists in recent years to capture data about living organisms at an unprecedented level of detail in time and space. Using this data, biologist can identify each single cell of a complex organism and investigate how it develops over time. With additional highlighting procedures, complementary information such as active genes can be recorded simultaneously to link the acquired information to processes on even finer biological scales. This wealth of information opens up new roads to study complex, yet fundamental processes in evolutionary and developmental biology (informally called evo-devo), such as: How does a single cell evolve into a complex organism? How do internal regulatory processes cause (nearly) identical cells to perform different tasks? How do genomic differences relate to differences in physiological structure? How do neurons connect to form rich and powerful structures within the brain?

Due to the wealth of evo-devo data and the relevant information therein, robust automatic preprocessing and a sophisticated visualization framework are central requirements to allow for future advances in this field. The fundamental data processing pipeline consists of three major steps:

1. *Data acquisition and storage:* commonly from digital microscopy
2. *Data preprocessing:* data fusion, image processing, and feature extraction
3. *Visualization and data analysis:* rendering of large-scale spatio-temporal data and deployment of application-focused interaction and data-mining techniques

In the following sections, we will investigate each step and outline relevant aspects for data analysis in general and data visualization in particular.

22.4.1 Data Acquisition and Storage

One of the most widely used data acquisition techniques in developmental biology is digital microscopy. Numerous specialized microscopes developed over the last decades support a large variety of applications. Most techniques enable the user to record 2D and 3D data, either from a single, volumetric specimen, or from cross-sections of one. Additional techniques exist to record volumetric time-series of living specimen (laser scanning microscopy) or data with a very high spatial resolution (electron microscopy). A variety of experimental preprocessing steps, such as selective dyes or fluorescent markers [19, 47], enhance data quality by highlighting structures and processes of interest. More details on experimental methods can be found in an excellent summary of microscopy usage in biology [49].

A lack of standardization in storage for microscopy data has resulted in a large variety of file formats, which poses a major challenge for generalizing processing pipelines. Most data, however, is stored as image files using, for example, the JPEG or TIFF file format. Depending on the application and the microscope manufacturer, the images have varying color depths and meta information, adding to the problem of data format diversity. To address this issue, the BioFormats group has developed a Java library for the reading and writing of a large variety of life science image file formats (http://www.loci.wisc.edu/software/bio-formats).

22.4.2 Data Preprocessing

The raw data coming off of a microscope often suffers from poor contrast, unregistered images, or the lack of precise information about individual structures in the image data. The following preprocessing steps are commonly employed to prepare the data for subsequent analysis:

1. *Data enhancement:* Due to limitations inherent in the microscope and the preparation of the specimen, the data often suffers from low contrast and high signal-to-noise ratios. To obtain better results during subsequent steps, the raw data is improved using image processing filters such as contrast adjustment or equalization of lighting. Deconvolution [26] is commonly used to decrease blurring.
2. *Data fusion:* Large specimens are often recorded in multiple passes that must be fused. This is commonly achieved using image stitching [45] and/or image registration [11]. Likewise, multiple recorded channels, e.g., from cameras recording parts of the sample that fluoresce with different wavelengths, must be registered. Additional problems arise in time-series data if a living specimen is recorded that can grow and/or move during recording. Such data not only requires the registration of individual time-steps, but also the application of drift correction algorithms.
3. *Image segmentation:* The image data is commonly too large and complex to be analyzed manually, thus causing automatic feature extraction algorithms to be mandatory. In a first step this means data segmentation [43], i.e., the extraction of relevant structures, such as cell boundaries, from the data. Segmentation and object classification is one of the major challenges in computer vision [44] and most microscopy images pose additional challenges as the algorithms have to cope with large variations in data intensity, morphological complexity and diversity, and varying signal-to-noise ratios. Despite the development of numerous, specialized segmentation algorithms, many applications still require manual or semi-automatic segmentation.
4. *Computation of metadata:* Many subsequent analysis steps, especially if time-series or multimodal data is recorded, require the computation of additional metadata. Simple examples of metadata include the size and shape of structures, textural properties, and image statistics of segmented regions. More advanced metadata is necessary for time series data where numerous cells are captured at individual time-steps. For the analysis of the morphology and morphodynamics, the cells need to be tracked over time. Here, cell tracking becomes a crucial part

of the preprocessing step. A large variety of metadata can be derived from the raw data, the inclusion of which depends on the specific application and scientific question.

Visualization is necessary in the first three preprocessing steps for data validation and quality assessment. Most automatic techniques come with errors and uncertainties which need to be conveyed to the user to help them judge the quality of their data and improve existing algorithms. Computation of metadata is commonly an important part of the visualization process itself and will be discussed in more detail in Sect. 22.4.3.

22.4.3 Visualization and Data Analysis in Evo-devo

A large number of evo-devo frameworks and techniques are being developed to support comprehensive analysis tasks. Most efforts currently concentrate on data preprocessing such as feature segmentation and the subsequent mathematical analysis where statistics are a widely applied tool [49]. Visualization is often reduced to the depiction of the data, which in itself is already rather difficult due to the size of the data and its time-dependent and multivariate nature. Tools widely used on the application side have a similar feature spectrum. Many of them support image processing tasks such as data registration and segmentation, feature extraction and analysis, and interactive rendering of three-dimensional data—even time-series data is supported by most applications. Examples of such tools are: (*commercial*) MetaMorph, Imaris, Volocity, Amira; and (*open-source*) ImageJ, Fiji, BioImageXD, V3D.

As the needs of specific applications diversify, specialized tools are being developed such as Cellenger (automated image segmentation and feature analysis: http://tiny.cc/rARky/), the CellProfiler project (automated image segmentation, feature analysis, data mining and visualization: http://cellprofiler.org), and various commercial products from microscope manufacturers. Recent research has also concentrated on the interactive validation of preprocessing steps applied to 3D spatio-temporal data. Examples for the validation of segmentation results are a combination of different rendering modes to overlay raw with segmented data [51] and the visualization of automatically quantified uncertainty [15, 20, 37]. A second direction is feature tracking, which has long been an important area of research in visualization [34], and is a crucial task in biological data analysis. Two major distinctions can be made for tracking algorithms operating on volumetric time-series data. One direction of research focuses on digital image processing approaches such as optical flow and operates directly on the scalar data [12, 30]. The other direction is based on tracking structures that have been segmented beforehand [25, 27]. If individual cells are tracked over time, they form cell lineages, i.e., binary trees that encode patterns of cell migration and division. Several browsers for the investigation of cell lineage data have recently been proposed [5, 32, 55], which combine volume rendering of the scalar data and graph drawing for the depiction of the lineage tree.

22.4.4 Challenges for Visualization

Evo-devo data presents several challenges for visualization, the largest of which is validation of the complex preprocessing pipeline. Assuming, however, that the data is perfectly preprocessed, the following visualization challenges remain for supporting data analysis:

- *Large-scale volumetric data*: The recorded data is typically quite large (up to several terabytes for a single specimen). Volume visualization needs to bridge different levels of detail to provide an overview in conjunction with detailed information about fine-scale structures.
- *Multimodal and time-series data:* Combining multimodal and temporal data is an open-research problem in visualization.
- *Wealth of data:* Many datasets consist of terabytes of raw data with numerous small features in long time series. Adequate abstraction mechanisms are therefore mandatory to help the user browse and investigate the data.
- *Combination of InfoVis and SciVis necessary:* Due to the versatile nature of the input data, techniques from both information and scientific visualization need to be combined in interactive frameworks.
- *Sophisticated data analysis required:* Feature selection and extraction is a crucial part of the visualization pipeline in evo-devo and requires knowledge in many related subjects such as computer vision, machine learning, and statistics.

22.4.5 Case Study: Visualization in Developmental Biology

As previously stated, visualization in evo-devo requires expertise in different areas of research. In this case study, we summarize the on-going work of our interdisciplinary team that studies the biological processes that occur during embryonic development, with the goal of creating a digital embryo model. The team consists of experts from: *developmental biology* who research the embryonic development of fish and flies; *microscopy research* who work towards better image acquisition techniques; *multidimensional image processing* who develop algorithms for the automatic segmentation and analysis of 3D spatio-temporal images; and *scientific visualization* who concentrate on data validation, feature extraction, and interactive data visualization and analysis. As the scientific questions related to embryonic development are very diverse and often change and expand over time, we decided to develop a versatile visualization framework (*Scifer*—http://scifer.info) with additional dedicated algorithms to address specific biological questions. *Scifer*, shown in Fig. 22.3, is a multi-window environment with linked views that combines interactive information and scientific visualization algorithms.

The primary GUI consists of three windows. The *main window* (Fig. 22.3a) is used to control the workflow. In this window, the user selects from a list of preprocessing algorithms, after which an interactive dialog box opens where additional parameters

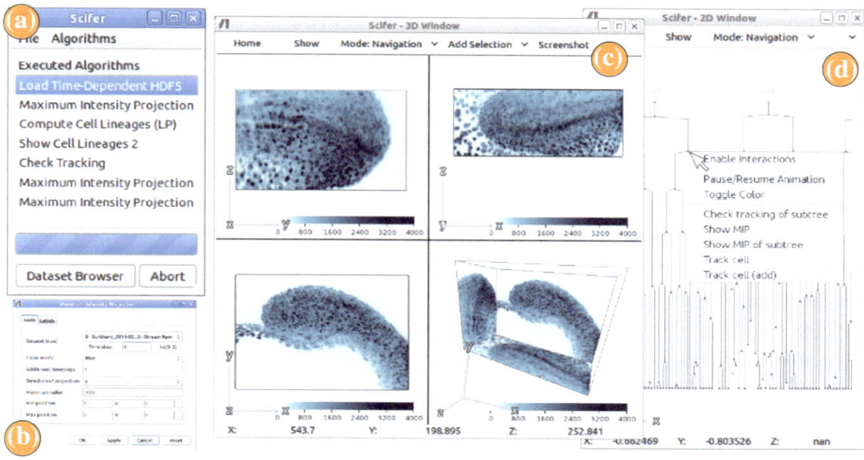

Fig. 22.3 Visualization in developmental biology using *Scifer*: **a** Main window, **b** dialog of an algorithm, **c** 3D window with MIP projection of part of a zebrafish embryo, and **d** 2D window with an interactive lineage depiction

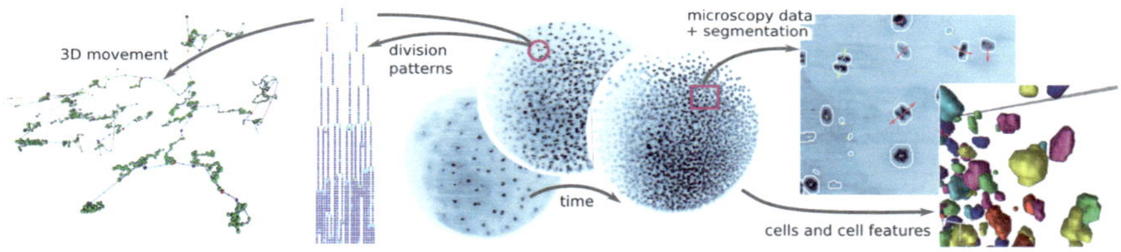

Fig. 22.4 Visualization techniques for data analysis in evo-devo (*left* to *right*): Cell lineage tree visualization in 3D, cell lineage tree visualization in 2D, volume visualization of microscopy data, combined rendering of microscopy and segmented data, isosurfaces of segmented cell nuclei

and user preferences are entered. Executed algorithms are listed in a separate widget where they can be altered and restarted. A progress bar and an abort button support the user while running computationally intensive algorithms. The two graphics windows support rendering and interaction with graphical primitives. The *2D window* (Fig. 22.3d) is mainly used for information visualization, such as plotting interactive scatterplots, heat maps, or cell lineages. Graphical primitives in the 2D window are selected by the user for further visualization in the native 3D space of the data, which are rendered in the *3D window* (Fig. 22.3c). Additional windows are added on demand to display supplementary information.

The system currently consists of several algorithms, the selection of which is driven by our interdisciplinary collaboration [22]. A selection of the interactive rendering techniques provided in *Scifer* is depicted in Fig. 22.4. We currently support the following functionalities: (i) *Data rendering*: We provide volume rendering, maximum intensity projection, and isosurfacing of the raw and segmented data to support validation and understanding of the measured and preprocessed data. (ii) *Feature computation and visualization*: Algorithms to compute additional cell features, such as cell volume, mean intensity, and texture properties, are provided to further investigate properties of individual cells and their temporal evolution. For the visualization and interactive exploration of this data, we provide suitable visual representations such as histograms, scatterplots, and heatmaps that are interactively linked to the 3D window, where selected cells are highlighted. (iii) *Lineage computation and rendering*: To visualize the data over time, we developed a robust cell tracking algorithm. The resulting cell lineages are rendered in the 2D window using standard graph drawing algorithms. Again, interaction is a central feature that adds additional information upon selecting a cell lineage or an individual cell, e.g., display of cell features or highlighting of the selected cell in the 3D window.

This initial set of algorithms forms the basis for exploring and analyzing spatio-temporal evo-devo data. We anticipate the ongoing development of novel visualization techniques within this framework to answer more specific biological questions as our collaboration continues.

22.5 Conclusions

Biology poses a wealth of interesting, and evolving, visualization challenges. The data is continually becoming larger and more complex, requiring novel solutions for processing and making sense of the patterns therein. Many biologists have already realized that pure mathematical data analysis does not suffice to answer their questions, or more precisely, help them formulate the right questions to understand the structures and processes they are interested in. The inherent relationships are often very diverse and complex, and visualization can help biologists successively gain deeper insight into the processes at hand. While dedicated visualization software has already helped a lot for specific problems, there are still many open research problems in visualization in biology that will not only advance biology but also visualization research.

We have found our collaborations with biologists to be truly symbiotic and that our contributions as visualization researchers goes beyond the development of a new tool. Our discussions with biologists help them to clarify, and sometimes change, their analysis needs and approaches. And the biological problems provide us with the motivation and inspiration to develop new visual representations and systems, as well as to refine our methods and processes. In short, there is a huge amount of interesting work at the intersection of visualization and biology.

From a visualization point of view, major areas of future research will have to be dedicated to *volume visualization* of large time-dependent datasets (terabytes per dataset) with multiple relevant levels of detail and varying levels of noise, integration of techniques from *scientific visualization, information visualization, and visual analytics*, and the visualization of *multi-input datasets*, i.e., data that consists of multiple information sources such as volumetric records with multiple channels and varying levels of detail, gene expression measurements, network data describing multiple correlated processes, mathematical simulations, or previous research/knowledge stored in databases.

References

1. Alm, E., Huang, K., Price, M., Koche, R., Keller, K., Dubchak, I., Arkin, A.: The MicrobesOnline web site for comparative genomics. Genome. Res. **15**(7), 1015–1022 (2005). doi:10.1101/gr.3844805
2. Barsky, A., Munzner, T., Gardy, J., Kincaid, R.: Cerebral: Visualizing multiple experimental conditions on a graph with biological context. IEEE Trans. Vis. Comput. Graph. **14**(6), 1253–1260 (2008) (Proc. InfoVis 2008). http://doi.ieeecomputersociety.org/10.1109/TVCG.2008.117
3. Bourqui, R., Westenberg, M.A.: Visualizing temporal dynamics at the genomic and metabolic level. In: IV '09: Proceedings of the 2009 13th International Conference Information Visualisation, pp. 317–322. IEEE Computer Society, Washington, DC, USA (2009)
4. Caspi, R., Foerster, H., Fulcher, C., Kaipa, P., Krummenacker, M., Latendresse, M., Paley, S., Rhee, S., Shearer, A., Tissier, C., Walk, T., Zhang, P., Karp, P.: The MetaCyc database of metabolic pathways and enzymes and the BioCyc collection of pathway/genome databases. Nucleic Acids Res. **36**(Database-Issue), 623–631 (2008)
5. Cedilnik, A., Baumes, J., Ibanez, L., Megason, S., Wylie, B.: Integration of information and volume visualization for analysis of cell lineage and gene expression during embryogenesis. In: Proceedings of SPIE, vol. 6809 (2007)
6. Clamp, M., Andrews, D., Barker, D., Bevan, P., Cameron, G., Chen, Y., Clark, L., Cox, T., Cuff, J., Curwen, V., Down, T., Durbin, R., Eyras, E., Gilbert, J., Hammond, M., Hubbard, T., Kasprzyk, A., Keefe, D., Lehvaslaiho, H., Iyer, V., Melsopp, C., Mongin, E., Pettett, R., Potter, S., Rust, A., Schmidt, E., Searle, S., Slater, G., Smith, J., Spooner, W., Stabenau, A., Stalker, J., Stupka, E., Ureta-Vidal, A., Vastrik, I., Birney, E.: Ensembl 2002: accommodating comparative genomics. Nucleic Acids Res. **31**(1), 38–42 (2003)
7. Cleveland, W.S., McGill, R.: Graphical perception: theory, experimentation, and application to the development of graphical methods. J. Am. Stat. Assoc. **79**(387), 531–554 (1984)
8. Darling, A.C., et al.: Mauve: multiple alignment of conserved genomic sequence with rearrangements. Genome Res. **14**(7), 1394–1403 (2004)
9. Eisen, M.B., Spellman, P.T., Brown, P.O., Botstein, D.: Cluster analysis and display of genome-wide expression patterns. Proc. Natl. Acad. Sci. **95**(25), 14,863–14,868 (1998)
10. Gehlenborg, N., O'Donoghue, S.I., Baliga, N.S., Goesmann, A., Hibbs, M.A., Kitano, H., Kohlbacher, O., Neuweger, H., Schneider, R., Tenenbaum, D., Gavin, A.C.: Visualization of omics data for systems biology. Nat. Methods **7**, S56–S68 (2010)
11. Goshtasby, A.: 2-D and 3-D Image Registration for Medical, Remote Sensing, and Industrial Applications. Wiley Press, Hoboken (2005)
12. Hand, A.J., Sun, T., Barber, D.C., Hose, D.R., MacNeil, S.: Automated tracking of migrating cells in phase-contrast video microscopy sequences using image registration. J. Microsc. **234**, 62–79 (2009)
13. Holten, D.: Hierarchical edge bundles: Visualization of adjacency relations in hierarchical data. IEEE Trans. Vis. Comput. Graph. (Proc. InfoVis 08), **12**(5), 741–748 (2006)
14. Hunt, E., et al.: The visual language of synteny. OMICS J. Integr. Biol. **8**(4), 289–305 (2004)
15. Janoos, F., Nouansengsy, B., Xu, X., Machiraju, R., Wong, S.T.: Classification and uncertainty visualization of dendritic spines from optical microscopy imaging. Comput. Graph. Forum **27**(3), 879–886 (2008)
16. Junker, B.H., Klukas, C., Schreiber, F.: VANTED: a system for advanced data analysis and visualization in the context of biological networks. BMC Bioinf. **7**, 109 (2006)

17. Kanehisa, M., Araki, M., Goto, S., Hattori, M., Hirakawa, M., Itoh, M., Katayama, T., Kawashima, S., Okuda, S., Tokimatsu, T., Yamanishi, Y.: KEGG for linking genomes to life and the environment. Nucleic Acids Res. **36**(Database-Issue), 480–484 (2008)
18. Karp, P., Paley, S., Romero, P.: The pathway tools software. Bioinformatics **18**, S225–S232 (2002)
19. Khairy, K., Keller, P.J.: Reconstructing embryonic development. Genesis **49**(7), 488–513 (2011). doi:10.1002/dvg.20698
20. Kniss, J., Van Uitert, R., Stephens, A., Li, G.S., Tasdizen, T., Hansen, C.: Statistically quantitative volume visualization. In: Visualization, 2005. VIS 05. IEEE, pp. 287–294 (2005)
21. Kurtz, S., et al.: Versatile and open software for comparing large genomes. Genome Biol. **5**(2), R12 (2004)
22. Leitte, H., Fangerau, J., Lou, X., Höckendorf, B., Lemke, S., Maizel, A., Wittbrodt, J.: Visualization software for 3D video microscopy: a design study. In: M. Meyer, T. Weinkauf (eds.) EuroVis 2012—Short Papers, pp. 73–77. Eurographics Association, Vienna, Austria (2012). doi:10.2312/PE/EuroVisShort/EuroVisShort2012/073-077
23. Letunic, I., Yamada, T., Kanehisa, M., Bork, P.: iPath: interactive exploration of biochemical pathways and networks. Trends Biochem. Sci. 33(3), 101–103 (2008). doi:10.1016/j.tibs.2008.01.001
24. Lewis, S., Searle, S., Harris, N., Gibson, M., Iyer, V., Richter, J., Wiel, C., Bayraktaroglu, L., Birney, E., Crosby, M., Kaminker, J., Matthews, B., Prochnik, S., Smith, C., Tupy, J., Rubin, G., Misra, S., Mungall, C., Clamp, M.: Apollo: a sequence annotation editor. Genome Biol. **3**(12), research0082.1–0082.14 (2002)
25. Lou, X., Kaster, F.O., Lindner, M.S., Kausler, B.X., Köthe, U., Höckendorf, B., Wittbrodt, J., Jänicke, H., Hamprecht, F.A.: Deltr: Digital embryo lineage tree reconstructor. In: ISBI 2011: IEEE International Symposium on Biomedical Imaging (2011)
26. Lucy, L.B.: An iterative technique for the rectification of observed distributions. Astron. J. **79**(6), 745–754 (1974)
27. Meijering, E., Dzyubachyk, O., Smal, I., van Cappellen, W.A.: Tracking in cell and developmental biology. Semin. Cell Dev. Biol. **20**(8), 894–902 (2009)
28. Meyer, M., Munzner, T., Pfister, H.: MizBee: a multiscale synteny browser. IEEE Trans. Vis. Comput. Graph. **15**(6), 897–904 (2009) (Proc. InfoVis 2009). http://dx.doi.org/10.1109/TVCG.2009.167
29. Meyer, M., Wong, B., Styczynski, M., Munzner, T., Pfister, H.: Pathline: A tool for comparative functional genomics. Comput. Graph. Forum **29**(3), 1043–1052 (2010) (Proc. EuroVis 10)
30. Miura, K.: Tracking movement in cell biology. In: Rietdorf, J. (ed.) Microscopy Techniques, Advances in Biochemical Engineering/Biotechnology, vol. 95, pp. 1304–1307. Springer, Berlin (2005)
31. Mlecnik, B., Scheideler, M., Hackl, H., Hartler, J., Sanchez-Cabo, F., Trajanoski, Z.: PathwayExplorer: web service for visualizing high-throughput expression data on biological pathways. Nucleic Acids Res. **33**(Web-Server-Issue), 633–637 (2005)
32. Olivier, N., Luengo-Oroz, M.A., Duloquin, L., Faure, E., Savy, T., Veilleux, I., Solinas, X., Débarre, D., Bourgine, P., Santos, A., Peyriéras, N., Beaurepaire, E.: Cell lineage reconstruction of early zebrafish embryos using label-free nonlinear microscopy. Science **329**(5994), 967–971 (2010)
33. Pan, X., Stein, L., Brendel, V.: SynBrowse: a synteny browser for comparative sequence analysis. Bioinformatics **21**(17), 3461–3468 (2005)
34. Post, F.H., Vrolijk, B., Hauser, H., Laramee, R.S., Doleisch, H.: The state of the art in flow visualisation: feature extraction and tracking. Comput. Graph. Forum (CGF) **22**(4), 775–792 (2003)
35. Rasko, D., Myers, G., Ravel, J.: Visualization of comparative genomic analyses by BLAST score ratio. BMC Bioinf. **6**(1), 2 (2005)
36. Roberts, J.C.: State of the art: Coordinated & multiple views in exploratory visualization. In: Proceedings of International Conference on Coordinated and Multiple Views in Exploratory Visualization (CMV), pp. 61–71. IEEE Computer Society (2007)

37. Saad, A., Hamarneh, G., Möller, T.: Exploration and visualization of segmentation uncertainty using shape and appearance prior information. IEEE Trans. Vis. Comput. Graph. **16**(6), 1366–1375 (2010)
38. Saldanha, A.J.: Java Treeview—extensible visualization of microarray data. Bioinformatics **20**(17), 3246–3248 (2004). http://dx.doi.org/10.1093/bioinformatics/bth349
39. Seo, J., Shneiderman, B.: Interactively exploring hierarchical clustering results. Computer **35**(7), 80–86 (2002). http://dx.doi.org/10.1109/MC.2002.1016905
40. Shannon, P., Markiel, A., Ozier, O., Baliga, N., Wang, J., Ramage, D., Amin, N., Schwikowski, B., Ideker, T.: Cytoscape: a software environment for integrated models of biomolecular interaction networks. Genome Res. **13**(11), 2498–2504 (2003). doi:10.1101/gr.1239303
41. Shneiderman, B.: The eyes have it: A task by data type taxonomy for information visualizations. In: Proceedings of IEEE Symposium on Visual Languages (VL), p. 336. IEEE Computer Society, Washington, DC, USA (1996)
42. Sinha, A., Meller, J.: Cinteny: flexible analysis and visualization of synteny and genome rearrangements in multiple organisms. BMC Bioinf. **8**(1), 82 (2007)
43. Suri, J.S., Wilson, D., Laxminarayan, S. (eds.): Handbook of Biomedical Image Analysis: Volume 2: Segmentation Models Part B. Springer, Berlin (2005)
44. Szekely, G., Gerig, G.: Model-based segmentation of radiological images. Kuenstliche Intelligenz **14**(3), 18–23 (2000)
45. Szeliski, R.: Image alignment and stitching: a tutorial. Technical Report MSR-TR-2004-92, Microsoft Research (2004)
46. TIGR (The Institute for Genomic Research): Sybil: Web-based software for comparative genomics (2009). Accessed 1 Mar 2009. http://sybil.sourceforge.net
47. Timpson, P., McGhee, E.J., Anderson, K.I.: Imaging molecular dynamics in vivo—from cell biology to animal models. J. Cell Sci. **124**, 2877–2890 (2011)
48. Tufte, E.R.: The Visual Display of Quantitative Information, 2 edn. Graphics Press (2001)
49. Walter, T., Shattuck, D.W., Baldock, R., Bastin, M.E., Carpenter, A.E., Duce, S., Ellenberg, J., Fraser, A., Hamilton, N., Pieper, S., Ragan, M.A., Schneider, J.E., Tomancak, P., Hériché, J.K.: Visualization of image data from cells to organisms. In: Visualizing Biological Data, vol. 7, pp. S26–S41. Nature Methods (2010)
50. Ware, C.: Information visualization: perception for design. Morgan Kaufmann , Burlington (2000) (Chapter 4)
51. Weber, G.H., Luengo, C.L.H., Keränen, S.V.E., Dillard, S.E., Ju, D., Sudar, D., Hamann, B.: Visualization for validation and improvement of three-dimensional segmentation algorithms. In: Data Visualization 2005 (Proceedings of the EUROGRAPHICS—IEEE VGTC Symposium on Visualization 2005). Eurographics Association (2005)
52. Weinstein, J.N.: A postgenomic visual icon. Science **319**(5871), 1772–1773 (2008). doi:10.1126/science.1151888
53. Wilkinson, L., Friendly, M.: The history of the cluster heat map. Am. Stat. **63**(2), 179–184 (2009)
54. Wong, B.: Points of view: color coding. Nat Methods **7**(8), 573 (2010)
55. Wylie, B., Baumes, J.: A unified toolkit for information and scientific visualization. Visual Data Analytics (2009)

Chapter 23
From Individual to Population: Challenges in Medical Visualization

C.P. Botha, B. Preim, A.E. Kaufman, S. Takahashi and A. Ynnerman

Abstract Due to continuing advances in medical imaging technology, and in medicine itself, techniques for visualizing medical image data have become increasingly important. In this chapter, we present a brief overview of the past 30 years of developments in medical visualization, after which we discuss the research challenges that we foresee for the coming decade.

23.1 Introduction

Since the advent of magnetic resonance imaging (MRI) and computed tomography (CT) scanners around the early seventies, and the consequent ubiquitousness of medical volume data, medical visualization has undergone significant development and is now a primary branch of Visualization. It finds application in diagnosis, for example virtual colonoscopy, in treatment, for example surgical planning and guidance, and in medical research, for example visualization of diffusion tensor imaging data. Although the field of medical visualization only established itself with this name in the late eighties [57], we shall see in the next section that already in the seventies

C.P. Botha (✉)
vxlabs, Somerset West, SA
e-mail: cpbotha@vxlabs.com

B. Preim
Department of Simulation and Graphics, University of Magdeburg, Magdeburg, Germany
e-mail: preim@isg.cs.uni-magdeburg.de

A.E. Kaufman
Department of Computer Science, Stony Brook University, New York, USA
e-mail: ari@cs.stonybrook.edu

S. Takahashi
Graduate School of Frontier Sciences, The University of Tokyo, Tokyo, Japan
e-mail: takahashi@acm.org

A. Ynnerman
Norrköping Visualization and Interaction Studio, Linköping University, Linköping, Sweden
e-mail: anders.ynnerman@itn.liu.se

C.D. Hansen et al. (eds.), *Scientific Visualization*, Mathematics and Visualization,
DOI 10.1007/978-1-4471-6497-5_23

there were published examples of computer-generated images, based on medical data and used for medical applications.

During the past decades, medical image acquisition technology has undergone continuous and rapid development. It is now possible to acquire much more complex data than ever before. For example, in High Angular Resolution Diffusion Imaging (HARDI), forty or more diffusion-weighted volumes are acquired in order to calculate and visualize water diffusion and, indirectly, structural neural connections in the brain [70]. In fMRI-based full brain connectivity, time-based correlation of neural activity is indirectly measured between all pairs of voxels in the brain, thus giving insight into the functional neural network [24]. Moreover, the questions that users attempt to answer using medical visualization have also become significantly more complex.

In this paper, we first give a high-level overview of medical visualization development over the past 30 years, focusing on key developments and the *trends* that they represent. During this discussion, we will refer to a number of key papers that we have also arranged on the medical visualization research timeline shown in Fig. 23.1. Based on the overview and our observations of the field, we then identify and discuss the medical visualization research challenges that we foresee for the coming decade.

23.2 Thirty-year Overview of Medical Visualization

Already in 1978, Sunguroff and Greenberg published their work on the visualization of 3D surfaces from CT data for diagnosis, as well as a visual radiotherapy planning system, also based on CT data [64]. Five years later, Vannier et al. published their results developing a system for the computer-based pre-operative planning of craniofacial surgery [71]. By this time, they had already used and evaluated their surgical planning system in treating 200 patients. The system was based on the extraction and visualization of 3D hard and soft tissue surfaces from CT data. Through the integration of an industrial CAD application, it was also possible to perform detailed 3D measurements on the extracted surfaces.

23.2.1 Practical and Multi-modal Volume Visualization

In 1986, Hohne and Bernstein [26] proposed using the gray-level gradient to perform shading of surfaces rendered from 3D CT data. In 1987, Lorensen and Cline published the now famous Marching Cubes isosurface extraction algorithm, which enabled the fast and practical extraction of 3D isosurfaces from real-world medical data. In the year thereafter, Levoy [47] introduced the idea of volume raycasting in May, and Drebin et al. [19] in August. Although medical volume visualization was possible before these publications, as witnessed by a number of publications, previous techniques were either not as fast or yielded less convincing results. With

1978 - Sunguroff and Greenberg: CT and 3D surfaces for diagnosis and radiotherapy planning [64].

1983 - Vannier et al.: 3D surfaces from CT for planning of craniofacial surgery [71].

1986 - Hohne and Bernstein: Shading 3D Images from CT using gray-level gradients [26].

1987 - Lorensen and Cline: Marching Cubes [49].

1988 - Levoy publishes Direct Volume Rendering paper in May [47], Drebin et al. in August [19].
Multi-modal volume rendering by Höhne et al. [27].

1993 - Altobelli et al.: Predictive simulation in surgical planning [1]. Gerig et al.: Vessel visualization [23].

1994 - Basser et al.: Diffusion Tensor Imaging [5].

1995 - Hong et al.: 3D Virtual Colonoscopy [28].

1996 - Behrens et al. et al.: Visualization of Dynamic Contrast-Enhanced MRI mammography data (time-varying) [7].

1998 - Basser et al.: DTI Tractography [4, 6].

2000 - Ebert and Rheingans - Medical Volume Illustration [20].

2001 - Tory et al.: Multi-timepoint MRI [69].

2003 - Krüger and Westermann: GPU raycasting [45].

2007 - Blaas et al.: Multi-Field Medical Visualization [9].

2008 - Wang et al.: LifeLines2 - multi-subject electronic health records [73].

2010 - Steenwijk et al.: Cohort studies - multi-subject imaging and metadata [63].

Fig. 23.1 Timeline with a subset of medical visualization papers showing the progression from scalar volume datasets through time-dependent data to multi-field and finally multi-subject datasets. This timeline is by no means complete, instead attempting to show a representative sample of papers that represent various trends in the development of the field

the introduction of Marching Cubes and volume raycasting, volume visualization became a core business of visualization and medical visualization for the years to come.

Up to this point, research had focused on uni-modality data, primarily CT. However, already in 1988 the first multi-modal volume rendering paper was published by Höhne et al., in which they demonstrated the registration and combined visualization of CT and MRI. A great deal of work has been done since then on the theory and applications of multi-modal volume visualization. The first notable example is the work of Cai and Sakas in 1999 where they classified voxel-voxel multi-modal volume rendering techniques according to the volume rendering pipeline stage where they take place [13]. The three classes are image level, where two volume renderings are combined pixel-by-pixel, accumulation level, where looked up samples along the ray are combined, and illumination model level, where the illumination model is adapted to process two volume samples directly. The second example we mention is a convincing application of multi-modal volume rendering for the planning of neurosurgical interventions, where MRI, CT, fMRI, PET and DSA data are all combined in an interactive but high quality visualization for the planning of brain tumor resection [8].

23.2.2 Therapy Planning, Predictive Simulation, and Diagnosis

Therapy planning was one of the first real applications of medical visualization and remains important to this day. In 1993, Altobelli et al. [1] published their work on using CT data to visualize the possible outcome of complicated craniofacial surgery. By manually repositioning soft tissue fragments based on the bony surfaces under them, in certain cases taking into account bone-skin motion ratios from literature, the expected outcome of a craniofacial procedure could be visualized. Although still rudimentary, this could be considered one of the earliest cases of *predictive or outcome simulation* integrated with visualization for surgical planning. The idea of predictive simulation, or predictive medicine, was further explored by Taylor et al. [66] for cardiovascular surgery.

With the introduction of virtual colonoscopy (VC) in 1995 [28], medical visualization also gained diagnosis as an important medical application, namely screening for colon cancer. VC combines CT scanning and volume visualization technologies. The patient's abdomen is imaged in a few seconds by a multi-slice CT scanner. A 3D model of the colon is then reconstructed from the scan by automatically segmenting the colon and employing "electronic cleansing" of the colon for computer-based removal of the residual material. The physician then interactively navigates through the volume rendered virtual colon employing camera control mechanisms, customized tools for 3D measurements, "virtual biopsy" to interrogate suspicious regions, and "painting" to support 100 % inspection of the colon surface [29]. VC is rapidly gaining popularity and is poised to become the procedure of choice in lieu of the conventional optical colonoscopy for mass screening for colon polyps—the precursor of colorectal cancer. Unlike optical colonoscopy, VC is patient friendly, fast, non-invasive, more accurate, and a more cost-effective procedure for mass screening for colorectal cancer.

VC technologies gave rise to the computer-aided detection (CAD) of polyps, where polyps are detected automatically by integrating volume rendering, conformal colon flattening, clustering, and "virtual biopsy" analysis. Along with the reviewing physician, CAD provides a second pair of "eyes" for locating polyps [30]. This work was also the basis for many other virtual endoscopy systems, such as virtual bronchoscopy, virtual cystoscopy, and virtual angioscopy. A careful integration of image analysis (e.g., segmentation, skeletonization), with efficient rendering (e.g., occlusion culling) and interaction (e.g., camera control based on predefined paths) are major ingredients of such systems [3, 29].

23.2.3 Multi-field Data

Diffusion Tensor Imaging, or DTI, is an MRI-based acquisition modality, introduced in 1994 by Basser et al., that yields 3×3 symmetric diffusion tensors as its native measurement quantity [5]. The tensors represent the local diffusion of water molecules, and hence indirectly indicate the presence and orientation of fibrous structures, such as neural fiber bundles or muscle fibers. Already in this first paper, the authors employed 3D glyphs to visualize the eigensystems represented by the tensors.

Basser and his colleagues were also some of the first to extract and visualize fiber-tract trajectories from DTI data of the brain [4, 6], thus linking together the point diffusion measurements to get an impression of the global connective structures in the brain. With DTI it was encouraging to see that the first visualization efforts were initiated by the scientists developing this new scanning modality themselves. Early work by the visualization community includes tensor lines for tractography [76] and direct volume rendering of DTI data [38, 39].

Importantly, DTI serves as one of the first examples of natively multi-field medical data, that is medical data with multiple parameters defined over the same spatio-temporal domain. The advent of DTI initiated a whole body of medical visualization research dedicated to the question of how best to visually represent and interact with diffusion tensor data in particular and multi-field medical data in general. The 2007 paper by Blaas et al. presented a visual analysis-inspired solution to this problem based on linked physical and feature space views [9].

23.2.4 Time-Varying Data

Time-varying medical volume data visualization made its entrance in 1996 with work by Behrens et al. [7] on supporting the examination of Dynamic Contrast-Enhanced MRI mammography data with the display of parameter maps, the selection of regions of interest (ROIs), the calculation of time-intensity curves (TICs), and the quantitative analysis of these curves. In 2001, Tory et al. [69] presented methods for visualizing multi-timepoint (1 month interval) MRI data of a multiple sclerosis (MS) patient, where the goal was to study the evolution of brain white matter lesions over time. Methods used included glyphs, multiple isosurfaces, direct volume rendering and animation. Coto et al. [16] applied multiple coupled views, including linked cursors and brushing on 3D renderings and scatterplots, to dynamic contrast-enhanced MRI (DCE-MRI) mammography data.

23.2.5 Illustrative Visualization

Illustrative visualization is primarily motivated by the attempt to create renditions that consider the *perceptual capabilities* of humans. As an example, humans infer information about shapes not only from realistic shading but also from appropriate hatching and from outlines that support the mental separation of nearby objects rendered in similar colours.

Illustrative visualization is related to the term *Non-Photorealistic Rendering* in computer graphics, or NPR for short. The term NPR was used since around 1990 when the seminal paper of Saito et al. clearly illustrated that complex 3D shapes could be rendered more *comprehensible* by using certain *feature lines* [61]. Compared to NPR, illustrative visualization is the more focused term that covers rendering techniques serving clear visualization goals, namely to convey shape information efficiently. In medical visualization, either surfaces or volume data are rendered in illustrative styles. For illustrative volume rendering, the term *volume illustration* was introduced by Ebert et al. in 2000 [20]. Boundary enhancement based on gradient approximation [17] and curvature-based transfer functions [40] are landmarks in illustrative medical visualization. Tietjen et al. applied silhouettes and other feature lines for various scenarios in liver surgery planning [68]. Besides silhouettes, stippling and, probably even more, hatching, have great potential to reveal details of shapes [32].

Later, Bruckner et al. [10–12] made a number of important contributions that support depth and shape perception with adapted transfer functions. In particular, they considered the peculiarities of interactive exploration of 3D datasets and elaborated on the idea of preserving essential context information. These and later refinements are integrated in the VolumeShop-system that is publicly available and used by several research groups.

23.2.6 Multi-subject Data

Medical visualization has also started to work on the problem of dealing with multi-subject data. These are datasets that include measurements, including imaging, of more than one subject. The goal is to be able to extract patterns that affect sub-groups of the whole collection, for example to explore which aspects of the data correlate with a specific disease outcome. An example of this type of work includes LifeLines2, an information visualization approach to visualize and compare multiple patient histories or electronic medical records [73]. More recently, work has been done on the interactive visualization of the multi-subject and mixed modality datasets acquired by medical cohort studies [63]. In these studies, mixed modality data, including imaging, genetics, blood measurements and so on, are acquired from a group of subjects in order to understand, diagnose or predict the clinical outcome of that group. Steenwijk et al. demonstrated that it was possible to create a highly interactive coupled view visualization interface, integrating both information and scientific visualization techniques, with which patterns, and also hypotheses, could be extracted from the whole data collection.

23.3 Challenges in Medical Visualization

23.3.1 Advances in Data Acquisition

Toshiba's 320-slice CT scanner, the Aquilion One, was introduced in 2007. It is able to acquire five 320 slice volumes *per second* [31] and can thus image a beating heart. Rapid and continuous advances in the dynamic nature and sheer magnitude of data in this and other mainstream medical imaging necessitates improvements to existing techniques in terms of computational and perceptual scalability.

High Angular Resolution Diffusion Imaging (HARDI) [70] and Diffusion Spectrum Imaging (DSI) [25] datasets contain hundreds of diffusion-weighted volumes describing the diffusion of water molecules and hence indirectly the orientation of directed structures such as neural fiber bundles or muscle fibers. This is a rather extreme example of multi-field medical data that is becoming more relevant in both medical research and clinical application. Completely new visual metaphors are required to cope with the highly multi-variate and three-dimensional data of diffusion weighted imaging in particular and many other new imaging modalities in general.

Molecular imaging enables the in vivo imaging of biochemical processes at the macroscopic level, meaning that, for example, pathological processes can be studied and followed over time in the same subject long before large-scale anatomical changes occur. Examples are bioluminescence (BLI) and fluorescence (FLI) imaging, two molecular imaging modalities that enable the in vivo imaging of gene expression. Molecular imaging yields datasets that vary greatly in scale, sensitivity, spatial-temporal embedding and in the phenomena that can be imaged. Each permutation brings with it new domain-specific questions and visualization challenges. Up to now, most of the visualization research has been focused on small animal imaging [41, 42], but due to its great diagnostic potential, molecular imaging will see increased application in humans.

The integration of microscopy imaging is an essential task for the future, where data handling, interaction facilities but also more classical rendering tasks such as transfer function design become essential. With more and more large scale and 3D microscopy data available, there are many opportunities for visualization researchers. Recent examples include techniques for interactively visualizing large-scale biomedical image stacks demonstrated on datasets of up to 160 gigapixels [34] and tools for the interactive segmentation and visualization of large-scale 3D neuroscience datasets, demonstrated on a 43 GB electron microscopy volume dataset of the hippocampus [33].

With these examples, we hope to have illustrated that advances in image acquisition are continuous, and due to the increasing demands of modern society are accelerating. Each new advance in imaging brings potentially greater magnitudes and exotic new types of data, leading to new challenges for medical visualization.

23.3.2 Heterogeneous Display and Computing Devices

Mobile devices, in particular the APPLE products IPAD and IPHONE, are extremely popular among medical doctors and indeed solve some serious problems of desktop devices in routine clinical use. In particular, bedside use of patient data is an essential use case for medical doctors of various disciplines.

Meanwhile, several mobile devices are equipped with powerful graphics cards and, using the OpenGL ES (Embedded Systems) standard, they are able to provide high-quality interactive rendering. Although the performance still trails that of modern desktop devices, slicing medical volume data and 3D rendering is feasible [52].

The rapid and widespread use of mobile devices also made gesture input popular. In particular, multi-touch interaction is considered an intuitive interaction since many potential users know a variety of gestures from their everyday activities with smart phones. Therefore, multitouch interaction is also incorporated in large displays in medical use, e.g. the Digital Lightbox[1] by BrainLab and the multi-touch table of Lundström et al. [50].

23.3.3 Interactive Image Segmentation

Image segmentation is important in clinical practice, for example, in diagnosis and therapy planning, and also in image-based medical research. In these applications, segmentation is complicated by the great deal of variation in image acquisition, pathology and anatomy. Furthermore, in matters of diagnosis or therapy planning, the accuracy of the segmentation can be critical. It comes as no surprise that user interaction is often required to help ensure the quality of the results, by initializing an image processing method, checking the accuracy of the result or to correct a segmentation [55].

A well-known interactive segmentation technique is the live-wire or intelligent scissors [51]. These ideas were later extended and applied to medical images [21]. More recently, visualization has been applied to the challenge of explicitly dealing with the uncertainty inherent in interactive 3D segmentation [56, 60].

Medical visualization research often combines elements of image analysis, graphics and interaction, and is thus ideally equipped to address the challenge of developing and validating effective interactive segmentation approaches for widespread use in medical research and practice.

23.3.4 Topological Methods

Topological data representation has played an important role in medical visualization, since it can allow us to segment specific features such as human organs and bones

[1] http://www.brainlab.com/art/2841/4/surgical-pacs-access/

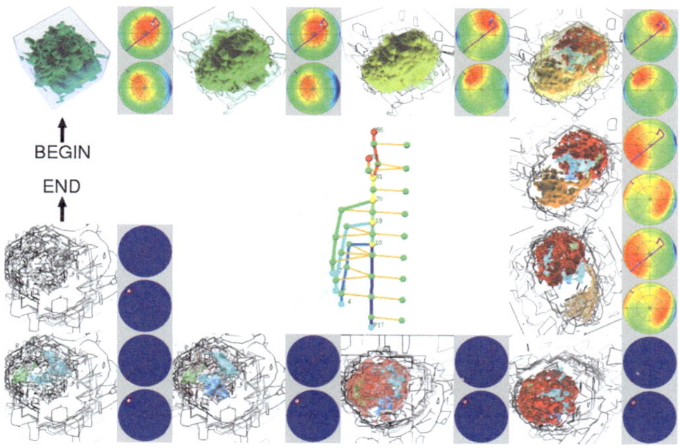

Fig. 23.2 Previewing a sheep heart volume along the optimal camera path. Image taken from Ref. [65]

from the input medical datasets systematically and identify their spatial relationships consistently. Actually, such topological concepts have been already introduced in the earlier stage of medical visualization. For example, the 3D surface of a human cochlea was reconstructed from a series of 2D CT cross-sectional images by identifying correct correspondence between the cross-sectional contours [62].

Topological approaches have also been extended to analyze 3D medical volume data. Contour trees [2] have been employed for designing transfer functions in order to illuminate human organs and bones systematically since the associated anatomical structure can be effectively captured as topological skeletons of isosurfaces [75]. Spatial relationships between bones and the position of an aneurysm were successfully extracted respectively from CT and angiographic datasets using a variant of contour trees [15]. Interesting features in medical volume data can be visually analyzed using an optimal camera path, which can be obtained by referring to the topological structure of human organs [65] (see Fig. 23.2). Topological methods are now being developed for visualizing multi-variate and high-dimensional datasets, and thus potentially for analyzing tensor fields obtained through DT-MRI, multi-subject data in group fMRI studies, and time-varying data measured by high-speed CTs.

23.3.5 Integration of Simulation Models

In their 1999 predictive medicine paper, Taylor et al. argued that surgical planning should not only address questions of surgical approach but also of the expected outcome, for example, predicted future states such as the efficacy of a treatment option or the performance of an implant [66]. Medical visualization approaches become significantly more valuable when enhanced with simulation models that help

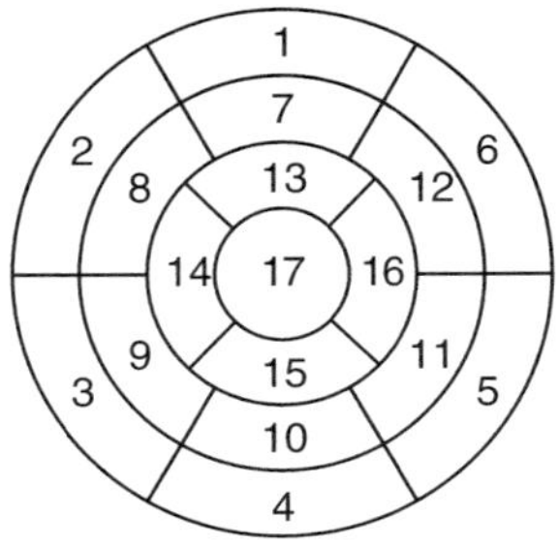

Fig. 23.3 The American Heart Association standardized 2D bull's eye plot (BEP) of the *left ventricle* of the heart (Courtesy of Konrad Mühler, inspired by Ref. [14]). Each numbered segment in 2D corresponds to a particular anatomical segment of the very much three dimensional heart muscle

to predict the outcome of a disease process or therapeutic procedure, or that enrich measured data with expected physiological phenomena. Examples besides the blood flow simulations of Taylor et al. include interactive skeletal range of motion [43] and biomechanical stress [18] simulation models for implant planning in orthopedics and nasal airflow simulation for reconstructive rhinosurgery [77].

The integration of these predictive models, although potentially valuable, brings with it new challenges. The addition of potentially complex and dynamic simulation output data to existing visualizations requires new visual representation techniques. Furthermore, for the simulation results to be maximally useful, the models should be tightly coupled to and steered by the user's interaction with the medical visualization. Finally, most simulations yield data with a certain degree of inherent uncertainty. The role of this uncertainty should be fully explored and it should be carefully but explicitly represented as an integral part of the visualization.

23.3.6 Mappings and Reformations

In 2002, the American Heart Association proposed a standardised segmentation and accompanying 2D bull's eye plot (see Fig. 23.3) of the myocardium, or heart muscle, of the left heart ventricle [14]. This 2D plot is a simple but great example of reducing complex 3D data to a standardized 2D representation that greatly facilitates the interpretation of that data. Another well-known example is that of curved planar reformation, or CPR, where volume data is sampled along a curved plane following the trajectory of a blood vessel or other tubular structure, thus enabling the study of the vessel and its surroundings with the minimum of interaction [36]. Other good examples of reformation can also be found in brain flattening [22] and colon unfolding [72].

Recently, the idea of intelligently reformatting or mapping 3D data was further explored by Neugebauer et al. [53] with aneurysm maps for the visualization of complex blood flow simulation data on the inside surfaces of aneurysms and by Rieder et al. [59] with their tumor maps for the post-operative assessment of radiofrequency ablation therapy. These types of reformations and mappings entail that more effort has to be put into carefully designing simplified, usually 2D, representations of complex 3D data, as opposed to, for example, the relatively straight-forward projection of volume data. The resultant visualization, if done right, requires less or no interaction and by definition avoids a number of problems inherent in 3D representations [74].

23.3.7 Illustrative Visualization in Medicine

For a long time, it was not possible to apply illustrative visualization techniques in practice due to performance constraints. With advances in graphics hardware and algorithms, such as GPU raycasting [45], it is now feasible from a computational standpoint. Now that computational problems have been largely solved, illustrative visualization approaches have to be finetuned and evaluated for diagnostic and treatment planning purposes. Recent examples of such work include the simulation of crepuscular rays for tumor accessibility planning [37] and multi-modal illustrative volume rendering for neurosurgical tumor treatment [58].

Illustrative medical visualization becomes increasingly important when visualizations become more complex and multi-modal, integrating functional (measured and simulated) information, anatomical information and, for example, surgical instruments. Illustration techniques enable visual representations to be simplified intelligently by the visualization designer, whilst still communicating as much information as possible. An example of this is the work of Zachow et al. [77] on the visualization of nasal air flow simulation where boundary enhancement was used as an illustrative technique to convey the simulated flow and the anatomy simultaneously.

23.3.8 Hyper-Realism

Analogous to the case of illustrative visualization, the rapid development in graphics hardware and algorithms has now enabled the *interactive* rendering of medical imaging datasets with physically-based lighting [44]. Figure 23.4 shows an example of such a visualization. These techniques make possible the simulation of an arbitrary number of arbitrarily shaped and textured lights, real shadows, a realistic camera model with lens and aperture, and so forth, all at interactive rates.

These techniques enable not only photo-realism, but also a technical form of *hyper-realism* in art, where it is possible to enhance visualizations with *additional* realistic detail in order to better convey information. Whilst there are strong indications that, for example, global illumination and shadows can have a positive effect on task performance in normal volume rendering [48], the possibilities and value of hyper-realistic effects in medical visualization need to be explored.

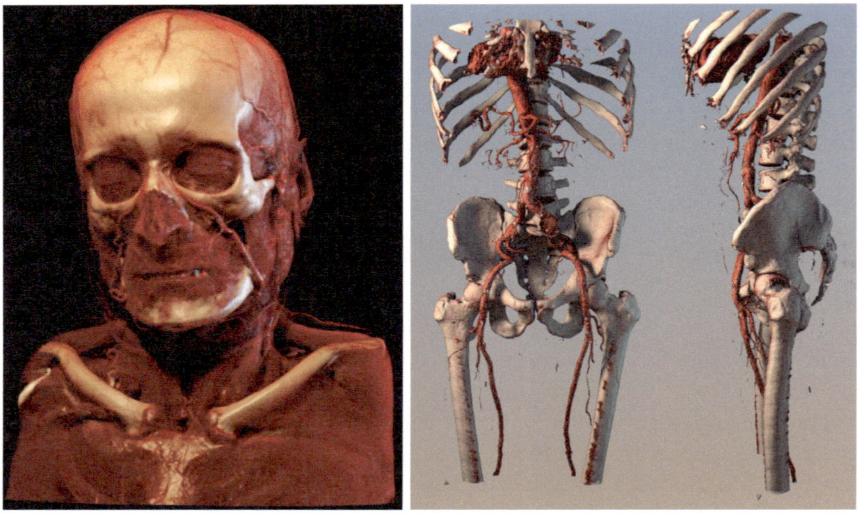

Fig. 23.4 Two examples of *interactive* visualizations made with the volume renderer of Kroes et al. [44]. Through the use of GPUs, physically-based lighting has become possible in an interactive volume rendering setting, enabling increased realism through soft shadows, depth of field and in this case mixed phase function and BRDF surface scattering

23.3.9 Visual Analysis in Healthcare

Visual analysis is becoming an essential component of medical visualization due to the rapidly growing role and availability of complex multi-dimensional, time-varying, mixed-modality, simulation and multi-subject datasets. In our view, the magnitude and especially the heterogeneity of the data necessitate the use of visual analysis techniques.

Existing examples involving time-varying data include the work of Coto et al. [16] on DCE-MRI mammography and Oeltze et al. [54] on perfusion data in general and MR perfusion of the brain in particular. Blaas et al. applied visual analysis techniques to multi-modal medical data, whilst Zachow et al. [77] focused on nasal airflow simulation data combined with anatomical information.

There is great potential for visual analysis in medical visualization, with clinical applications including advanced diagnosis and medical research and, even more importantly, treatment planning and evaluation, e.g. radio therapy planning and post-chemotherapy evaluation. The new Visual Analysis in Healthcare (VAHC) workshops that were held at IEEE VisWeek in 2010 and 2011 underline the emerging importance of this research direction.

23.3.10 Population Imaging

In population imaging, medical image data and other measurements are acquired of a large group of subjects, typically more than one thousand, over a longer period, typically years, in order to study the onset and progression of disease, general aging effects, and so forth in larger groups of people. Examples include the Rotterdam Scan Study focusing on neuro-degeneration [46] and the Study of Health In Pomerania (SHIP) focusing on general health [35].

This application domain is an extreme example of multi-subject medical visualization discussed in Sect. 23.2, integrating large quantities of heterogeneous, multi-modal and multi-timepoint data acquired of a large group of subjects. The scientists running these studies usually do not formulate strictly-defined hypotheses beforehand, instead opting for meticulous data acquisition, followed by an extended period of analysis in order to extract patterns and hypotheses from the data. Recently, Steenwijk et al. [63] set the first steps for the visualization of population imaging by applying visual analysis techniques to cohort study imaging data. The extreme heterogeneity and magnitude of the data, coupled with the explorative nature of the research, renders this a promising long-term application domain for visual analysis and medical visualization.

23.4 Conclusions

In this chapter, we gave a compact overview of the history of medical visualization research, spanning the past 30 years. Based on this history and on our own observations working in the field, we then identified and discussed the research challenges of the coming decade.

Our discussion of classic medical visualization problems related to efficient and high quality display of *one* static dataset was brief. We devoted more space to data that change over time, to the integration of anatomy with simulation and finally to cohort studies. We refer to problems where such time-dependent and high-dimensional data are employed as *MedVis 2.0* problems. While the classic problems are—from an application perspective—solved, there are many research opportunities in MedVis 2.0 problems. These data are significantly more difficult to analyze, to process and to visualize. Time-dependent MRI data, e.g., exhibit all artifacts of static MRI data but a number of additional artifacts, e.g. due to motion. Integrated analysis and visualization is a key feature of MedVis 2.0 solutions. In general, successful solutions to these problems require a considerably deeper understanding of the medical background and thus favor close collaborations with medical doctors over merely having access to medical data.

References

1. Altobelli, D.E., Kikinis, R., Mulliken, J.B., Cline, H., Lorensen, W., Jolesz, F.: Computer-assisted three-dimensional planning in craniofacial surgery. Plast. Reconstr Surg **92**(4), 576–585; discussion 586–587 (1993)
2. Bajaj, C.L., Pascucci, V., Schikore, D.R.: The contour spectrum. In: Proceedings of IEEE Conference on Visualization, 1997. Visualization'97, pp. 167–173. IEEE (1997). doi:10.1109/VISUAL.1997.663875
3. Bartz, D.: Virtual endoscopy in research and clinical practice. Comput. Graph. Forum **24**(1), 111–126 (2005). doi:10.1111/j.1467-8659.2005.00831.x
4. Basser, P.: Fiber-tractography via diffusion tensor MRI (DT-MRI). In: Proceedings of the 6th Annual Meeting ISMRM, Sydney, Australia, vol. 1226 (1998)
5. Basser, P., Mattiello, J., LeBihan, D.: MR diffusion tensor spectroscopy and imaging. Biophys. J. **66**(1), 259–267 (1994). doi:10.1016/S0006-3495(94)80775-1
6. Basser, P.J., Pajevic, S., Pierpaoli, C., Duda, J., Aldroubi, A.: In vivo fiber tractography using DT-MRI data. Magn. Resonance Med. **44**(4), 625–632 (2000). (Official Journal of the Society of Magnetic Resonance in Medicine / Society of Magnetic Resonance in Medicine)
7. Behrens, U., Teubner, J., Evertsz, C., Walz, M., Jürgens, H., Peitgen, H.O.: Computer-assisted dynamic evaluation of contrast-enhanced-MRI. In: Proceedings of Computer Assisted Radiology, pp. 362–367 (1996)
8. Beyer, J., Hadwiger, M., Wolfsberger, S., Buhler, K.: High-quality multimodal volume rendering for preoperative planning of neurosurgical interventions. IEEE Trans. Vis. Comput. Graph. **13**(6), 1696–1703 (2007). doi:10.1109/TVCG.2007.70560
9. Blaas, J., Botha, C.P., Post, F.H.: Interactive visualization of multi-field medical data using linked physical and feature-space views. In: K. Museth, A. Ynnerman, T. Möller (eds.) Proceedings of Eurographics / IEEE-VGTC EuroVis, pp. 123–130 (2007)
10. Bruckner, S., Grimm, S., Kanitsar, A., Gröller, M.: Illustrative context-preserving volume rendering. Proc. EUROVIS **2005**, 69–76 (2005)
11. Bruckner, S., Grimm, S., Kanitsar, A., Gröller, M.E.: Illustrative context-preserving exploration of volume data. IEEE Trans. Vis. Comput. Graph. **12**(6), 1559–1569 (2006). doi:http://doi.ieeecomputersociety.org/10.1109/TVCG.2006.96
12. Bruckner, S., Gröller, E.: Enhancing depth-perception with flexible volumetric halos. IEEE Trans. Vis. Comput. Graph. **13**(6), 1344–1351 (2007). doi:http://doi.ieeecomputersociety.org/10.1109/TVCG.2007.70555
13. Cai, W., Sakas, G.: Data intermixing and multi-volume rendering. Comput. Graph. Forum **18**(3), 359–368 (1999). doi:10.1111/1467-8659.00356
14. Cerqueira, M.D., Weissman, N.J., Dilsizian, V., Jacobs, A.K., Kaul, S., Laskey, W.K., Pennell, D.J., Rumberger, J.A., Ryan, T., Verani, M.S.: Standardized myocardial segmentation and nomenclature for tomographic imaging of the heart. Circulation **105**(4), 539–542 (2002). doi:10.1161/hc0402.102975
15. Chan, M.Y., Qu, H., Chung, K.K., Mak, W.H., Wu, Y.: Relation-aware volume exploration pipeline. IEEE Trans. Vis. Comput. Graph. **14**(6), 1683–1690 (2008). doi:10.1109/TVCG.2008.159
16. Coto, E., Grimm, S., Bruckner, S., Gröller, E., Kanitsar, A., Rodriguez, O.: MammoExplorer: an advanced CAD application for breast DCE-MRI. In: Proceedings of Vision, Modelling, and Visualization (VMV), pp. 91–98 (2005)
17. Csébfalvi, B., Mroz, L., Hauser, H., König, A., Gröller, E.: Fast visualization of object contours by Non-Photorealistic volume rendering. Comput. Graph. Forum **20**(3), 452–460 (2001). doi:10.1111/1467-8659.00538
18. Dick, C., Georgii, J., Burgkart, R., Westermann, R.: Stress tensor field visualization for implant planning in orthopedics. IEEE Trans. Vis. Comput. Graph. **15**(6), 1399–1406 (2009). doi:10.1109/TVCG.2009.184
19. Drebin, R.A., Carpenter, L., Hanrahan, P.: Volume rendering. SIGGRAPH Comput. Graph. **22**(4), 65–74 (1988). doi:10.1145/378456.378484

20. Ebert, D., Rheingans, P.: Volume illustration: non-photorealistic rendering of volume models. In: Proceedings of the conference on visualization '00. VIS '00, pp. 195–202. IEEE Computer Society Press, Los Alamitos, CA, USA (2000)
21. Falcão, A.X., Udupa, J.K., Samarasekera, S., Sharma, S., Hirsch, B.E., Lotufo, R.D.A.: User-steered image segmentation paradigms: Live wire and live lane. Graph. Models Image Process. **60**(4), 233–260 (1998). doi:10.1006/gmip.1998.0475
22. Fischl, B., Sereno, M.I., Dale, A.M.: Cortical surface-based analysis: II: inflation, flattening, and a surface-based coordinate system. NeuroImage **9**(2), 195–207 (1999). doi:10.1006/nimg.1998.0396
23. Gerig, G., Koller, T., Székely, G., Brechbühler, C., Kübler, O.: Symbolic description of 3-d structures applied to cerebral vessel tree obtained from MR angiography volume data. In: Barrett, H.H., Gmitro, A.F. (eds.) Inf. Process. Med. Imaging, vol. 687, pp. 94–111. Springer, Berlin (1993)
24. Greicius, M.D., Krasnow, B., Reiss, A.L., Menon, V.: Functional connectivity in the resting brain: A network analysis of the default mode hypothesis. Proc. Natl. Acad. Sci. **100**(1), 253–258 (2003). doi:10.1073/pnas.0135058100
25. Hagmann, P., Jonasson, L., Maeder, P., Thiran, J.P., Wedeen, V.J., Meuli, R.: Understanding diffusion MR imaging techniques: from scalar diffusion-weighted imaging to diffusion tensor imaging and beyond. Radiograph. Rev. Publ. Radiol. Soc. North Am. Inc. **26**(Suppl 1), S205–S223 (2006). doi:10.1148/rg.26si065510
26. Höhne, K.H., Bernstein, R.: Shading 3D-Images from CT using gray-level gradients. IEEE Trans. Med. Imaging **5**, 45–47 (1986). doi:10.1109/TMI.1986.4307738
27. Höhne, K.H., Bomans, M., Tiede, U., Riemer, M.: Display of multiple 3D-objects using the generalized voxel-model. In: Medical Imaging II, Part B, Proceedings of SPIE 914, pp. 850–854. Newport Beach (1988)
28. Hong, L., Kaufman, A., Wei, Y.C., Viswambharan, A., Wax, M., Liang, Z.: 3D virtual colonoscopy. In: Proceedings of Biomedical Visualization, 1995, pp. 26–32, 83. IEEE (1995). doi:10.1109/BIOVIS.1995.528702
29. Hong, L., Muraki, S., Kaufman, A., Bartz, D., He, T.: Virtual voyage: interactive navigation in the human colon. In: Proceedings of the 24th Annual Conference on Computer Graphics and Interactive Techniques, SIGGRAPH'97, pp. 27–34. ACM Press/Addison-Wesley Publishing Co., New York, NY, USA (1997). doi:10.1145/258734.258750
30. Hong, W., Qiu, F., Kaufman, A.: A pipeline for computer aided polyp detection. IEEE Trans. Vis. Comput. Graph. **12**(5), 861–868 (2006). doi:10.1109/TVCG.2006.112
31. Hsiao, E.M., Rybicki, F.J., Steigner, M.: CT coronary angiography: 256-slice and 320-detector row scanners. Current Cardiol. Reports **12**(1), 68–75 (2010). doi:10.1007/s11886-009-0075-z.PMID:20425186PMCID:2893879
32. Interrante, V., Fuchs, H., Pizer, S.: Enhancing transparent skin surfaces with ridge and valley lines. In: IEEE Conference on Visualization, 1995. Proceedings of Visualization'95, pp. 52–59, 438 (1995). doi:10.1109/VISUAL.1995.480795
33. Jeong, W., Beyer, J., Hadwiger, M., Blue, R., Law, C., Vazquez, A., Reid, C., Lichtman, J., Pfister, H.: SSECRETT and NeuroTrace: interactive visualization and anaysis tools for large-scale neuroscience datasets. IEEE Comput. Graph. Appl. **99**, 1–1 (2011). doi:10.1109/MCG.2011.33
34. Jeong, W.K., Schneider, J., Turney, S.G., Faulkner-Jones, B.E., Meyer, D., Westermann, R., Reid, R.C., Lichtman, J., Pfister, H.: Interactive histology of large-scale biomedical image stacks. IEEE Trans. Vis. Comput. Graph. **16**(6), 1386–1395 (2010). doi:10.1109/TVCG.2010.168
35. John, U., Hensel, E., Lüdemann, J., Piek, M., Sauer, S., Adam, C., Born, G., Alte, D., Greiser, E., Haertel, U., Hense, H.W., Haerting, J., Willich, S., Kessler, C.: Study of health in pomerania (SHIP): a health examination survey in an east german region: objectives and design. Soc. Preventive Med. **46**, 186–194 (2001). doi:10.1007/BF01324255
36. Kanitsar, A., Fleischmann, D., Wegenkittl, R., Felkel, P., Gröller, M.E.: CPR: curved planar reformation. In: Proceedings of the Conference on Visualization '02. VIS'02, pp. 37–44. IEEE Computer Society, Washington, DC, USA (2002)

37. Khlebnikov, R., Kainz, B., Muehl, J., Schmalstieg, D.: Crepuscular rays for tumor accessibility planning. IEEE Trans. Vis. Comput. Graph. **17**(12), 2163–2172 (2011). doi:10.1109/TVCG.2011.184
38. Kindlmann, G., Weinstein, D.: Hue-balls and lit-tensors for direct volume rendering of diffusion tensor fields. In: Proceedings of the Conference on Visualization '99: Celebrating Ten Years. VIS '99, pp. 183–189. IEEE Computer Society Press, Los Alamitos, CA, USA (1999)
39. Kindlmann, G., Weinstein, D., Hart, D.: Strategies for direct volume rendering of diffusion tensor fields. IEEE Trans. Vis. Comput. Graph. **6**(2), 124–138 (2000). doi:10.1109/2945.856994
40. Kindlmann, G., Whitaker, R., Tasdizen, T., Moller, T.: Curvature-based transfer functions for direct volume rendering: Methods and applications. In: Proceedings of the 14th IEEE Visualization 2003 (VIS'03), VIS '03, p. 67. IEEE Computer Society, Washington, DC, USA (2003). doi:10.1109/VISUAL.2003.1250414
41. Kok, P., Baiker, M., Hendriks, E.A., Post, F.H., Dijkstra, J., Löwik, C.W., Lelieveldt, B.P., Botha, C.P.: Articulated planar reformation for change visualization in small animal imaging. IEEE Trans. Vis. Comput. Graph. **16**(6), 1396–1404 (2010). doi:10.1109/TVCG.2010.134
42. Kok, P., Dijkstra, J., Botha, C.P., Post, F.H., Kaijzel, E., Que, I., Lowik, C., Reiber, J., Lelieveldt, B.P.F.: Integrated visualization of multi-angle bioluminescence imaging and micro CT. In: Cleary, K.R., Miga, M.I. (eds.) Proceedings of SPIE Medical Imaging 2007, vol. 6509 (2007)
43. Krekel, P.R., Botha, C.P., Valstar, E.R., Bruin, P.W.d., Rozing, P.M., Post, F.H.: Interactive simulation and comparative visualisation of the bone-determined range of motion of the human shoulder. In: T. Schulze, G. Horton, B. Preim, S. Schlechtweg (eds.) Proceedings of Simulation and Visualization, pp. 275–288. SCS Publishing House Erlangen (2006). Best Paper Award
44. Kroes, T., Post, F.H., Botha, C.P.: Exposure render: an interactive photo-realistic volume rendering framework. PLoS ONE **7**, e38586 (2012)
45. Krüger, J., Westermann, R.: Acceleration techniques for GPU-based volume rendering. In: IEEE Visualization Conference, p. 38. IEEE Computer Society, Los Alamitos, CA, USA (2003). doi:http://doi.ieeecomputersociety.org/10.1109/VIS.2003.10001
46. de Leeuw, F.E., de Groot, J.C., Achten, E., Oudkerk, M., Ramos, L.M.P., Heijboer, R., Hofman, A., Jolles, J., van Gijn, J., Breteler, M.M.B.: Prevalence of cerebral white matter lesions in elderly people: a population based magnetic resonance imaging study. the rotterdam scan study. J. Neurol. Neurosurg. Psychiatry **70**(1), 9–14 (2001). doi:10.1136/jnnp.70.1.9
47. Levoy, M.: Display of surfaces from volume data. IEEE Comput. Graph. Appl. **8**(3), 29–37 (1988). doi:10.1109/38.511
48. Lindemann, F., Ropinski, T.: About the influence of illumination models on image comprehension in direct volume rendering. IEEE Trans. Vis. Comput. Graph. **17**(12), 1922–1931 (2011). doi:10.1109/TVCG.2011.161
49. Lorensen, W.E., Cline, H.E.: Marching cubes: a high resolution 3D surface construction algorithm. SIGGRAPH Comput. Graph. **21**(4), 163–169 (1987). doi:10.1145/37402.37422
50. Lundström, C., Rydell, T., Forsell, C., Persson, A., Ynnerman, A.: Multi-touch table system for medical visualization: application to orthopedic surgery planning. IEEE Trans. Vis. Comput. Graph. **17**(12), 1775–1784 (2011). doi:10.1109/TVCG.2011.224
51. Mortensen, E.N., Barrett, W.A.: Intelligent scissors for image composition. In: Proceedings of the 22nd Annual Conference on Computer graphics and Interactive Techniques, SIGGRAPH'95, pp. 191–198. ACM, New York, NY, USA (1995). doi:10.1145/218380.218442
52. Moser, M.,Weiskopf, D.: Interactive volume rendering on mobile devices. In: Proceedings of Vision, Modeling, and Visualization 2008, 8–10 Oct 2008, Konstanz, Germany, p. 217 (2008)
53. Neugebauer, M., Gasteiger, R., Beuing, O., Diehl, V., Skalej, M., Preim, B.: Map displays for the analysis of scalar data on cerebral aneurysm surfaces. Comput. Graph. Forum **28**(3), 895–902 (2009). doi:10.1111/j.1467-8659.2009.01459.x
54. Oeltze, S., Doleisch, H., Hauser, H., Muigg, P., Preim, B.: Interactive visual analysis of perfusion data. IEEE Trans. Vis. Comput. Graph. **13**(6), 1392–1399 (2007). doi:10.1109/TVCG.2007.70569
55. Olabarriaga, S., Smeulders, A.: Interaction in the segmentation of medical images: a survey. Med. Image Anal. **5**(2), 127–142 (2001). doi:10.1016/S1361-8415(00)00041-4

56. Prassni, J.S., Ropinski, T., Hinrichs, K.: Uncertainty-aware guided volume segmentation. IEEE Trans. Vis. Comput. Graph. **16**(6), 1358–1365 (2010). doi:10.1109/TVCG.2010.208
57. Preim, B., Bartz, D.: Visualization in Medicine. Morgan Kaufmann, Burlington (2007)
58. Rieder, C., Ritter, F., Raspe, M., Peitgen, H.: Interactive visualization of multimodal volume data for neurosurgical tumor treatment. Comput. Graph. Forum **27**, 1055–1062 (2008)
59. Rieder, C., Weihusen, A., Schumann, C., Zidowitz, S., Peitgen, H.: Visual support for interactive Post-Interventional assessment of radiofrequency ablation therapy. Comput. Graph. Forum **29**(3), 1093–1102 (2010). doi:10.1111/j.1467-8659.2009.01665.x
60. Saad, A., Hamarneh, G., Möller, T.: Exploration and visualization of segmentation uncertainty using shape and appearance prior information. IEEE Trans. Vis. Comput. Graph. **16**(6), 1366–1375 (2010). doi:10.1109/TVCG.2010.152
61. Saito, T., Takahashi, T.: Comprehensible rendering of 3-d shapes. SIGGRAPH Comput. Graph. **24**(4), 197–206 (1990). doi:10.1145/97880.97901
62. Shinagawa, Y., Kunii, T.L.: Constructing a Reeb graph automatically from cross sections. IEEE Comput. Graph. Appl. **11**(6), 44–51 (1991). doi:10.1109/38.103393
63. Steenwijk, M.D., Milles, J., Buchem, M.A., Reiber, J.H., Botha, C.P.: Integrated visual analysis for heterogeneous datasets in cohort studies. In: IEEE VisWeek Workshop on Visual Analytics in Health Care (2010)
64. Sunguroff, A., Greenberg, D.: Computer generated images for medical applications. In: Proceedings of the 5th Annual Conference on Computer Graphics and Interactive Techniques, SIGGRAPH'78, pp. 196–202. ACM, New York, NY, USA (1978). doi:10.1145/800248.807390
65. Takahashi, S., Fujishiro, I., Takeshima, Y., Bi, C.: Previewing volume decomposition through optimal viewpoints. In: Scientific Visualization: Interactions, Features, Metaphors (2011)
66. Taylor, C., Draney, M., Ku, J., Parker, D., Steele, B., Wang, K., Zarins, C.: Predictive medicine: computational techniques in therapeutic decision-making. Comput. Aided Surg. **4**(5), 231–247 (1999)
67. Termeer, M., Bescos, J.O., Breeuwer, M., Vilanova, A., Gerritsen, F., Gröller, M.E.: CoViCAD: comprehensive visualization of coronary artery disease. IEEE Trans. Vis. Comput. Graph. **13**(6), 1632–1639 (2007). doi:10.1109/TVCG.2007.70550
68. Tietjen, C., Isenberg, T., Preim, B.: Combining silhouettes, surface, and volume rendering for surgery education and planning. In: IEEE/Eurographics Symposium on Visualization (EuroVis), pp. 303–310 (2005)
69. Tory, M., Röber, N., Möller, T., Celler, A., Atkins, M.S.: 4D space-time techniques: a medical imaging case study. In: Proceedings of IEEE Visualization 2001, pp. 473–476. IEEE Computer Society, Washington, DC, USA (2001)
70. Tuch, D.S., Reese, T.G., Wiegell, M.R., Makris, N., Belliveau, J.W., Wedeen, V.J.: High angular resolution diffusion imaging reveals intravoxel white matter fiber heterogeneity. Magn. Reson. Med. **48**(4), 577–582 (2002). doi:10.1002/mrm.10268
71. Vannier, M.W., Marsh, J.L., Warren, J.O.: Three dimensional computer graphics for craniofacial surgical planning and evaluation. SIGGRAPH Comput. Graph. **17**(3), 263–273 (1983). doi:10.1145/964967.801157
72. Vilanova, A., Wegenkittl, R., König, A., Gröller, E.: Nonlinear virtual colon unfolding. In: Proceedings of the Conference on Visualization'01. VIS '01, pp. 411–420. IEEE Computer Society, Washington, DC, USA (2001)
73. Wang, T.D., Plaisant, C., Quinn, A.J., Stanchak, R., Murphy, S., Shneiderman, B.: Aligning temporal data by sentinel events: discovering patterns in electronic health records. In: Proceedings of the Twenty-Sixth Annual SIGCHI Conference on Human Factors in Computing Systems, CHI '08, pp. 457–466. ACM, New York, NY, USA (2008). doi:10.1145/1357054.1357129
74. Ware, C.: Designing with a 2 1/2-d attitude. Inf. Des. J. **10**(3), 258–265 (2001). doi:10.1075/idj.10.3.07war
75. Weber, G.H., Dillard, S.E., Carr, H., Pascucci, V., Hamann, B.: Topology-controlled volume rendering. IEEE Trans. Vis. Comput. Graph. **13**(2), 330–341 (2007). doi:10.1109/TVCG.2007.47

76. Weinstein, D., Kindlmann, G., Lundberg, E.: Tensorlines: advection-diffusion based propagation through diffusion tensor fields. In: Proceedings of the Conference on Visualization '99: Celebrating Ten Years. VIS '99, pp. 249–253. IEEE Computer Society Press, Los Alamitos, CA, USA (1999)
77. Zachow, S., Muigg, P., Hildebrandt, T., Doleisch, H., Hege, H.C.: Visual exploration of nasal airflow. IEEE Trans. Vis. Comput. Graph. **15**(6), 1407–1414 (2009). doi:http://doi.ieeecomputersociety.org/10.1109/TVCG.2009.198

Chapter 24
The Ultrasound Visualization Pipeline

Åsmund Birkeland, Veronika Šoltészová, Dieter Hönigmann, Odd Helge Gilja, Svein Brekke, Timo Ropinski and Ivan Viola

Abstract Radiology is one of the main tools in modern medicine. A numerous set of deceases, ailments and treatments utilize accurate images of the patient. Ultrasound is one of the most frequently used imaging modality in medicine. The high spatial resolution, its interactive nature and non-invasiveness makes it the first choice in many examinations. Image interpretation is one of ultrasound's main challenges. Much training is required to obtain a confident skill level in ultrasound-based diagnostics. State-of-the-art graphics techniques is needed to provide meaningful visualizations of ultrasound in real-time. In this paper we present the process-pipeline for ultrasound visualization, including an overview of the tasks performed in the specific steps. To provide an insight into the trends of ultrasound visualization research, we have selected a set of significant publications and divided them into a technique-based taxonomy covering the topics pre-processing, segmentation, registration, rendering and augmented reality. For the different technique types we discuss the difference between ultrasound-based techniques and techniques for other modalities.

Å. Birkeland (✉) · V. Šoltészová · S. Brekke · I. Viola
University of Bergen, Bergen, Norway
e-mail: asmund.birkeland@uib.no

D. Hönigmann
N22 Research and Technology Transfer, Wiener Neustadt, Austria

O. H. Gilja
Haukeland University Hospital, Bergen, Norway

S. Brekke
Archer—The Well Company, Bergen, Norway

T. Ropinski
Linköping University, Linköping, Sweden

V.Šoltészová
Christian Michelsen Research, Bergen, Norway

I. Viola
Vienna University of Technology,Vienna, Austria

C.D. Hansen et al. (eds.), *Scientific Visualization*, Mathematics and Visualization,
DOI 10.1007/978-1-4471-6497-5_24

24.1 Introduction

Medical ultrasound has a strong impact on clinical decision making and its high significance in patient management is well established [49, 50]. Ultrasonography (US) has in comparison with CT, MRI, SPECT and PET scanning very favorable cost, great availability world-wide, high flexibility, and extraordinary patient friendliness. In addition to these factors, ultrasonography stands out as the imaging method with the highest temporal resolution and also often the best spatial resolution. Furthermore, ultrasonography is a clinical method that easily can be applied bedside, even using mobile, hand-carried scanners [23] and even pocket sized scanners [19], thus expanding the field of applications considerably. However, a low signal-to-noise ratio, "shadowing" and the relatively small scan sector make ultrasound images very difficult to interpret. Accordingly, improved visualization of the broad spectrum of ultrasound images has a great potential to further increase the impact of ultrasonography in medicine.

As advancement of technology is fertilizing and stimulating medical development, there is a continuous need for research and new applications in visualization. Visualization methods have the capacity to transform complex data into graphical representations that enhance the perception and meaning of the data [22]. Ordinary ultrasound scanning produces real-time 2D slices of data, and these dynamic sequences pose in itself a challenge to visualization methods. One example is functional ultrasonography (f-US), i.e. ultrasound imaging of (patho)physiology and/or organ function, in contrast to conventional imaging of anatomic structures. Using f-US, information on motility, biomechanics, flow, perfusion, organ filling and emptying can be obtained non-invasively [24, 57]. Moreover, the 2D images can be aligned to form 3D data sets. In such cases, 3D visualization provides added value in terms of a more holistic understanding of the data. Typical examples are demonstration of complex anatomy and pathology, pre-operative surgical planning or virtual training of medical students. Furthermore, there are now matrix 3D probes on the market that allow real-time 3D acquisition. To benefit from the high temporal resolution, advanced graphics techniques are required in ultrasound visualization, preventing the visualization technique from being the performance *bottleneck*. This opens up new challenges to the visualization community to develop fast and efficient algorithms for rendering on-the-fly.

In addition, co-registration techniques enable use of multi-modal data sets. Fusion imaging, where ultrasound is combined with either CT, MRI, or PET images, allows for more precise navigation in ultrasound-guided interventions. This challenging new arena demands advanced visualization research to enlighten how different data types can be combined and presented in novel ways.

The diversity of the ultrasound imaging technology provides a great tool for medical diagnostics, but the nature of the data can make it challenging to process. Techniques which work well for other modalities are being adapted to suit the special characteristics of ultrasound. In this chapter, we present an overview of the pipeline for advanced visualization specific to ultrasound data. The chapter is divided into the chosen taxonomy, in essence each step of the visualization pipeline; pre-processing, entation, registration, rendering and augmented reality.

24.2 Taxonomy

Techniques for ultrasound visualization can be categorized in a variety of ways, e.g., when they where developed, which types of data modalities were utilized, which anatomy the technique was focused on, etc. The development of new ultrasound technology leads to different visualization techniques. The step from 2D ultrasound images to 3D freehand ultrasound (2D ultrasound with position information) revealed new challenges as spatial information could be included to generate volumetric data. The recent development of 2D matrix probes provided again a new challenge of 3D + time (4D) data visualization. Karadayi et al. published a survey regarding 3D ultrasound [32]. This chapter has a greater focus on data acquisition and volume handling, but also gives a brief overview of visualization of 3D ultrasound data.

Another taxonomic scheme for ultrasound visualization is based on the different types of data the technique utilized. 3D freehand and 4D ultrasound pose very different challenges compared to 2D ultrasound or when handling multiple modalities. Blending B-mode ultrasound for tissue and Doppler ultrasound for blood flow can be challenging enough in 2D and even more in 3D. An example image is shown in Fig. 24.1d. In addition to the ultrasound input, the combination of other medical imaging modalities, such as CT or MRI with ultrasound, provide more information, but also more challenges to the visualization researcher.

Different anatomic regions have different characteristics in ultrasound images, as can be seen in Fig. 24.1. For instance, in a liver scan one might look for tumors using a high-resolution abdominal 2D probe. For heart infarctions, the doctor might need to examine the strain in the heart muscle to detect defective muscle tissue. The large difference between tissue and pathology leads to anatomically specific visualization techniques.

In this survey we categorized around 60 papers and from the different categories we generated a parallel-coordinate plot, shown in Fig. 24.2. Looking at the graph, we

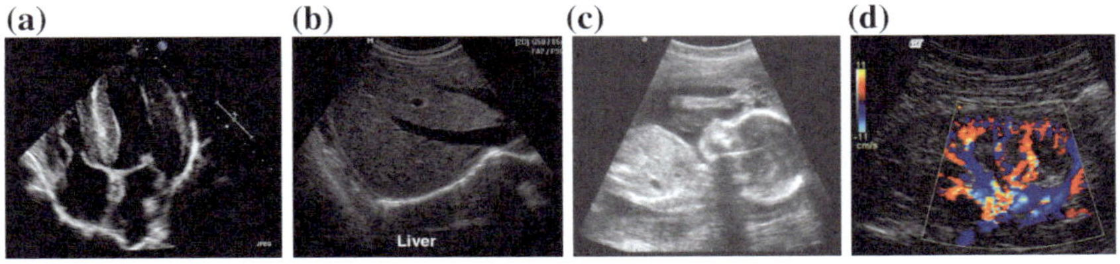

Fig. 24.1 Example ultrasound images from the cardiac (**a**), gastric (**b**), fetal (**c**) and Blood flow (**d**) domain

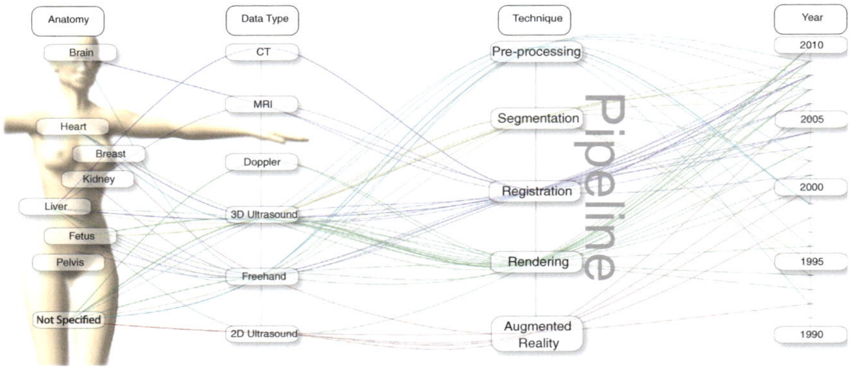

Fig. 24.2 The different classifications shown in a parallel-coordinate plot. The *colors* depict which technique a publication has given the most weight

see an increase in rendering techniques for 3D ultrasound in the last five years. Volume rendering is often considered to be a *solved* problem. However, our study shows that much work dealt with volumetric ultrasound data. Yet, 3D ultrasound rendering can still not be considered a solved problem. The high presence of noise, shadows from hyper-echoic areas and inconsistent data values provide a great challenge to make 3D ultrasound a more easy-to-use tool for examiners.

We also see an absence of augmented reality techniques for 3D ultrasound. Yet another trend is the neglection of 2D ultrasound from the visualization community. 2D ultrasound is the most used modality by physicians and while presenting the signal data onto the screen is straight forward, understanding what you see is not so trivial. Increasing the readability of 2D ultrasound is mostly worked on in the commercial domain, aiming to give a company an edge over its rivals.

In Fig. 24.2 we see the categorized papers in a parallel coordinate plot where each axis corresponds to the different taxonomy classification. The third axis (the pipeline axis) is selected as the classification for this survey. Five categories where chosen based on what we recognize as the essential parts in the visualization pipeline for ultrasound data:

- *Pre-processing*: Processing ultrasound data prior to segmentation, registration or rendering.
- *Segmentation*: Extracting features from ultrasound data.
- *Registration*: Combining ultrasound data with other types of medical imaging modalities.
- *Rendering*: Presenting ultrasound data.
- *Augmented Reality*: Mixing ultrasound rendering with natural vision.

In the following sections we motivate the need for each of the major topics, focusing on significant techniques and how they deal with the characteristics of ultrasound data.

24.3 Pre-processing

3D ultrasound is often employed in clinical diagnostic imaging. If a dedicated 3D probe is unavailable, 3D volumes can be acquired using freehand ultrasound systems; a 2D probe with an attached tracker which places and orients the 2D images in 3D space. Volume compounding consists of two levels: acquisition and reconstruction. Precise reconstruction requires calibration of the tracking system and correction of pressure-induced artifacts from the probe onto the skin.

Ultrasound allows for extracting more information, such as tissue strain. Strain is a tissue-deformation property and can be used to detect functional deficiencies, e.g., from myocardial infarction. Strain determination via tissue tracking is a complex task and can be done by using tissue Doppler [27]. Deprez et al. advanced in 3D strain estimation by providing a better out-of-plane motion estimation [12]. Visualization of strain has, however, stagnated compared to the development of technology and is mostly depicted by elementary color coding.

For freehand ultrasound systems, it is necessary to calibrate the position and orientation of the 2D image with respect to the tracking sensor. Wein and Khamene proposed to make two perpendicular sweeps through tissue containing well-visible structures [79]. They used an optimization strategy to maximize the similarity between two volumes reconstructed from each sweep.

To achieve the best possible quality of scans, the clinician presses the probe against the body. However, the human factor causes a non-constant pressure and different deformations of underlying structures in the body. Prager et al. correlated images in the sequence and used a rigid translation in the x and y directions followed by a non-rigid shift in depth z [58].

Ultrasound acquisition takes place in polar coordinates (ϕ, R) for 2D or (ϕ, ψ, R) for 3D. The angles ϕ and ψ correspond to the azimuth and elevation angles of the beam and R is the depth of the tissue boundary which has reflected the echo. In order to use off-the-shelf 3D volume rendering techniques, the grid must be scan-converted to a Cartesian lattice. This can be done as a preprocessing step or on-the-fly directly at the rendering stage.

This section is dedicated to selected methods for *volume reconstruction* from scan-converted freehand ultrasound and for *data enhancement* tailored for ultrasound volumes, which in the pipeline typically follow the reconstruction stage.

24.3.1 Reconstruction

Volume reconstruction from a set of 2D images needs to solve several important problems. Each image must be inserted precisely into the right spatial context. Space-filling between individual images is also crucial and the high framerate of 2D ultrasound implies speed requirements.

A detailed categorization of reconstruction algorithms was done by Rohling et al. [61] and Solberg et al. [70]. We adopt the categorization by Solberg et al. into *voxel-*, *pixel-* and *function-based* methods and complete it with recent works.

Voxel-based methods, i.e., *backward compounding*, run through the voxel grid and assign each of them a value estimated by an interpolation method such as the Stradx system [59]. It allows for real-time visualization of freehand ultrasound, including plane re-slicing based on nearest-neighbor interpolation and later also for direct volume rendering [58]. They blend images generated by re-slicing as described in their previous work. Gee et al. also used nearest neighbor interpolation for direct plane re-slicing [21]. The reconstructed plane is intended for direct viewing—implying only one re-sampling stage. Linear, bilinear and trilinear interpolation methods have also been used [6, 74]. Recent developments by Wein et al. improve both quality and performance by applying a backward-warping paradigm implemented on dedicated graphics hardware [80].

Karamalis et al. used interpolation on the GPU for high-quality volume reconstruction [33]. They select an optimal orientation of reconstruction slices based on the orientation of the scans and reconstruct the volume by following this direction. Each sampling layer is reconstructed from scans which intersect this layer by interpolating intensity values between the intersections. The visualization pipeline includes two re-sampling steps: one during the reconstruction and one while volume rendering.

Pixel-based methods, i.e., *forward compounding*, traverse each pixel of all acquired 2D images and update the value of one or several voxels of the target grid. Gobbi and Peters used splatting as a high-quality interpolation method and described a technique that operates in real-time while the data is captured [25].

Function-based methods employ a specific function to interpolate between voxels. In most applications, the shape of the underlying data is not considered. Rohling et al. investigated the quality of interpolation using splines, which is a polynomial function [60]. They compared this technique with other standard methods and showed that it produces more accurate reconstructions.

Tetrahedron-based methods reconstruct a 3D model built from tetrahedra using an iterative subdivision of an initial tetrahedron instead of a regular grid [63]. The subdivision terminates if all tetrahedra contain one data point. Each point is assigned a value which corresponds to the barycentric coordinates of the data point in this tetrahedron. This strategy is adaptive; the model adapts as new data is streamed in.

We listed selected algorithms in categories based on how they were implemented. If choosing a specific algorithm, one must choose between speed and quality. Solberg et al. compared the performance of some of the algorithms [70]. From all listed methods, the radial-based function reconstruction by Rohling et al. [61] delivers reconstructions of the best quality but it is also the most computationally expensive. However, the increasingly powerful dedicated graphics hardware for computational acceleration solves this problem.

24.3.2 Data Enhancement

Ultrasound is a challenging modality for visualization due to its natural properties such as low dynamic range, noisiness and speckle patterns [64]. Also, the geometric resolution varies with depth and the tissue boundaries can be several pixels wide depending on their orientation. Tissue boundaries can even disappear if they are parallel to the ultrasound beam. 2D images are preferred without filtering and enhancement. Speckle patterns refer to the texture of the tissue boundary, which is valuable information for clinicians. However, speckle in 3D brings no added value to the visualization and is considered as an artifact the same as noise. Therefore, prior to the rendering stage, the 3D data is filtered to enhance its quality.

For a review on early speckle reduction techniques, refer to the survey of Forsberg et al. [16]. Belohlavek et al. [5] use the *eight hull* algorithm with a geometric filter [10]. Recent techniques are based on region growing [9], adaptive filtering [67], compression techniques [26] and anisotropic diffusion filters [38].

Systems usually employ a blend of image-processing techniques to enhance the data. Sakas et al. listed techniques with a good trade-off between loss of information and quality [64]. They employed Gaussian filters for noise reduction, speckle-removal methods for contour smoothing and median filters for gap closing and noise reduction. Median filters remove small surface artifacts and preserve the sharpness of boundaries. There exist fast implementations where a histogram can be used to keep track of values [29]. Still, they require a more advanced memory management, making them less parallelizable than the evaluation of fast Gaussian filters. Lizzi and Feleppa described a technique to increase the axial resolution by processing the signal in the frequency domain. This resolution gain is especially valuable in opthalmology when visualizing thin layers within the cornea [45].

24.4 Segmentation

Selecting interesting features to be visualized is important to be able to root out the occluding elements from large datasets. For most modalities, segmentation can be performed by extracting regions with similar data values. For instance, because of the physical properties of x-rays, the data values in a CT scan are recorded into Hounsfield units which provide a good basis for binary thresholding techniques for certain tissue types. Early work indicated that binary thresholding techniques are not very well suited for ultrasound data [72]. More sophisticated techniques are required for satisfactory segmentation. An extensive survey on ultrasound image segmentation was presented by Noble and Boukerroui [48] in 2006. In this section we have focused on significant publications from recent years.

To increase robustness of the ultrasound segmentation, the various approaches are usually tailored for specific anatomies. Carneiro et al. have developed an automatic technique for segmenting the brain of a fetus [8]. By first detecting the cerebellum,

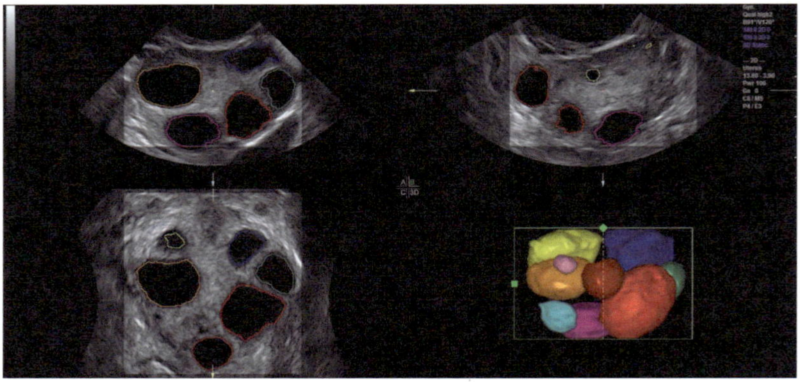

Fig. 24.3 Automatic segmentation of the ovarian follicles [13]

the system can *narrow down* the search for other features. On the other hand, segmentation is an extremely critical procedure which may obscure diagnostically relevant aspects of the anatomy under examination. Consequently, fully automatic segmentation techniques have not been implemented in clinical systems so far, with the exception of a method for follicle volumetry [13], as shown in Fig. 24.3.

A great challenge with ultrasound segmentation is that the data is dependent on many factors. For one, different positions and orientations of the probe, while looking at the same anatomical part, can provide very different images. Hyper-echoic regions cast shadows onto the tissue behind it according to the probe position. This alone, makes ultrasound segmentation data highly uncertain. Most segmentation techniques return a model with no indication of the uncertainty of the result. To compensate for the fuzzy nature of the ultrasound data, Petersch et al. developed a soft segmentation technique for 3D ultrasound data [56]. This technique calculates a probability map for 3D ultrasound data, which in turn can be used to create *soft* representations of the features extracted.

24.4.1 Clipping

Feature extraction can be computationally costly. In-vivo 3D ultrasound examination cannot always afford the extra time necessary to extract the interesting structures. Therefore, clipping is a commonly used tool in live visualization of 3D ultrasound. Interactively removing regions which are not interesting, the user gets a clear view of the features normally occluded. Sakas et al. developed a clipping tool in their ultrasound visualization system [64] which is nowadays a standard feature in commercial 3D ultrasound systems. The user can segment the dataset in-vivo using three approaches: drawing on one of the three axially-aligned slices, selecting everything along the current axis, and within the sketch. Another tool is based on sketching

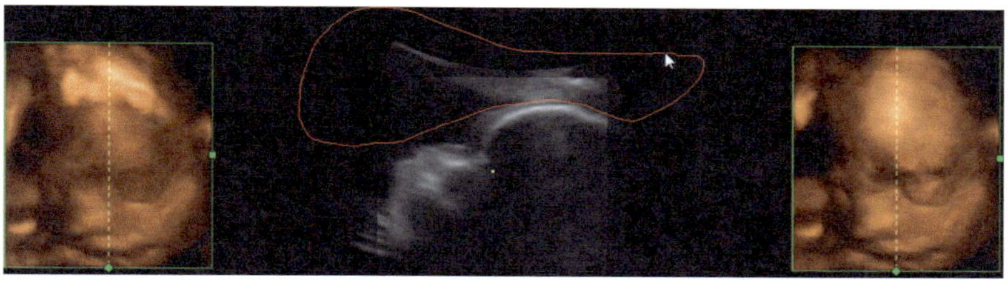

Fig. 24.4 Using MagiCut to Clip the volume, generating a clear view to the desired structure [1]

directly on the 3D rendered scene. Each voxel is then projected onto the screen and removed if it lies within the marked area. The third clipping tool is based on the distance from a single mouse-click on the view-plane. A hemispherical wave front is propagated from the seed-point and stops when the voxels reach a user-specified threshold. Figure 24.4 shows an example of clipping implemented in the GE Voluson machines [1].

24.5 Registration

Merging ultrasound with other modalities can be very beneficial. While ultrasound provides high resolution images at a high frame-rate, other modalities, such as MRI or CT can provide information complimentary to the ultrasound images. Data registration is the process of transforming different modalities into the same reference frame to achieve as much comprehensive information about the underlying structure as possible. While CT and MRI are typically pre-operative imaging techniques, ultrasound can easily be performed live during surgery. For instance, the radiation from CT is dangerous and the large electro magnets in an MRI scanner require that everything in the room is non-magnetic. Recently Curiel et al. built a non-magnetic ultrasound scanner for proper simultaneous intra-operative imaging [11]. Though there was some electric interference between the two modalities, the technique is promising, although availability will most likely be very low.

Nikas et al. published an evaluation of the application of co-registered 2D ultrasound and MRI for intra-operative navigation [47]. Ultrasound based navigation shows promising results due to live acquisition at high frame rates and easy portability. For prostate brachytherapy a combination of ultrasound and co-registered CT can be used, as shown by Fuller et al. [17]. Existing commercial products apply optical tracking for intra-operative navigation during neurosurgery [71]. Figure 24.5 shows how ultrasound and CT can be blended together into a single reference frame [7].

Registration can be divided into two different types: Rigid and non-rigid. Rigid registration can be used to quickly obtain a registration between two modalities and is suitable for rigid anatomies such as the skull. A common approach to register two images is to search for the transformation which minimizes a difference function,

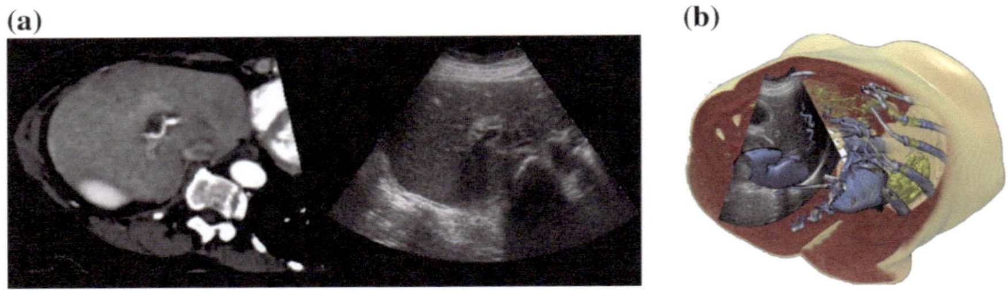

Fig. 24.5 Registering ultrasound and CT **a**: Slice-view of a CT scan co-registered with 2D ultrasound. **b**: Cut-away view of a CT scan co-registered with 2D ultrasound [7]

for instance sum-of-square-differences. Direct image based registration between ultrasound and CT or MRI can be difficult due to the different nature of the imaging techniques and usually some pre-processing, such as filtering, is required. For instance, an approach presented by Leroy et al. used a gradient-preserving speckle filter and then looked for the similarity in the gradients.

Penney et al. proposed a technique for registering MRI and ultrasound. The system calculates a probability map of each element being a part of a liver-vessel [52]. Later Penney et al. extended their technique for CT-ultrasound registration of the pelvis and femur [53]. The system was validated using cadavers, showing that the registration was accurate to a 1.6 mm root-mean-square error on average. A similar technique for the cardiovascular domain was proposed later by Zhang et al. [83].

Combining segmentation with registration, King et al. presented a technique for registering pre-segmented models with ultrasound [35]. The technique predicts the probability that the ultrasound image was produced by the segmented anatomy.

In addition to a rigid transformation, affine registration includes non-uniform scaling which sometimes needs to be applied in order to get a more correct registration. Wein et al. developed an automatic affine-registration technique between CT and ultrasound [78]. To provide a better similarity of the ultrasound and CT, the system creates a simulated ultrasound image out of the CT scan based on the tracked probe position. The simulated ultrasound image is generated using a ray-traced approach to calculate the ultrasound wave reflection and attenuation in the tissue. To simulate tissue specific echogeneity, they apply an angle-independent polynomial function based on which tissue the region corresponds to.

External pressure or different laying positions of the patient when acquiring the images are influential factors. To account for local deformations while imaging soft tissue, a more complex registration is required. Papenberg et al. proposed two approaches for CT ultrasound registration [51] given a set of paired landmarks in both the CT and ultrasound data set. One approach uses the landmarks as hard constraints and in the other, the landmarks are considered as soft constraints and are combined with intensity value information, in this case the normalized gradient field. The paper shows a non-rigid registration between the liver vascular structures. The latter technique was later evaluated by Lange et al. [40].

24.6 Rendering

Visual presentation of the data is the last stage of the pipeline before involving the user. The basic B-mode ultrasound images can be depicted on a screen in a straightforward manner as varying pixel intensities according to the echo amplitude. Doppler information can be included as well with color-encoded blood-flow direction. Other data, such as tissue strain, can also be included into 2D as overlays. Another example of overlays is the *CycleStack Plot* which superimposes the respiratory signal onto a selected feature of interest in the ultrasound image [41]. Doctors use this information to account for the respiration-caused motion of the tumor in order to minimize the damage done by certain tumor treatments.

Freehand ultrasound In Sect. 24.3.1, we discussed how freehand ultrasound systems can be used to create large volumes by putting images into 3D spatial context. Garrett et al. presented a technique for correct visibility ordering of images using a binary positioning tree [18]. Visualization of large volumes leads to visual clutter. Therefore, Gee et al. extended existing re-slicing tools to create narrow-band volumes which contain less elements and are easier to present [20].

3D ultrasound is not as trivial to present due to its natural properties. In an early work, Nelson and Elvis discussed the effect of existing techniques for presenting 3D ultrasound data, such as surface fitting and volume rendering [46]. Later, seven ultrasound-dedicated volume projection techniques were evaluated by Steen and Olstad [72]. They included maximum intensity projection (MIP), average intensity projection (AIP) and gradient magnitude projection (GMP). The techniques were applied to 3D fetal data, where GMP was valued to give the best detail and robustness towards viewing parameters.

Data definition in the polar coordinate system is another challenge for ultrasound volume rendering. Kuo et al. presented a technique for quick on-the-fly scan-conversion [39]. To reduce the costs of the functional evaluation of $\tan(\phi)$ and $\tan(\psi)$, the functional values were pre-calculated and stored in a texture as a look-up-table.

Surface Rendering is a common tool for many imaging modalities. In ultrasound, the low signal-to-noise ratio and parallel tissue boundary discontinuities make defining smooth surfaces difficult. Smoothing of a surface can be performed at the rendering stage. Fattal et al. presented an approach to render smooth surfaces from 3D ultrasound [15]. The surface is extracted based on the variational principle. Fuzzy surface rendering is done by a technique called oriented splatting. Oriented splatting creates triangles aligned with the gradient of the surface function. The triangle is then colored with a Gaussian function and rendered in a back-to-front order. Wang et al. proposed an improved surface rendering technique for 3D ultrasound data of fetuses [77]. To remove the noise and to preserve edges, a modified anisotropic diffusion is first applied to the dataset. To enhance low intensities which appear due to signal loss as the sound wave propagates through the tissue, a light absorption function based on the distance from a point is applied to the data. Finally, a texture-based surface rendering is used, where the texture is extracted from images of infants. The textures are warped and blended with the surface of the fetus face. To create smooth surfaces and remove unimportant noise in direct volume rendering, Lim et al. proposed a filtering technique in their GPU based ultrasound rendering framework [43]. This technique employs different sized filters to smooth out the noise.

24.6.1 Transfer Function Design

For direct volume rendering, *transfer functions* map ultrasound data, i.e., voxel echogenicity in B-mode imaging and frequency information in Doppler imaging, onto colors and opacities. Usually, this mapping is based on look-up tables. In color Doppler imaging the commonly used red-to-blue color transfer function encodes direction and velocity of flow, whereas a variety of predefined color maps is in use for B-mode volume rendering. Custom color map editors are available, but hardly ever used. Overall, there is a well-established set of color-maps used in clinical practice.

Different from color transfer functions, where the selection largely depends on the preferences of the sonographer, the proper design of an appropriate *opacity transfer function* (OTF) is crucial: When designing OTFs, the goal is to assign a high opacity to voxels of structures of interest, while mapping all other samples to low opacities, thus avoiding any occlusion of the target structure. Whereas computed tomography allows classification of tissue based on voxel intensities, tissue classification-based transfer functions do not work in B-mode imaging due to the completely different data characteristics. Generally, a high signal intensity arises at a transition *between* tissues of different acoustic properties. Thus, at least in the case of soft tissue structures, we will measure high signal intensity at transitional areas and lower intensity signals within homogeneous tissue. This is the reason for applying monotonically increasing opacity transfer functions in DVR of ultrasound data. The aim is to opacify the tissue transitions in the hope of obtaining a visualization of an entire target structure.

The most commonly used OTF in volume rendering of B-mode data assigns voxels to one of three classes depending on their echogenicity, namely invisible, transparent, and opaque. The corresponding piecewise linear OTF is modified manually by means of two parameters, namely a threshold intensity I_{thresh} and a transparency value α controlling the increase of opacity for intensities above I_{thresh}. The effect of modifying I_{thresh} is depicted visually on MPR images, see Fig. 24.6.

The parameters of the OTF affect the rendered image in a substantial way. The lower the I_{thresh} value, the lower the rendered image's brightness, due to an increasing number of hypoechoic voxels contributing to the image. Furthermore, the OTF affects depth contrast, i.e., the contrast arising from a spatial discontinuity in the target structure, and tissue contrast, i.e., contrast due to different echogenicity of adjacent tissue. See Ref. [28] for an evaluation of these effects on linear and parabolic OTFs. On the other hand, any modification of fundamental acquisition parameters, such as, overall gain, or depth gain compensation, and any change of the position of the transducer or the target structure, changes the echogenicity distribution and thus requires modifying the OTF for an optimal visualization. For a real time imaging modality, incessant modification is not feasible. Hence, in clinical practice sonographers use

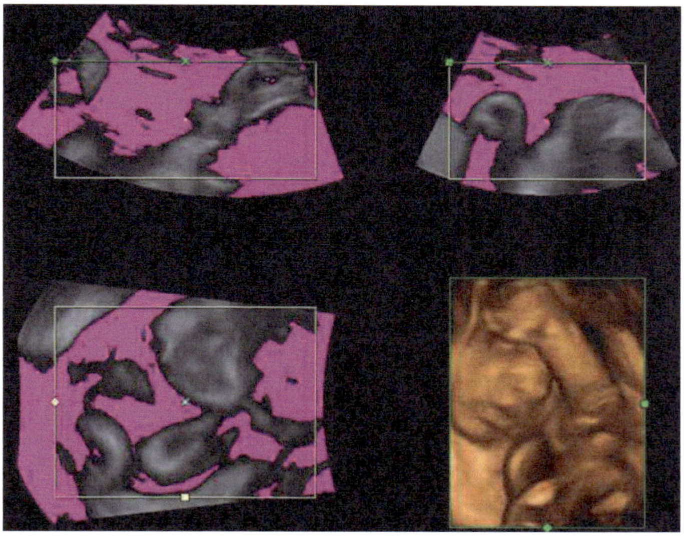

Fig. 24.6 The parameter I_{thresh} determines which echo intensity values to render transparently. A user control with immediate feedback, indicating transparent regions in pink, is essential

a default OTF providing reasonable visualization in the majority of cases, and hardly ever touch the OTF control panel.

Therefore, there is a need for automatic determination of an optimal OTF for every single acquisition. Due to the distinct characteristics and the real-time nature of ultrasound imaging, most conventional approaches for transfer function design have proven inadequate or require substantial modification in order to be applicable to ultrasound volume imaging. Among the most important advances in transfer function design for CT data is the work by Kindlmann et al. [34] and subsequent work by Kniss et al. [36], introducing the concept of histogram volumes for semi-automatic generation of OTFs for datasets where the regions of interest are boundaries between materials of relatively constant value. In [30], von Jan et al. adapt this approach to ultrasound data and apply it successfully to 3D freehand acquired volumes of hyperechoic structures.

Hönigmann et al. suggest an approach dedicated to the specific problem of rendering hyperechoic structures embedded in hypoechoic fluid [28]. By analyzing so called *tube cores*, they yield an estimate for the position of the most prominent tissue transition in the rendering direction. Voxel intensities *prior to* and *at* the detected interface steer the extent of modification of an initial, parabolic OTF in a multiplicative way. In a subsequent publication the authors assess the temporal coherence of the tube core method and conclude that it is sufficiently efficient and robust for online-computation of OTFs for an entire sequence of acquired volumes, if smoothing in the temporal domain is employed [54].

24.6.2 Multi-modal Rendering

Additional challenges arise when it comes to DVR of multiple datasets. Multi-modal rendering is meant to bring two or more data sets of the same object into a single image. Having two or more datasets in the same scene creates a challenge to keep the cluttering of less interesting regions to a minimum from the datasets. For ultrasound, 3D Doppler data can be acquired simultaneously with 3D B-mode data. Jones et al. discuss several approaches to explore and visualize 4D Doppler data [31]. Multi-planar rendering, showing several slices at once, provides a surface fitting of the Doppler data based on the YCbCr color scheme values to improve separation between Doppler data and B-mode data. An approach is presented to blend multi-planar slice rendering into a DVR scene. The DVR is shown highly transparent and the slices provide better detail along the perspective. A different way of combining B-mode with Doppler data was presented by Petersch and Hönigmann [55]. They propose a one level composite rendering approach allowing for blending flow and tissue information arbitrarily, using silhouette rendering for the B-Mode and a mix of Phong shaded DVR and silhouette rendering on color Doppler.

A new technique for blending Doppler and B-mode was introduced by Yoo et al. [82]. Instead of blending two 2D rendered images (post fusion), or a blending the two volumes while rendering (composite fusion), they proposes a way to do both called progressive fusion (PGF). Post fusion has a problem with depth blending and composite fusion will get a too early ray termination. PGF compensates for this by using an if-clause to adjust the alpha-out value in the ray-caster to composite either the Doppler-signal or the B-mode-signal.

Burns et al. applied illustrative cut-aways combined with 3D freehand ultrasound and CT [7]. This provides a better spatial overview for the ultrasound images. To add more information onto the 2D ultrasound image, Viola et al. proposed an approach to enhance the ultrasound image by overlaying higher order semantics [75], in this case in the form of Couinaud segmentation. The segmentation is pre-defined in a CT dataset and visually verified using exploded views. To combine it with ultrasound images, the CT dataset is co-registered with the ultrasound using rigid transformation according to user defined landmarks. The different segments are superimposed onto the ultrasound image enabling the user to directly see which segments are being imaged. To improve ultrasound video analysis, Angelelli et al. used a degree-of-interest (DOI) distribution superimposed on the image [3]. The video sequence was presented as a function of time (x-axis), where the y-axis was defined by the amount the current ultrasound image covered the DOI-function.

24.6.3 Shading and Illumination

Light is an indispensable part of scenes we see in real life. Also in computer graphics, light sources and light transport models have to be taken into account when rendering realistic scenes. In volume graphics, the problem of illumination and light transport has been tackled by a handful of researchers as well.

We distinguish between local and global illumination models. Local illumination models use gradients of the volumetric function instead of surface normals to evaluate the diffuse and specular terms of the Phong illumination model [42]. While local illumination models already reveal structures, global illumination methods result in a more realistic appearance, which further supports spatial perception. While gradient-based local illumination methods are faster to evaluate, gradient computation is sensitive to noise and high frequencies, which are natural properties of ultrasound data.

Recent works show that global illumination models based on gradient-free methods are suitable for rendering ultrasound volumes [62, 76]. Ropinski et al. described a volumetric lighting model which simulates scattering and shadowing [62]. They use slice-based volume rendering from the view of the light source to calculate a light volume and raycasting to render the final image (see Fig. 24.7b). A perceptual evaluation of the generated images indicates that the proposed model yields stronger depth cues than gradient-based shading. Šoltészová et al. presented a single-pass method for the simulation of light scattering in volumes [76]. Light transport is approximated using a tilted cone-shaped function which leaves elliptic footprints in the opacity buffer during slice-based volume rendering. They use a slice-based renderer with an additional opacity buffer. This buffer is incrementally blurred with an elliptical kernel, and the algorithm generates a high-quality soft-shadowing effect (see Fig. 24.7c). The light position and direction can be interactively modified. While these two techniques have been explicitly applied to 3D US data, the application of other volumetric illumination models potentially also improves the visual interpretation of 3D US data. Figure 24.8 shows a comparison of six different shading techniques as applied to a 3D US scan of a human heart. While the first row of Fig. 24.8 shows examples for the already addressed shading techniques, the second row shows three alternative approaches. Figure 24.8d incorporates scattering of light in volume data, as proposed by Kniss et al. [37]. Their slicing technique allows textured slices to be rendered from both light and viewing direction simultaneously. By sampling the incident light from multiple directions while updating the light's attenuation map, they

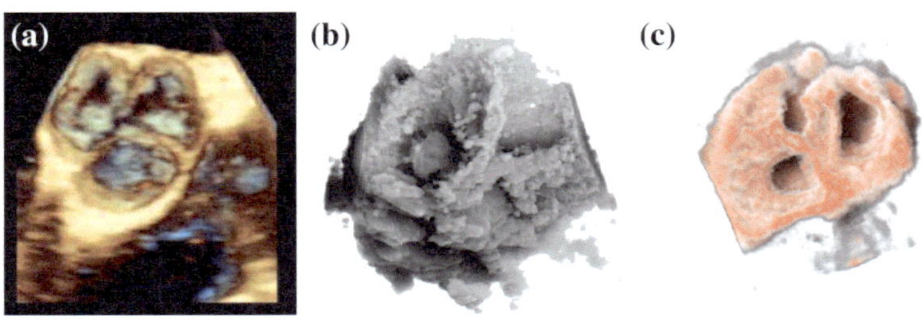

Fig. 24.7 **a** Diastole of the aortic valve on a modern ultrasound workstation using color-coding based on depth. **b** Rendering of 3D ultrasound of a human heart with shadowing from the work of Ropinski et al. [62] and **c** rendered using the technique presented in the work of Šoltészová et al. [76] © IEEE Reprinted, with permission, from IEEE Pacific Visualization Symposium (PacificVis), 2010

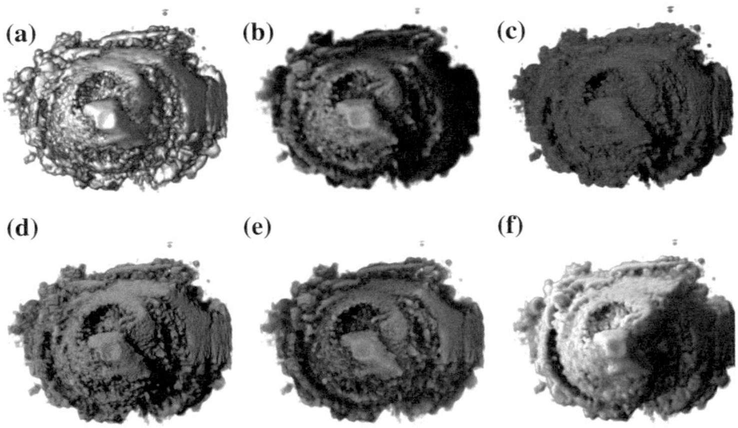

Fig. 24.8 Comparison of six volume shading models as applied to a 3D US scan of a human heart. **a** Phong, **b** [62], **c** [76], **d** [37], **e** [66], **f** [44]

account for scattering effects in slice-based volume rendering. Figure 24.8e shows the application of the directional occlusion shading technique [66]. This technique constrains the light source position to coincide with the view point. Finally, Fig. 24.8f shows the application of a technique based on spherical harmonic lighting [44].

Advanced illumination techniques are now being implemented in the commercial ultrasound workstations. Some workstations use additional color coding based on depth. Deeper tissues are colored with cold tones such as blue while close regions have red and orange tones. This effect has been firstly described by Einthoven [14] and is also referred to as chromostereopsis [2]. Figure 24.7a shows a chromatic depth-encoding rendering of a 3D human heart in a modern ultrasound workstation.

24.7 Ultrasound and Augmented Reality

Ultrasound is commonly viewed on a separate monitor. Therefore, it is difficult to comprehend the spatial relationship between what you see on the monitor and where it is located in the patient's body. Augmented reality can aid the user by, for instance, super-imposing the ultrasound image onto the body where the ultrasound probe is positioned. Bajura et al. presented a system which linked 3D freehand ultrasound with a head-mounted display (HMD) [4]. The HMD contains a camera, tracker and two displays, one for each eye. The system can then project the tracked ultrasound image onto the tracked camera feed so the user can see where in the body the image is actually positioned.

Combining segmentation, surface rendering and augmented reality, Sato et al. aimed to aid surgeons during breast tumor removal for minimizing risk and maximizing breast preservation [65] by projecting a segmented tumor onto a video feed. The tumor is segmented using a minimal intensity projection based selection of the volume of interest. In the final stage, the tumor is surface rendered and superimposed on the video image.

Stetten et al. show how tomographic reflection can provide a superimposed image onto the body without any tracking systems [73]. The ultrasound probe carries a half-silvered mirror. The mirror reflects the ultrasound image which is shown on a flat panel monitor mounted on the probe. This technique was extended in the Sonic Flashlight [68]. The tomographic reflection was shown to increase the localization perception compared to conventional ultrasound [81].

Augmented reality shows a great potential benefit in medical ultrasound imaging. Yet, there is a lag from technology development to the actual integration into every day usage. Sielhorst et al. published a detailed review for advanced medical displays in 2008 [69]. This chapter discuss the potential benefit and the increasing use for augmented reality in medical imaging in general. They state that improvements in both technologies are needed to be able to create a seamless integration into the workflow of physicians and surgeons.

24.8 Summary and Discussion

In this chapter, we have categorized several of the most important works in what constitute the ultrasound visualization pipeline. The pipeline is defined as the five major categories in data processing and rendering. The five categories are pre-processing, segmentation, registration, rendering and augmented reality.

Medical ultrasound data is very different compared to other medical imaging modalities. Techniques for the individual steps in the visualization pipeline are tailored to suit the special nature of the data. For instance, techniques meant for in-vivo use have strong performance requirements to handle the high frame rate of ultrasound images. Segmentation and registration becomes very challenging, due to inconsistent data values for similar tissue. Still, ultrasound remains as one of the most used imaging modalities in medicine. Research in advanced ultrasound visualization techniques focuses greatly on 3D ultrasound, but the trend in diagnostics is mostly 2D due to higher frame-rates, high resolution and a minimal requirement for interaction. The temporal and spatial resolution for ultrasound is approaching the physical limits of the speed of sound. It is very important to explore what strengths and weaknesses the different modalities possess and combine the strengths into the natural work flow of medical personnel.

Acknowledgments This work has been carried out within the IllustraSound research project (#193170), which is funded by the VERDIKT program of the Norwegian Research Council with support of the MedViz network in Bergen, Norway (PK1760-5897-Project 11). We would also like to thank Helwig Hauser for invaluable help and fruitful discussions.

References

1. Ag, K.: 3d ultrasound: a dedicated system. Eur. Radiol. **9**, S331–S333 (1999). doi:10.1007/PL00014067
2. Allen, R.C., Rubin, M.L.: Chromostereopsis. Surv. Ophthalmology **26**(1), 22–27 (1981)
3. Angelelli, P., Viola, I., Nylund, K., Gilja, O.H., Hauser, H.: Guided visualization of ultrasound image sequences. In: Proceedings of Visual Computing for Biomedicine, pp. 125–132 (2010)
4. Bajura, M., Fuchs, H., Ohbuchi, R.: Merging virtual objects with the real world: seeing ultrasound imagery within the patient. Proc. ACM SIGGRAPH **26**(2), 203–210 (1992)
5. Belohlavek, M., Dutt, V., Greenleaf, J., Foley, D.A., Gerber, T.C., Seward, J.B.: Multidimensional ultrasonic visualization in cardiology. Proc. IEEE Ultrason. Symp. **2**, 1137–1145 (1992)
6. Berg, S., Torp, H., Martens, D., Steen, E., Samstad, S., Hik, I., Olstad, B.: Dynamic three-dimensional freehand echocardiography using raw digital ultrasound data. Ultrasound Med. Biol. **25**(5), 745–753 (1999)
7. Burns, M., Haidacher, M., Wein, W., Viola, I., Gröller, E.: Feature emphasis and contextual cutaways for multimodal medical visualization. In: Proceedings of Eurographics/IEEE VGTC Symposium on Visualization (EuroVis 2007), pp. 275–282 (2007)
8. Carneiro, G., Amat, F., Georgescu, B., Good, S., Comaniciu, D.: Semantic-based indexing of fetal anatomies from 3-D ultrasound data using global/semi-local context and sequential sampling. In: Proceedings of Computer Vision and Pattern Recognition, pp. 1–8 (2008)
9. Chen, Y., Yin, R., Flynn, P., Broschat, S.: Aggressive region growing for speckle reduction in ultrasound images. Pattern Recogn. Lett. **24**(4–5), 677–691 (2003)
10. Crimmins, T.R.: The geometric filter for speckle reduction. Appl. Optics **24**(10), 1438–1443 (1985)
11. Curiel, L., Chopra, R., Hynynen, K.: Progress in multimodality imaging: truly simultaneous ultrasound and magnetic resonance imaging. IEEE Trans. Med. Imaging **26**(12), 1740–1746 (2007)
12. Deprez, J., Brusseau, E., Schmitt, C., Cloutier, G., Basset, O.: 3d estimation of soft biological tissue deformation from radio-frequency ultrasound volume acquisitions. Med. Image Anal. **13**(1), 116–127 (2009)
13. Deutch, T.D., Joergner, I., Matson, D.O., Oehninger, S., Bocca, S., Hönigmann, D., Abuhamad, A.: Automated assessment of ovarian follicles using a novel three-dimensional ultrasound software. Fertility Steril. **92**(5), 1562–1568 (2009)
14. Einthoven, W.: Stereoscopie durch Farbendifferenz. Von Graefes. Archives **19**, 211–238 (1885)
15. Fattal, R., Lischinski, D.: Variational classification for visualization of 3D ultrasound data. In: Proceedings of IEEE Visualization, pp. 403–410 (2001)
16. Forsberg, F., Healey, A.J., Leeman, S., Jensen, J.A.: Assessment of hybrid speckle reduction algorithms. Phys. Med. Biol. **36**(11), 1539–1549 (1991)
17. Fuller, D.B., Jin, H., Koziol, J.A., Feng, A.C.: CT-ultrasound fusion prostate brachytherapy: a dynamic dosimetry feedback and improvement method. A report of 54 consecutive cases. Brachytherapy **4**(3), 207–216 (2005)
18. Garrett, W.F., Fuchs, H., Whitton, M.C., State, A.: Real-time incremental visualization of dynamic ultrasound volumes using parallel BSP trees. In: Proceedings of IEEE Visualization, pp. 235–240 (1996)
19. GE Healthcare: A closer look at GE's pocket-sized vscan ultrasound. http://www.gereports.com/a-closer-look-at-ges-pocket-sized-vscan-ultrasound/. Accessed 13 Sept 2010
20. Gee, A., Prager, R., Treece, G., Berman, L.: Narrow-band volume rendering for freehand 3D ultrasound. Comput. Graph. **26**(3), 463–476 (2002)
21. Gee, A., Prager, R., Treece, G., Cash, C., Berman, L.: Processing and visualizing three-dimensional ultrasound data. Br. J. Radiol. **77**, 186–193 (2004)
22. Gilja, O.H., Hatlebakk, J.G., Ødegaard, S., Berstad, A., Viola, I., Giertsen, C., Hausken, T., Gregersen, H.: Advanced imaging and visualization in gastrointestinal disorders. World J. Gastroenterol. **13**, 1408–1421 (2007)

23. Gilja, O.H., Hausken, T., Ødegaard, S., Wendelbo, Ø., Thierley, M.: Mobile ultrasonography in a medical department. Tidskr. Nor. Lægeforen. **19**, 270–285 (2003)
24. Gilja, O.H., Heimdal, A., Hausken, T., Gregersen, H., Matre, K., Berstad, A., Ødegaard, S.: Strain during gastric contractions can be measured using doppler ultrasonography. Ultrasound Med. Biol. **28**(11–12), 1457–1465 (2002)
25. Gobbi, D., Peters, T.: Interactive intra-operative 3D ultrasound reconstruction and visualization. In: Proceedings of Medical Image Computing and Computer-Assisted Intervention, pp. 156–163 (2002)
26. Gupta, N., Swamy, M.N.S., Plotkin, E.: Despeckling of medical ultrasound images using data and rate adaptive lossy compression. IEEE Trans. Med. Imaging **24**(6), 743–754 (2005)
27. Heimdal, A., Sten, A., Torp, H., Skje, T.: Real-time strain rate imaging of the left ventricle by ultrasound. J. Am. Soc. Echocardiogr. **11**(11), 1013–1019 (1998)
28. Hönigmann, D., Ruisz, J., Haider, C.: Adaptive design of a global opacity transfer function for direct volume rendering of ultrasound data. In: IEEE Transactions on Ultrasonics, Ferroelectrics, and Frequency Control, pp. 489–496 (2003)
29. Huang, T., Yang, G., Tang, G.: A fast two-dimensional median filtering algorithm. IEEE Trans. Acoust. Speech Signal Process. **27**(1), 13–18 (1979)
30. Jan, U., Sandker, D., Rauberger, M., Matthies, H.K., Overhoff, H.M.: Enhanced visualization of ultrasound volumes for diagnostic and therapeutic purposes. In: Proceedings of 4th European Conference of the International Federation for Medical and Biological Engineering, IFMBE Proceedings, pp. 689–692, IFMBE (2009)
31. Jones, M.G., Shipley, J.A., Robinson, T.M.: Visualisation of 4-D colour and power Doppler data. Ultrasound Med. Biol. **29**(12), 1735–1747 (2003)
32. Karadayi, K., Managuli, R., Kim, Y.: Three-dimensional ultrasound: from acquisition to visualization and from algorithms to systems. IEEE Rev. Biomed. Eng. **2**, 23–39 (2009)
33. Karamalis, A., Wein, W., Kutter, O., Navab, N.: Fast hybrid freehand ultrasound volume reconstruction. In: Proceedings of SPIE Medical Imaging. vol. 7261, pp. 726114–726118 (2009)
34. Kindlmann, G., Durkin, J.W.: Semi-automatic generation of transfer functions for direct volume rendering. In: Proceedings of IEEE Symposium on Volume Visualization, pp. 79–86 (1998)
35. King, A.P., Ma, Y., Yao, C., Jansen, C., Razavi, R., Rhode, K.S., Penney, G.P.: Image-to-physical registration for image-guided interventions using 3-d ultrasound and an ultrasound imaging model. In: Proceedings of International Conference on Information Processing in Medical Imaging, pp. 188–201. Springer, Berlin (2009)
36. Kniss, J., Kindlmann, G., Hansen, C.: Multidimensional transfer functions for interactive volume rendering. Trans. Visual. Comput. Graph. **8**(3), 270–285 (2002)
37. Kniss, J., Premoze, S., Hansen, C., Ebert, D.: Interactive translucent volume rendering and procedural modeling. In: Proceedings of IEEE Visualization, pp. 109–116 (2002)
38. Krissian, K., Westin, C.F., Kikinis, R., Vosburgh, K.G.: Oriented speckle reducing anisotropic diffusion. IEEE Trans. Image Process. **16**(5), 1412–1424 (2007)
39. Kuo, J., Bredthauer, G.R., Castellucci, J.B., von Ramm, O.T.: Interactive volume rendering of real-time three-dimensional ultrasound images. IEEE Trans. Ultrason. Ferroelectr. Freq. Control **54**(2), 313–318 (2007)
40. Lange, T., Papenberg, N., Heldmann, S., Modersitzki, J., Fischer, B., Lamecker, H., Schlag, P.: 3D ultrasound-CT registration of the liver using combined landmark-intensity information. Int. J. Comput. Assist. Radiol. Surg. **4**, 79–88 (2008)
41. Lee, T., Chaudhuri, A., Porikli, F., Shen, H.: Cyclestack: Inferring periodic behavior via temporal sequence visualization in ultrasound video. In: Proceedings of IEEE Pacific Visualization, pp. 89–96, IEEE (2010)
42. Levoy, M.: Display of surfaces from volume data. IEEE Comput. Graph. Appl. **8**(3), 29–37 (1988)
43. Lim, S., Kwon, K., Shin, B.S.: GPU-based interactive visualization framework for ultrasound datasets. Computer Anim. Virtual Worlds **20**(1), 11–23 (2008)
44. Lindemann, F., Ropinski, T.: Advanced light material interaction for direct volume rendering. In: Proceedings of IEEE/EG International Symposium on Volume Graphics, pp. 101–108 (2010)

45. Lizzi, F.L., Feleppa, E.J.: Image processing and pre-processing for medical ultrasound. In: Proceedings of Imagery Pattern Recognition Workshop, p. 187 (2000)
46. Nelson, T.R., Elvins, T.T.: Visualization of 3D ultrasound data. IEEE Comput. Graph. Appl. **13**(6), 50–57 (1993)
47. Nikas, D.C., Hartov, A., Lunn, K., Rick, K., Paulsen, K., Roberts, D.W.: Coregistered intraoperative ultrasonography in resection of malignant glioma. Neurosurg. Focus **14**(2), e6 (2003)
48. Noble, J.A., Boukerroui, D.: Ultrasound image segmentation: a survey. IEEE Trans. Med. Imaging **25**(8), 987–1010 (2006)
49. Ødegaard, S., Gilja, O.H., Gregersen, H.: Basic and New Aspects of Gastrointestinal Ultrasonography. Advanced Series in Biomechanics. World Scientific, Singapore (2005)
50. Ødegaard, S., Nesje, L.B., Gilja, O.H.: Atlas of Endoscopic Ultrasonography. Fagbokforlaget, Bergen (2007)
51. Papenberg, N., Lange, T., Modersitzki, J., Schlag, P.M., Fischer, B.: Image registration for CT and intra-operative ultrasound data of the liver. In: SPIE Medical Imaging, vol. 6918, p. 691808. SPIE (2008)
52. Penney, G., Blackall, J., Hamady, M., Sabharwal, Y., Adam, A., Hawkes, D.: Registration of freehand 3D ultrasound and magnetic resonance liver images. Med. Image Anal. **8**(1), 81–91 (2004)
53. Penney, G.P., Barratt, D.C., Chan, C.S.K., Slomczykowski, M., Carter, T.J., Edwards, P.J., Hawkes, D.J.: Cadaver validation of intensity-based ultrasound to CT registration. Med. Image Anal. **10**(3), 385–395 (2006)
54. Petersch, B., Hadwiger, M., Hauser, H., Hönigmann, D.: Real time computation and temporal coherence of opacity transfer functions for direct volume rendering of ultrasound data. Comput. Med. Imaging Graph. **29**(1), 53–63 (2005)
55. Petersch, B., Hönigmann, D.: Blood flow in its context: combining 3D B-mode and color Doppler ultrasonic data. Trans. Visual. Comput. Graph. **13**(4), 748–757 (2007)
56. Petersch, B., Serrano-Serrano, O., Hönigmann, D.: 3D soft segmentation and visualization of medical data based on nonlinear diffusion and distance functions. In: Proceesings of Eurographics/IEEE VGTC Symposium on Visualization (Eurovis 2006), pp. 331–338 (2006)
57. Postema, M., Gilja, O.H.: Ultrasound-directed drug delivery. Curr. Pharm. Biotechnol. **8**, 355–361(7) (2007)
58. Prager, R., Gee, A., Treece, G., Berman, L.: Freehand 3D ultrasound without voxels: volume measurement and visualisation using the Stradx system. Ultrasonics **40**(1–8), 109–115 (2002)
59. Prager, R.W., Gee, A., Berman, L.: Stradx: real-time acquisition and visualization of freehand three-dimensional ultrasound. Med. Image Anal. **3**(2), 129–140 (1999)
60. Rohling, R., Gee, A., Berman, L.: A comparison of freehand three-dimensional ultrasound reconstruction techniques. Med. Image Anal. **3**(4), 339–359 (1999)
61. Rohling, R., Gee, A., Berman, L., Treece, G.: Radial basis function interpolation for 3D ultrasound. Lect. Notes Comput. Sci. **1613**, 478–483 (1999)
62. Ropinski, T., Döring, C., Rezk-Salama, C.: Interactive volumetric lighting simulating scattering and shadowing. In: Proceedings of IEEE Pacific Visualization, pp. 169–176 (2010)
63. Roxborough, T., Nielson, G.M.: Tetrahedron based, least squares, progressive volume models with application to freehand ultrasound data. In: Proceedings of IEEE Visualization, pp. 93–100 (2000)
64. Sakas, G., Schreyer, L., Grimm, M.: Preprocessing and volume rendering of 3D ultrasonic data. IEEE Comput. Graph. Appl. **15**(4), 47–54 (1995)
65. Sato, Y., Nakamoto, M., Tamaki, Y., Sasama, T., Sakita, I., Nakajima, Y., Monden, M., Tamura, S.: Image guidance of breast cancer surgery using 3-D ultrasound images and augmented reality visualization. IEEE Trans. Med. Imaging **17**(5), 681–693 (1998)
66. Schott, M., Pegoraro, V., Hansen, C., Boulanger, K., Bouatouch, K.: A directional occlusion shading model for interactive direct volume rendering. In: Proceedings of Eurographics, pp. 855–862 (2009)
67. Shankar, P.M.: Speckle reduction in ultrasonic images through a maximum likelihood based adaptive filter. Phys. Med. Biol. **51**(21), 5591–5602 (2006)

68. Shelton, D., Stetten, G.D., Chang, W.: Ultrasound visualization with the sonic flashlight. In: Proceedings of ACM SIGGRAPH, p. 82 (2002)
69. Sielhorst, T., Feuerstein, M., Navab, N.: Advanced medical displays: a literature review of augmented reality. J. Disp. Technol. **4**(4), 451–467 (2008)
70. Solberg, O.V., Lindseth, F., Torp, H., Blake, R.E., Hernes, T.A.N.: Freehand 3d ultrasound reconstruction algorithms: a review. Ultrasound Med. Biol. **33**(7), 991–1009 (2007)
71. SonoWand: Sonowand—product description. http://www.sonowand.com/ (2011)
72. Steen, E., Olstad, B.: Volume rendering of 3D medical ultrasound data using direct feature mapping. IEEE Trans. Med. Imaging **13**(3), 517–525 (1994)
73. Stetten, G.D., Chib, V.S., Tamburo, R.J.: Tomographic reflection to merge ultrasound images with direct vision. In: Proceedings of 29th Applied Imagery Pattern Recognition Workshop, pp. 200–205 (2000)
74. Thune, N., Gilja, O.H., Hausken, T., Matre, K.: A practical method for estimating enclosed volumes using 3D ultrasound. Eur. J. Ultrasound **3**(1), 83–92 (1996)
75. Viola, I., Nylund, K., Øye, O.K., Ulvang, D.M., Gilja, O.H.: Illustrated ultrasound for multimodal data interpretation of liver examinations. In: Proceedings of Eurographics Workshop on Visual Computing in Biomedicine, pp. 125–133 (2008)
76. Šoltészová, V., Patel, D., Bruckner, S.: A multidirectional occlusion shading model for direct volume rendering. Comput. Graph. Forum **29**(3), 883–891 (2010)
77. Wang, S., Sun, Y., Chang, F.: Artifact removal and texture-based rendering for visualization of 3D fetal ultrasound images. Med. Biol. Eng. Comput. **46**(6), 575–588 (2008)
78. Wein, W., Brunke, S., Khamene, A., Callstrom, M.R., Navab, N.: Automatic CT-ultrasound registration for diagnostic imaging and image-guided intervention. Med. Image Anal. **12**(5), 577–585 (2008)
79. Wein, W., Khamene, A.: Image-based method for in-vivo freehand ultrasound calibration. In: SPIE Medical Imaging, vol. 6920, p. 69200K (2008)
80. Wein, W., Pache, F., Roeper, B., Navab, N.: Backward-warping ultrasound reconstruction for improving diagnostic value and registration. In: Proceedings of Medical Image Computing and Computer-Assisted Intervention. Lecture Notes in Computer Science, pp. 750–757. Springer (2006)
81. Wu, B., Klatzky, R.L., Shelton, D., Stetten, G.D.: Psychophysical evaluation of in-situ ultrasound visualization. Trans. Visual. Comput. Graph. **11**(6), 684–693 (2005)
82. Yoo, Y.M., Managuli, R., Kim, Y.: New multi-volume rendering technique for three-dimensional power Doppler imaging. Ultrasonics **46**(4), 313–322 (2007)
83. Zhang, W., Noble, J.A., Brady, J.M.: Real time 3-D ultrasound to MR cardiovascular image registration using a phase-based approach. In: Proceedings of IEEE International Symposium on Biomedical Imaging: Macro to Nano, pp. 666–669 (2006)

Chapter 25
Visual Exploration of Simulated and Measured Blood Flow

A. Vilanova, Bernhard Preim, Roy van Pelt, Rocco Gasteiger, Mathias Neugebauer and Thomas Wischgoll

Abstract Morphology of cardiovascular tissue is influenced by the unsteady behavior of the blood flow and vice versa. Therefore, the pathogenesis of several cardiovascular diseases is directly affected by the blood-flow dynamics. Understanding flow behavior is of vital importance to understand the cardiovascular system and potentially harbors a considerable value for both diagnosis and risk assessment. The analysis of hemodynamic characteristics involves qualitative and quantitative inspection of the blood-flow field. Visualization plays an important role in the qualitative exploration, as well as the definition of relevant quantitative measures and its validation. There are two main approaches to obtain information about the blood flow: simulation by computational fluid dynamics, and in-vivo measurements. Although research on blood flow simulation has been performed for decades, many open problems remain concerning accuracy and patient-specific solutions. Possibilities for real measurement of blood flow have recently increased considerably by new developments in magnetic resonance imaging which enable the acquisition of 3D quantitative

A. Vilanova (✉) · R. van Pelt
Computer Graphics and Visualization, EEMCS - Delft University of Technology,
Room : HB 11.270 Building 36 - Mekelweg 4, 2628 CD Delft Postbox 5, 2600 AA Delft, The Netherlands
e-mail: A.Vilanova@tudelft.nl

R. van Pelt
e-mail: r.f.p.v.pelt@tue.nl

B. Preim, R. Gasteiger · M. Neugebauer
Institut Für Simulation Und Graphik, Otto-von-Guericke-Universität Magdeburg, Magdeburg, Germany
e-mail: bernhard.preim@ovgu.de

R. Gasteiger
e-mail: rocco.gasteiger@ovgu.de

M. Neugebauer
e-mail: mathias.neugebauer@ovgu.de

T. Wischgoll
Computer Science and Engineering, Wright State University, Dayton USA
e-mail: thomas.wischgoll@wright.edu

C.D. Hansen et al. (eds.), *Scientific Visualization*, Mathematics and Visualization,
DOI 10.1007/978-1-4471-6497-5_25

measurements of blood-flow velocity fields. This chapter presents the visualization challenges for both simulation and real measurements of unsteady blood-flow fields.

25.1 Introduction

Cardiovascular disease (CVD) is a class of conditions affecting the heart and blood vessels, with an estimated overall prevalence of over thirty percent of the American population [1], and is currently the leading cause of death worldwide [53].

Diagnosis of CVD typically involves an evaluation of both the anatomical structure and function, while the behavior of blood flow is still rarely inspected. The flow behavior is, nevertheless, of vital importance to the cardiovascular system. Morphology of cardiovascular tissue is significantly influenced by the unsteady behavior of flowing blood and vice versa. Therefore, blood flow analysis potentially harbors a considerable value for both diagnosis and risk assessment. A wide range of pre-clinical research indicates that flow behavior directly relates to medical conditions [13, 28].

In particular, congenital heart diseases imply anomalous hemodynamics that strongly influence the progression and treatment of the innate defects. For the adult case, a noteworthy application is the aortic dissection, which is caused by a tear in the inner aortic wall. This allows blood to flow between the disintegrated layers of the vessel wall, resulting in a high risk of rupture. Again, the blood flow behavior plays a predominant role in the course of the condition. Decision support in case of cerebral aneurysms is one of the main applications of blood flow analysis. Blood flow is essential for the assessment of risk of rupture, urgency of treatment in case of multiple aneurysms, selection of treatment strategy (e.g., coiling/stenting, neurosurgical clipping).

The analysis of hemodynamic characteristics involves qualitative and quantitative inspection of the blood flow field. Physicians in clinical research investigate both the spatiotemporal flow behavior, as well as derived measures, such as the mean flux or cardiac output. The analysis of the blood flow data often requires complex mental reconstruction processes by the physician. Visualization plays an important role in the qualitative exploration, as well as the definition of relevant quantitative measures and its validation.

There are two main approaches to obtain information about the blood flow: simulations (i.e., computational fluid dynamics) and in-vivo measurements. Both of these methodologies can obtain information about the unsteady blood flow characteristics, where each has different advantages and disadvantages.

Although research on simulations of blood flow has been active for several decades, still a lot of open problems remain concerning accuracy and patient-specific solutions. Recently, research around measurement of blood flow has increased considerably. Developments in magnetic resonance imaging (MRI) have made the acquisition of 3D quantitative measurements of blood flow velocities fields possible. Furthermore, several vendors have made essential postprocessing software to inspect the data clinically available. Therefore, clinical pilot studies have been possible and have shown the relevance and potential of this data. Furthermore, the MRI acquisition development towards 7 and 9 T machines have the potential to provide the required resolution and signal-to-noise ratios (SNR) to analyze blood flow in smaller vessels compared to the main arteries around the heart.

In this chapter, we will consider the visualization challenges for both simulations and in-vivo measurements of unsteady flow fields. We will present a review of the existing literature, the main challenges related to blood flow visualization and analysis, as well as the open issues.

25.2 Blood Flow Simulation

One way of determining blood flow within a vascular system is through simulation. This approach involves two major steps. First, the vascular structure needs to be segmented, and the geometry of the vessel boundary determined as accurately as possible. Next, a Computational Fluid Dynamics (CFD) model simulates the blood flow within the reconstructed geometry. The next sections will discuss these steps in more detail.

25.2.1 Grid Generation

In order to simulate blood flow with CFD, the boundary conditions of the underlying mathematical model have to be defined properly. In case of vascular flow, the boundary conditions are defined by two different components: the first one is the the geometric boundary of the vessels; the second one consists of the inflow and outflow characteristics as defined by the circulatory system.

There are different ways of identifying the vessel boundary. Typically, some imaging technique is used to generate a scan of the vascular structure for which the flow is supposed to be simulated, for example a Computed Tomography (CT) scan. In order to extract the vascular structure from such a volumetric image, the data needs to be segmented. Simple thresholding based on the intensity value can be used. However, this may not be sufficient for anatomical structures where significant perfusion and noise occurs, such as the heart. More sophisticated segmentation techniques are necessary, for example, gradient-based thresholding techniques tend to produce better results in those cases.

The segmentation process also has great influence on the overall accuracy of the simulation. Basic intensity thresholding techniques, for example, determine individual voxel locations as being part of the vessel boundary. However, it is unlikely that the vessel boundary is located precisely at such a voxel location, especially given that the volumetric data set only imposes an artificial grid on the organ at hand. There-

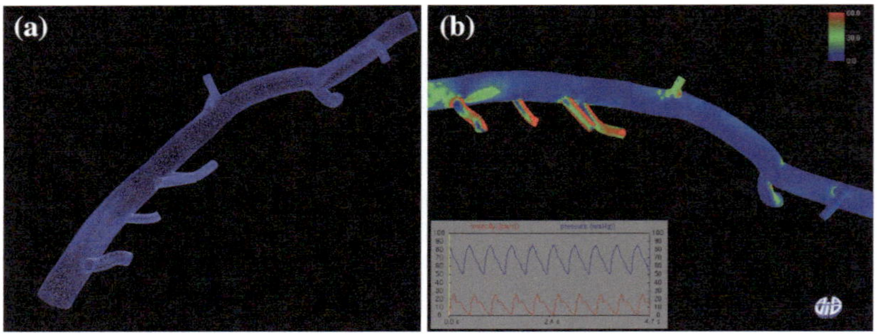

Fig. 25.1 An example of a vessel boundary based on a CT scan segmented using gradient-based thresholding with sub-voxel precision (**a**) and resulting simulation of blood flow visualized by color-coding the wall-shear stress (**b**)

fore, the accuracy can be improved by using segmentation techniques that operate at a sub-voxel level, e.g., Marching Cubes algorithm [25].

The downside of the Marching Cubes algorithm, however, is that it only operates with a fixed intensity threshold across the entire data set. Since typically a point spread function with a radius greater than one has to be assumed for most imaging techniques, this may cause errors in the boundary geometry, overestimating larger vessels and underestimating smaller vessels. Gradient-based approaches can achieve better results in these cases identifying the location where the maximal gradient value is assumed to find a more accurate estimate for the exact location of the vessel boundary [52]. The geometric model resulting from the segmentation step can then be further refined, for example by using smoothing or rounding off the transitions at vessel bifurcations [33], resulting in a vessel boundary that can be used for a CFD simulation. Figure 25.1a shows an example of such a vessel boundary generated based on a CT scan using gradient-based thresholding with sub-voxel precision.

25.2.2 Computational Fluid Dynamics Model

In addition to the geometric boundary of the vascular structure inflow, outflow, and wall boundary conditions have to be defined properly for the CFD simulation [5, 49]. Inflow and outflow conditions arise from the fact that the current vascular structure has to be isolated from the rest of the arterial system. In practice, the flow rate or the speed profile at the inlets and the pressure at the outlets are utilized for a cardiac cycle. These quantities are obtained based on experimental measurements or by time-resolved Phase-contrast MRI flow measuring from the patient (see subsequent section). boundary condition arises from the fact that the vessel wall is distensible, which may influence the local hemodynamic and vice versa [44]. However, typically no proper characterization of the arterial wall, such as modulus of elasticity, wall thickness, or pressure wave that form at the wall, is available or is difficult to measure noninvasively [5]. Thus, a rigid wall is assumed in most cases, which also decreases the numerical computation time. With all boundary conditions defined, the flow can be computed based on the 3D unsteady Navier-Stokes equations for an incompressible Newtonian fluid. Typically, common CFD solvers are used for this step, for example ANSYS Fluent or OpenFOAM. The resulting velocity and pressure of the blood flow can subsequently be used for further analysis. For example, the computation of the dimensionless Reynolds number [6]. The Reynolds (Re) number characterizes the local flow behaviour in terms of laminar ($Re < 300$) or turbulent ($Re > 300$). In addition to velocity and pressure values, other hemodynamic quantities are obtained during the simulation. An important quantity is the wall shear stress (*WSS*), which represents the tangential force produced by blood moving across the vessel surface. It is known that *WSS* has an influence on the tissue structure of the vessel wall and it is likely that *WSS* plays an important role in initiation, growth and rupture of cerebral aneurysms [32]. The *WSS* can be computed based on the velocity field and the geometry [6, 17]. Figure 25.1b shows the result of such a CFD simulation using the inlet pressure and velocity based on a typical heart rate. This simulation is based on 124 time steps. For each time step, a grid size of 500,000 cells was used to accurately represent the vascular structure resulting in close to 900 MB of data. CFD simulations give blood flow information at high resolution. However, CFD simulations are based on models with assumptions and simplifications which make it difficult to obtain patient-specific accurate results.

25.3 Blood Flow Measurement

25.3.1 Acquisition Methods

Measured blood flow information is mostly obtained by quantitative ultrasound (US) acquisition (see Chap. 5). US is a cost-effective modality, providing flow information at high spatiotemporal resolution. However, US acquisition requires a skilled operator, is generally subject to a substantial amount of noise, and volumetric measurements of the vector velocity field are not possible. Consequently, this modality is less suitable for challenging cardiovascular conditions. Alternatively, computed tomography provides a limited number of blood flow acquisition sequences while delivering better signal-to-noise ratios. CT has the drawback of not measuring flow directly and exposing the patient to harmful radiation, which is impermissible for young patients. Instead, we focus on non-invasive Phase-Contrast (PC) MRI acquisition, which is the only modality providing volumetric quantitative measurements of blood flow velocities throughout the cardiac cycle. A typical size of such a volumetric data is $150 \times 150 \times 50$ voxels with velocity vectors with a resolution of $2 \times 2 \times 2.5$ mm per voxel, and a time series of 20–25 steps per cardiac cycle.

Phase-contrast MRI sequences enable acquisition of flow data that is linearly related to the actual blood flow velocities, capturing both speed and direction. This linear relation is described by the velocity encoding (VENC) acquisition parameter, representing the largest speed that can be measured unambiguously and is typically

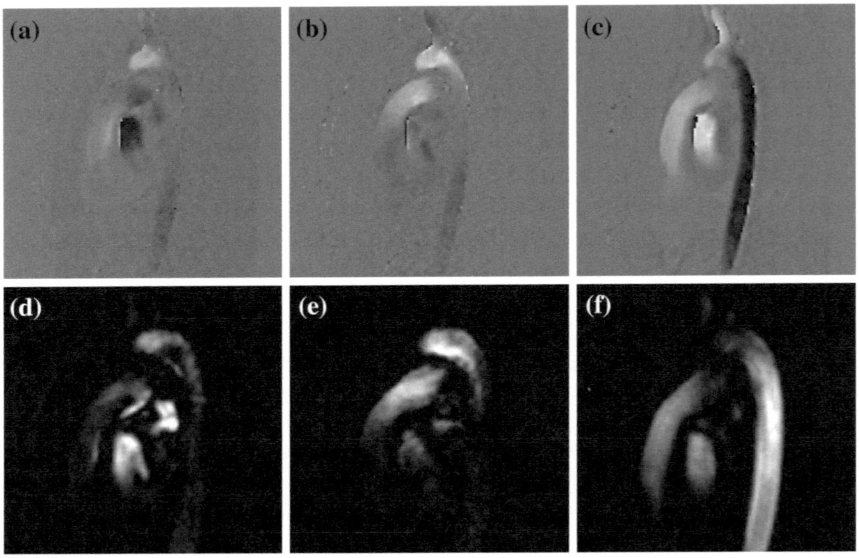

Fig. 25.2 The PC flow data set consists of 20 phases in time. Each phase in the series comprises a velocity vector volume with a resolution of $144 \times 144 \times 50$ voxels of $2.0 \times 2.0 \times 2.7$ mm. **a** PC-P right to left; **b** PC-P anterior to posterior; **c** PC-P head to feet; **d** PC-M right to left; **e** PC-M anterior to posterior; **f** PC-M head to feet © IEEE Reprinted, with permission, from IEEE transactions on visualization and computer graphics 16(6)

defined in centimeters per second. The range of the imposed speed limit, for example (−100 cm/s, 100 cm/s], corresponds to the phase extremities, i.e., $-\pi$ and π radians. If a suitable VENC is chosen, PC-MRI provides a data set with great correspondence to the actual blood flow velocity field [15]. As a consequence, the acquired data allows for quantitative analysis of the blood flow behavior. PC cine MRI sequences support the acquisition of volumetric blood flow data throughout the cardiac cycle, generating a 4D blood flow velocity field [27, 34]. There are two customary approaches to reconstruct the acquired raw data to the desired flow images [3]: phase (PC-P) and magnitude (PC-M) reconstruction. Figure 25.2 depicts a single slice of the reconstructed 4D flow data, at a certain point in time. The top row, Fig. 25.2a–c, represents the blood flow data in the three patient-oriented orthogonal directions, encoding both speed and directions of the blood flow quantitatively. This data is commonly referred to as the phase (PC-P) reconstruction. The bottom row, Fig. 25.2d–f, represents the blood flow data in three directions, encoding only speed. This data is commonly referred to as the complex difference or magnitude (PC-M) reconstruction. Even though the blood flow direction cannot be resolved from the PC-M reconstruction, the resulting data is inherently less prone to the uncorrelated noise that is typical for the PC-P reconstructed data.

25.3.2 Noise and Artifacts

Measurements come with imperfections which complicate the meaningful quantitative analysis of the flow data. In particular, measures derived from the data are sensitive to relatively small errors in the flow measurements. Inaccuracies are caused by a combination of many factors, associated with the MRI hardware, imaging sequences and their parametrization, and patient movement. For sensitive cardiac applications, the generally accepted objective is to acquire flow data with less than 10 % error [14].

The parametrization of the imaging sequence has a large influence on the accuracy of flow measurements. The parametrization directly influences the spatial and temporal resolution [15]. In particular quantitative analysis of small vessels (e.g., in the brain) becomes cumbersome at low resolution [2].

Besides user parametrization, motion is an important cause of imaging artifacts. There are three major causes of tissue displacement due to patient movement: motion artifacts by peristaltic motion, artifacts caused by contraction of the heart muscle, and respiration. Contraction of the heart muscle artifacts can be considerably reduced during acquisition. The impact of the respiratory motion can also be largely suppressed, by exploiting the relatively motionless period after exhalation.

In addition, flow measurements are subject to general MRI artifacts, largely due to hardware imperfections common in all MRI scans [4]. A relevant artifact for flow is due to the fast gradient switching which induces eddy currents in the electromagnetic field. This causes background phase errors in the image, which manifest as slowly varying image gradients in both the spatial and temporal domains. These effects are difficult to predict and therefore challenging to correct [14, 38]. The conventional MRI noise follows a Rician distribution. For flow imaging, it can be shown that the noise in flow regions depends on the velocity encoding speed and is inversely proportional to the SNR of the corresponding magnitude image [26]. Hence, the VENC parameter should be chosen as small as possible, while capturing the full dynamic range of the actual flow. The decision about the VENC value is often not easy to make.

There are additional artifacts that are specific to flow data. For instance, aliasing, or phase wrapping, erroneously introduces regions with opposite flow directions. Whenever the actual blood flow speed transcends the VENC value, a *phase wrap* occurs. Several methods have been devised to correct these artifacts caused by a single phase wrap through postprocessing [22, 54]. Another flow-specific artifact is misregistration where blood flow regions are shifted from the stationary tissue. This is due to the time between the phase encoding and frequency encoding gradients. These artifacts can be corrected by adding a bipolar gradient to each phase encoding gradient [51].

The flow imaging sequences are based on the assumption that the blood flow velocities are constant at the time of measurement. Hence, measurements of accelerated flows are less accurate and can cause undesirable artifacts. Accelerated flows can be found in pulsatile flows, stenotic flows, or jets that can cause so-called flow voids

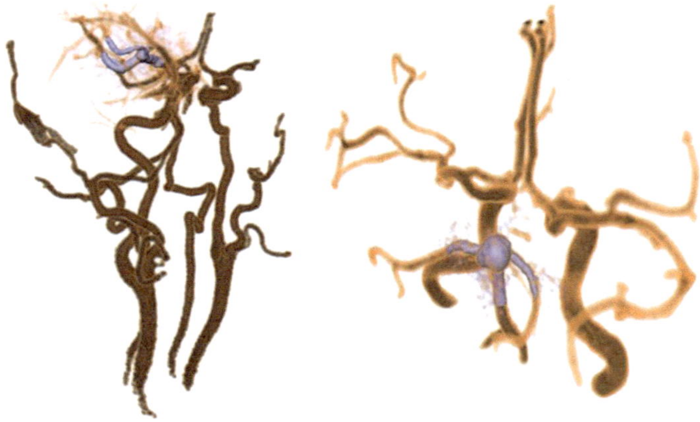

Fig. 25.3 The focus structures, the aneurysm and its immediate inflow and outflow are rendered opaquely with highly saturated colors. Context information is shown with a less striking colour with decreasing opacity for more distant vasculars structures

at relatively low spatial resolutions. To verify whether this behavior has occurred, black blood scans are often employed to inspect the vessel delineation.

25.4 Visual Exploration

25.4.1 Visualization of the Anatomical Context

The visual exploration of blood flow data is usually focused on a rather small anatomical region. In case of simulated blood flow data, this represents the domain where the simulation was performed. It may be necessary to present this focus region embedded in a somehow larger context to better understand the location of a pathology and the in- and outflow regions. Such a visualization goal may be achieved with a coordinated focus-and-context view, where the detail view presents only the target region and the context view provides the big picture with additional anatomical context. An integrated focus-and-context view is mentally easier to interpret. A reasonable strategy is to employ distance-based transfer functions [42], where the distance to the target anatomy is mapped to opacity in order to hide distant vascular structures. This strategy is illustrated in Fig. 25.3. The specific choice of colours and opacity as well as the amount of information to be displayed requires careful discussions with physicians [31]. Such a visualization may be a first step in a pipeline of exploration and analysis, as it presents an overview and needs to be followed by a more local analysis.

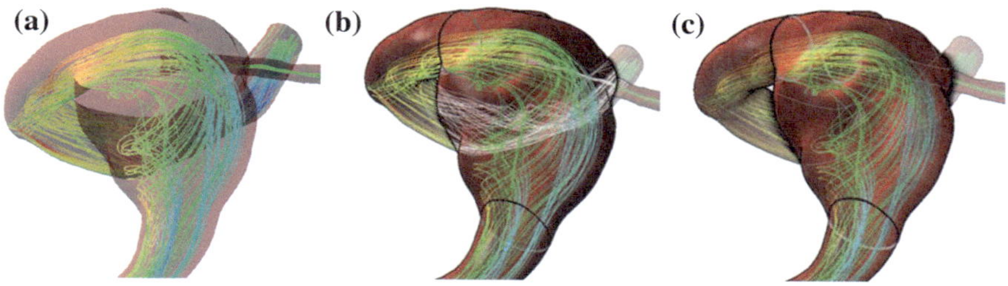

Fig. 25.4 Comparison between a semi-transparent visualization (**a**) and a ghosted view technique (**b**, **c**) applied to the enclosed vessel surface to show the internal flow. In **b** the hidden streamlines are depicted in grey and omitted in **c** to reduce visual clutter

The flow strongly depends on local variations of the enclosing vascular structures. Large changes in flow speed occur at stenotic regions, and turbulent flow occurs primarily at bifurcations or strongly curved areas. Thus, it is important to investigate the morphology of anatomical structures and the internal flow simultaneously. In case of simulated flow, such an integrated analysis may reveal that a significant flow feature is due to a small variation of the surface, which may result from an inaccuracy in the segmentation. The simplest idea to display flow and vascular anatomy at the same time is to render the vascular surface transparently. However, depending on the transparency level, either the vascular anatomy is hardly recognizable, or the internal flow is strongly obscured by the vessel wall.

As a remedy, smart visibility techniques [50], such as ghosted views, may be employed. The flow may be considered as an important object and the vessel walls transparency is modified to reveal flow lines. This idea has been realized by Gasteiger et al. [12]. The specific solution to provide ghosted view visualizations is based on a Fresnel reflection model [39], where the reflection term is replaced by opacity. In Fig. 25.4 a comparison of that technique with conventional semi-transparent rendering is presented. Gasteiger et al. refined their technique by an integration of landmarks described in the next section, and the ability to remove all hidden flow lines to further reduce visual clutter (see Fig. 25.4).

Van Pelt et al. [46] presented an anatomical context based on methods inspired by medical illustrations, where the detail is removed while the morphological information is preserved (see Fig. 25.5). To this end, they used cel-shaded silhouettes, combined with superimposed occluding contours. Hidden contours were visible during viewpoint interactions in order to resolve occlusion problems and to clarify spatial relations. Their user evaluation showed that these methods had a positive impact for the purpose of anatomical context representation of the flow.

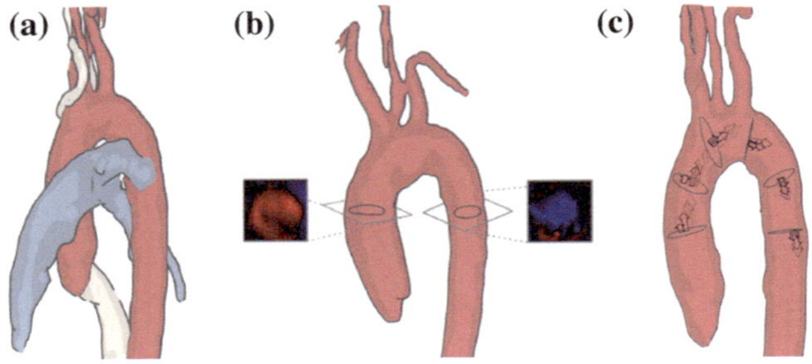

Fig. 25.5 Anatomical context visualization with cel shaded silhouettes and occluding contours. **a** Thoracic arteries. **b** MPR visualization exploded view of planes at cross-section positions. **c** Flow-rate arrow-trails arrows at 280 ms after the start of the cardiac cycle. [46] © IEEE Reprinted, with permission, from IEEE transactions on visualization and computer graphics 16(6)

25.4.2 Localization of Anatomical Landmarks

The exploration of vascular structures and the embedded flow benefits strongly from geometric descriptors that enable a carefully guided or constrained interaction. Such geometric descriptors may also be used to decompose the relevant anatomy in meaningful subregions to ease the exploration of complex flow data.

A widely used geometric descriptor is the vessel centerline, determined by a skeletonization algorithm (see, e.g., [19]). The vessel centerline is often used in order to move a cross-sectional plane that is always aligned perpendicular to the centerline, presenting the maximum-sized area. In conventional vessel analysis packages, the cross-sectional view displays the intensity values from the original image data, e.g., the CT Hounsfield values. In case of blood flow data, this strategy may be used to present any scalar value derived from the flow or the flow data itself, e.g., by using some glyph mapping. Van Pelt et al. [46] presented this cross-sectional visualization approach for the main arteries (see Fig. 25.5).

Better support for exploration tasks may be achieved by detecting and analyzing further anatomical landmarks of a particular region. Once these landmarks are identified, they may be used for labeling and for guiding movements in the complex 3D anatomy. The choice of such landmarks is specific for a particular anatomical region. We describe and illustrate this principle for the specific example of cerebral aneurysms. First, it is essential to understand *which* landmarks are actually important to characterize the local vessel anatomy. Neugebauer et al. [29] questioned a couple of neuroradiologists to draw cerebral aneurysms and extracted characteristic points commonly used by them. The following points were deemed essential (see Fig. 25.6).

- the *dome point* of an aneurysm,
- the (curved) *ostium plane*, where the blood enters the aneurysm, and
- a so-called *central aneurysm axis* (the closest connection between the parent vessel's centerline and the dome point)

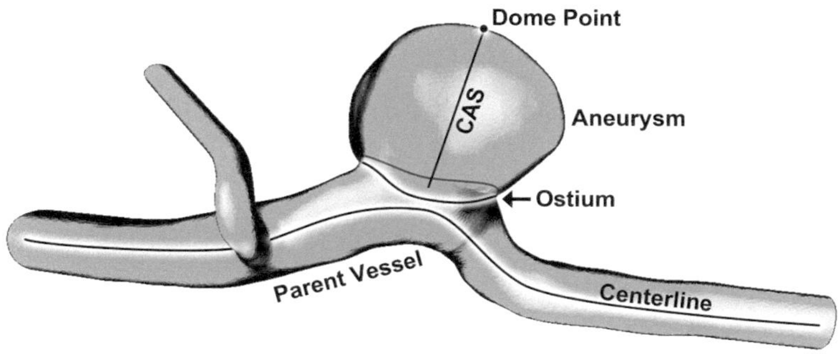

Fig. 25.6 Geometric descriptors for characterizing cerebral aneurysms

These landmarks may be utilized for example to move a plane along the central aneurysm axis or to employ the ostium plane as the seeding region, where streamline integration starts. The robust and precise detection is challenging due to the large variety of pathologic situations. Nevertheless even if it is successful only in about 90 percent of the cases, it provides a valuable support (see Neugebauer et al. [29] for a description of landmark extraction in saccular cerebral aneurysms). A similar landmark extraction process can be significant for other anatomical regions as well, since it provides a familiar reference frame for medical doctors. Constrained navigation does not necessarily mean that it is impossible to deviate from a predefined path. There are many variants to combine a preference direction with free exploration where the user is attracted to the predefined path, but may deviate. For a general discussion of constrained navigation techniques, see Hanson et al. [16] and more recently Elmqvist and Tudoreanu [8].

25.4.3 Exploration of Surface Flow Scalar Features

Blood flow simulations result in flow data as well as scalar flow features, such as pressure, speed and wall shear stress (WSS). It is known that WSS plays an essential role in the understanding of initiation and progression of vascular diseases. A simple solution to display scalar flow features is to show the surface of the relevant vascular region with a color-coded scalar flow feature.

The disadvantage of this simple solution is that only a small portion of the surface is visible at the same time. Map projections, which unfold an anatomical structure onto a plane, allow the visualization of the whole scalar information simultaneously. However, a map exhibits distortions (not all spatial relations, such as distances or angles can be preserved) and, even worse, a simple map is very hard to relate to the complex 3D anatomy of pathologic vessels. One promising approach is to combine a faithful 3D anatomy representation and a map view, where interaction in both views are synchronized. It is inspired by map views in other areas of medical diagnosis,

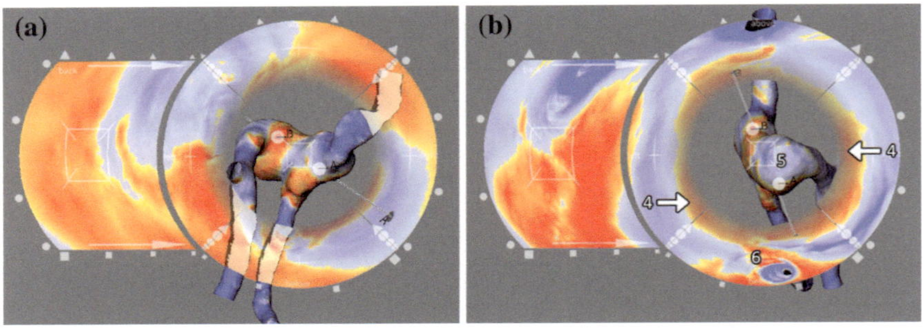

Fig. 25.7 A 3D model of the relevant vascular anatomy is surrounded by map views that display scalar flow features of five sides (features at the *left*, *right*, *bottom*, and *up* side are shown at the corresponding ring portions). Scalar features of the backside are shown at the most right display. The *lines* pointing from the map portions to the 3D view indicate correspondences, where scalar features are shown in both views. If the user drags a point, representing an interesting feature from a map view to the *center*, the anatomical model is rotated to make that region visible. All map views change accordingly

such as the Bull's eye plot in cardiology and stretched curved planar reformations in vessel diagnosis [18].

Neugebauer et al. [30] introduced a map display for scalar flow features, where the 3D anatomy model is presented in a central part and flow features of the surrounding sides are presented as flat regions of a map of the anatomical view. The map views and the 3D anatomy view are linked to depict positions of interest selected by the user (see Fig. 25.7). This enables a systematic exploration of all regions. Neuroradiologists emphasized that this technique enables a better exploration of scalar flow features.

Despite encouraging feedback obtained by Neugebauer et al., more evaluation and corresponding refinements are necessary to make this strategy broadly applicable. While in principle their approach is applicable to unsteady flows, it is likely that modifications are necessary if the scalar flow features change over time, leading to frequent changes of both views. Furthermore, flow information is volumetric. Although some measures are meaningful on the vessel wall, several flow features can only be analyzed through full volumetric visualization.

25.4.4 Blood Flow Probing

Time consuming segmentation is a necessary step in most tools for the inspection of measured flow data. Many tools incorporate functionality for local 2D segmentation of the vasculature [37]. This step is necessary not just to provide the anatomical context, as described in Sect. 25.4.1, but also to limit the domain where the vector field is considered valid. Due to the acquisition process, measured flow data presents values outside the vessel boundaries. This limits direct applicability of some visualization methods, such as integral lines or particle traces.

Segmentation techniques have been developed to directly segment measured 4D flow data. These methods are based on the assumption that the flow outside the vessel boundary exhibits incoherent behavior [7, 35, 40]. Van Pelt et al. [45] presented an extension of active surfaces to segment flow. Krishnan et al. [20] introduce a segmentation technique based on Finite-Time Lyaponov Exponents (FTLE). The main drawback of these methods is that they will fail in some pathologies due to the characteristics of the flow, e.g., areas with slow flow.

Most flow visualization techniques require seeding or region selection as initialization. The main reason for the selection is to avoid the clutter that visualizing the flow in the full domain supposes. The definition of the seeding region is usually done by probing in the volume domain, often with the help of segmentation.

Van Pelt et al. [46] presented a semi-automatic technique to probe cross-sections of anatomical data avoiding full segmentation (see Fig. 25.5). If anatomical data is not available an option for cross-sectional placement is to use the so called temporal maximum intensity projection (TMIP). For each voxel position of the TMIP scalar volume, the maximum speed is determined along the time axis of the 4D flow data. Hence, each voxel with a bright intensity indicates that a flow velocity with a substantial speed has occurred there at least once during the cardiac cycle. This probing method has several drawbacks: it assumes tubular structures, so it is only valid for vessels, and it does not consider the movement of the vessels during the heart cycle.

In later work, Van Pelt et al. [47] presented a probing technique to allow fast qualitative inspection, avoiding full segmentation. The user positions a 3D virtual probe on the viewing plane with common 2D interaction metaphors. An automatic fitting of the probe is provided for the third dimension, i.e., the viewing direction. Using the available velocity information of the measured blood flow field, the probe is aligned to this field. In particular, the automated fitting aligns the orientation of the long axis of the virtual probe to be tangential to the average local blood flow orientation. The probe is the basis for further visualizations (see Fig. 25.9).

Van Pelt et al. [46] also investigated different local seeding strategies based on the vessel center (e.g., radial or circular) concluding that fixed template seeding cannot accommodate flow variations. Krishnan et al. [20] presented a seeding strategy based on the segmentation of flow maps [41]. Flow maps are based on the end position of the particle after integration or advection. It is expected that this seeding strategy will adapt to real flow patterns.

In visualization, focus-and-context approaches are commonly used to avoid clutter. Gasteiger et al. [11] propose the FlowLens which is a user-defined 2D magic lens. This lens combines flow attributes by showing a different attribute and visualization within and outside the lens. Additionally, they incorporate a 2.5D lens to enable probing and slicing through the flow. To simplify the interface, they provide scopes which are task-based. Each scope consists of pairs of focus vs. context attributes, and propose visualization templates to represent each pair (see Fig. 25.8).

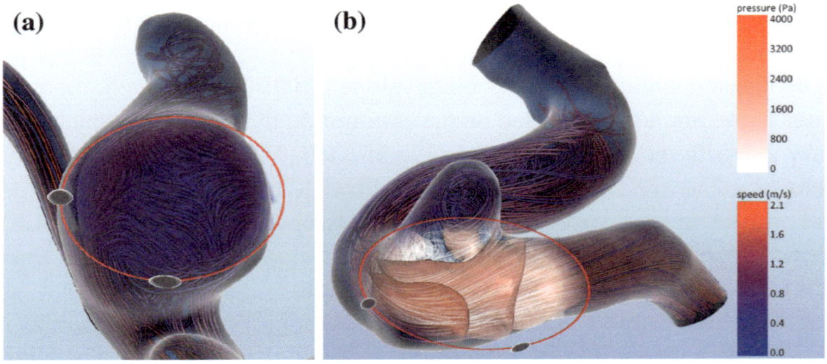

Fig. 25.8 Two examples of the FlowLens, which is depicted with a *red* contour line and two handles. Outside the lens the flow is visualized with illustrative and color-coded streamlines. Inside the lens a view-aligned probe plane with a LIC visualization for the investigation of the degree of vorticity (**a**) as well as the flow pressure as isosurfaces (**b**) are embedded

25.4.5 Blood Flow Visualization

Flow visualization has been an active field of research for decades. It has developed a large number of methods for inspecting flow data, and there are different review articles that define and classify these techniques [23, 36]. Although these techniques can be directly applied to blood flow fields, it is important to note that not all techniques are meaningful due to the characteristics of the data, e.g., measured data has low temporal resolution. Furthermore, the chosen visualization should be comprehensible to physicians and clinical researchers. In other words, the features shown should be linked to an intuitive understanding of the flow, and the pathology. In the remainder of this section, we will present the most common blood flow visualization techniques that have been proposed in literature, and their variations.

One of the most common ways to depict flow is using *integral curves*. There are two main approaches used for blood flow the so-called streamlines and pathlines. These integral curves represent the trajectory that a massless particle would follow through the vector field. Streamlines assume steady flow, so the vector field does not change in time. Pathlines are the extension of the streamlines that convey the temporal behavior of unsteady flow fields. Therefore, pathlines are the curves to depict particles trajectories in the vascular system. However, streamlines are still often used to depict instantaneous flow-field structure. In measured flow data, where the temporal resolution is low, streamlines can be informative, since error is accumulated at each integration step of the pathlines, and therefore the reliability of the lines decreases rapidly. Streaklines are another category of integral curves which have been used less often for blood flow. Streaklines are generated by a continuous seeding through time. Each point of the line corresponds to a seed that is continuously integrated through time.

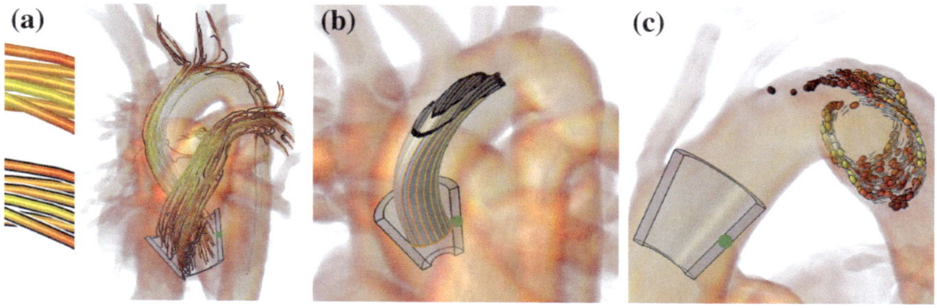

Fig. 25.9 GPU-generated long pathlines, statically seeded in the heart chambers to capture aorta and pulmonary arteries. **a** *shaded* imposter tuboids. The *left-hand* side shows a larger versions of the pathlines, respectively without and with halos and contours. **b** pathsurfaces presented as *tube-shaped* surfaces. The latter may be nested, approximately capturing the wave-front profile. A stripe pattern is employed to convey rotation [47] © IEEE. Reprinted, with permission, from IEEE transactions on visualization and computer graphics 17(12)

Integral curves are often rendered as illuminated lines or shaded tuboids. Perception of the spatial relations between pathlines is improved by means of halos [9]. Van Pelt et al. [47] additionally applied contours in order to enhance the structure of the pathlines (see Fig. 25.9a). Two types of seeding strategies are also proposed by Van Pelt et al. [46]. On the one hand, lines may be seeded statically from a fixed position in space and time. On the other hand, integration curves may be seeded dynamically, tracing the lines from a fixed spatial location, and varying seed time with the current time frame of the cardiac cycle. Dynamically seeded pathlines consist of comparatively short traces. Although the covered temporal range is relatively narrow, the pathlines are more reliable, and provide an approximative depiction of the pulse-wave in the cardiovascular system. The drawback is that it does not provide enough information on a large scale; it only provides sufficient local information.

Integral surfaces are a generalization of integral lines. Integral surfaces are formed by a continuum of integral curves. The surfaces enable surface shading techniques which improve the perception of 3D structures. Integral surfaces are initialized by a seeding curve which defines the topological connection between the integral curves. Depending on the integral curve used there exist: streamsurfaces, pathsurfaces and streaksurface. Integral surfaces have been recently studied for blood flow [20, 47].

The seeding curve used for initialization is crucial for the correct interpretation of the integral surface. Krishnan et al. [20] define the seeding curve as the boundary of segmented regions based on flow maps. Van Pelt et al. [47] presented cross-shaped and tube-shaped patterns of the integral surfaces (see Fig. 25.9b). Integral surfaces allow shading and texturing. For example, for tube-shaped surfaces stripes texturing emphasizes the rotational motion around the centerline. The color may convey various derived measures of the blood flow. In the user evaluation of Van Pelt et al. [47], the integral surfaces were considered valuable to explore the local rotational aspects of the flow.

Particle systems are readily applied to blood flow. Commonly, particles are depicted as spheres [48], or otherwise represented by small integral curves [37]. Both approaches convey blood flow speed through color, while direction information is captured by temporal cohesion of the animated particles. Integral curves additionally provide a short history of the particle trajectory. Both conventional approaches employ the available visual cues, such as color and shape, to capture only the blood flow velocity information. Van Pelt et al. [47] propose an illustrative particle-based approach that captures the velocity information by means of shape, keeping the color cue available for more elaborate blood flow characteristics. They mimic techniques often used to convey motion in comic books by deforming a ball at high-speed motion, and adding speed lines to improve the perception of direction (see Fig. 25.9c).

Visual clutter remains an important issue in 4D blood flow field visualizations. Usually, this clutter is avoided by user biased interaction methods which can miss important properties of the flow. Grouping vector field areas with meaningful similar characteristics, i.e., clustering, can help in developing techniques to improve the visual exploration and minimize the user bias. Some work exists in the clustering of static 3D vector fields [10, 21, 24, 43] while little research has been conducted to extend it to 3D unsteady flow fields [55]. Developing and extending these techniques to blood flow data is an interesting research direction. The main challenge is to provide a clustering that has a meaning for the user, and an adequate visualization technique that enables efficient exploration of the clusters.

25.5 Discussion and Open Issues

Simulated and measured blood flow data have been two distinct research fields that have developed in parallel. Simulation data is based on many assumptions, it is difficult to make it patient-specific, and also validation is a challenge. Measured flow data, on the other hand, represents the patient-specific flow, but it has a lot of limitation concerning resolution, artifacts, and noise in the data. An interesting direction is to combine both methods to strengthen each other. For example, measured data can be used as boundary conditions for a simulation, or simulation methods could be used to compensate for the lack of temporal resolution.

Blood flow data sets are considerably large data sets since they consist on a time series of vector-field volumes. The issue of dealing with large data will be of increasing importance given the improvements in spatial and/or temporal resolution that are expected.

Recently, new blood flow visualization techniques have been developed. Many decisions with respect to seeding, segmentation, integral curves, the use of illustration techniques are rather ad-hoc decisions based on intuition. It is important to link the decision to the users' needs. Additionally a more thorough exploration of the design space and comparisons of existing methods is needed. A major challenge is that this data is new to the domain experts, and it is challenging for them to identify the relevant features to visualize.

The existing methods are rather complex visual representations that overwhelm a considerable portion of the target user group. Future research should address simplifications of the blood flow by either clustering flow or by detecting and emphasizing relevant features. Existing techniques for flow field analysis may serve as orientation, but certainly need to be combined with the in-depth knowledge of domain experts regarding the relevance of certain blood flow characteristics. A better understanding of specific tasks, decisions and relevant information is necessary to support blood flow exploration with a guided workflow-based interaction.

Radiologists need to prepare reports where they summarize their findings verbally including relevant images. A better understanding of such reports may help to better support reporting, e.g., in case of cardiovascular diseases. While current applications are strongly focused on measured cardiac flow and simulated cerebral blood flow, advances in image acquisition will lead to further applications, e.g., where renal or liver flow is represented.

References

1. American Heart Association: Heart and stroke statistics (2010). http://www.americanheart.org/statistics
2. Arheden, H.k., Saeed, M., Törnqvist, E., Lund, G., Wendland, M.F., Higgins, C.B., Ståhlberg, F.: Accuracy of segmented MR velocity mapping to measure small vessel pulsatile flow in a phantom simulating cardiac motion. J. Magn. Reson. Imaging 13(5), 722–728 (2001)
3. Bernstein, M.A., Ikezaki, Y.: Comparison of phase-difference and complex-difference processing in phase-contrast MR angiography. Magn. Reson. Imaging 1(2), 725–729 (1991). http://www3.interscience.wiley.com/journal/112146770/abstract
4. Bernstein, M.A., Zhou, X.J., Polzin, J.A., King, K.F., Ganin, A., Pelc, N.J., Glover, G.H.: Concomitant gradient terms in phase contrast MR: analysis and correction. Magn. Reson. Med. **39**(2), 300–308 (1998)
5. Cebral, J.R., Castro, M.A., Appanaboyina, S., Putmann, C.M., Millan, D., Frangi, A.F.: Efficient pipeline for image-based patient-specific analysis of cebral aneurysm hemodynamics:technique and sensitivity. IEEE Trans. Med. Imaging **24**(4), 457–467 (2005)
6. Cebral, J.R., Putman, C.M., Alley, M.T., Hope, T., Bammer, R., Calamante, F.: Hemodynamics in normal cerebral arteries: qualitative comparison of 4d phase-contrast magnetic resonance and image-based computational fluid dynamics. J. Eng. Math. **64**(4), 367–378 (2009)
7. Chung, A.C.S., Noble, J.A., Summers, P.: Vascular segmentation of phase-contrast magnetic resonance angiograms based on statistical mixture modeling and local phase coherence. IEEE Trans. Med. Imaging **23**(12), 1490–1507 (2004)
8. Elmqvist, N., Tudoreanu, M.E., Tsigas, P.: Evaluating motion constraints for 3d wayfinding in immersive and desktop virtual environments. In: Proceedings of the ACM SIGCHI Conference on Human Factors in Computing Systems, CHI '08, pp. 1769–1778. ACM, New York, NY, USA (2008)
9. Everts, M.H., Bekker, H., Roerdink, J.B., Isenberg, T.: Depth-dependent halos: illustrative rendering of dense line data. IEEE Trans. Vis. Comput. Graph. 15, 1299–1306 (2009). http://doi.ieeecomputersociety.org/10.1109/TVCG.2009.138
10. Garcke, H.: Preuβer, T., Rumpf, M., Telea, A., Weikard, U., van Wijk, J.: A continuous clustering method for vector fields. In: Proceedings of the conference on Visualization '00. VIS'00, pp. 351–358. IEEE Computer Society Press, Los Alamitos, CA, USA (2000)

11. Gasteiger, R., Neugebauer, M., Beuing, O., Preim, B.: The FLOWLENS: A Focus-and-Context Visualization Approach for Exploration of Blood Flow in Cerebral Aneurysms. IEEE Trans. Vis. Comput. Graph. **17**, 2183–2192 (2011)
12. Gasteiger, R., Neugebauer, M., Kubisch, C., Preim, B.: Adapted surface visualization of cerebral aneurysms with embedded blood flow information. In: Eurographics Workshop on Visual Computing for Biology and Medicine (EG VCBM), pp. 25–32 (2010)
13. Gatehouse, P., Keegan, J., Crowe, L., Masood, S., Mohiaddin, R., Kreitner, K.F., Firmin, D.: Applications of phase-contrast flow and velocity imaging in cardiovascular MRI. Eur. Radiol. **15**, 2172–2184 (2005)
14. Gatehouse, P.D., Rolf, M.P., Graves, M.J., Hofman, M.B., Totman, J.J., Werner, B., Quest, R.A., Liu, Y., Spiczak von, J., Dieringer, M., Firmin, D.N., Rossum van, A., Lombardi, M., Schwitter, J., Schulz-Menger, J., Kilner, P.J.: Flow measurement by cardiovascular magnetic resonance: a multi-centre multi-vendor study of background phase offset errors that can compromise the accuracy of derived regurgitant or shunt flow measurements. J. Cardiovasc. Magn. Reson. **12**(5) (2010)
15. Greil, G., Geva, T., Maier, S.E., Powell, A.J.: Effect of acquisition parameters on the accuracy of velocity encoded cine magnetic resonance imaging blood flow measurements. Magn. Reson. Imaging **15**(1), 47–54 (2002). doi:10.1002/jmri.10029
16. Hanson, A.J., Wernert, E.A.: Constrained 3d navigation with 2d controllers. In: Proceedings of the 8th Conference on Visualization '97, VIS '97, pp. 175–182. IEEE Computer Society Press (1997)
17. Huo, Y., Kaimovitz, B., Lanir, Y., Wischgoll, T., Hoffman, J.I., Kassab, G.S.: Biophysical model of spatial heterogeneity of myocardial flow. Biophys. J. **96**(10), 4035–4043 (2009)
18. Kanitsar, A., Fleischmann, D., Wegenkittl, R., Felkl, P., Gröller, E.: CPR: curved planar reformation. pp. 37–44. IEEE Computer Society (2002)
19. Kirbas, C., Quek, F.: A review of vessel extraction techniques and algorithms. ACM Comput. Surv. **36**, 81–121 (2004)
20. Krishnan, H., Garth, C., Guhring, J., Gulsun, M., Greiser, A., Joy, K.: Analysis of time-dependent flow-sensitive PC-MRI data. IEEE Trans. Vis. Comput. Graph. **99**, 1 (2011). doi:10.1109/TVCG.2011.80
21. Kuhn, A., Lehmann, D.J., Gaststeiger, R., Neugebauer, M., Preim, B., Theisel, H.: A clustering-based visualization technique to emphasize meaningful regions of vector fields. In: Proceedings of Vision, Modeling, and Visualization (VMV 2011), pp. 191–198. Eurographics Assosciation (2011)
22. Langley, J., Zhao, Q.: Unwrapping magnetic resonance phase maps with chebyshev polynomials. J. Magn. Reson. Imaging **27**(9), 1293–1301 (2009)
23. Laramee, R.S., Hauser, H., Doleisch, H., Vrolijk, B., Post, F.H., Weiskopf, D.: The state of the art in flow visualization: dense and texture-based techniques. Comput. Graph. Forum **23**(2), 203–221 (2004)
24. Li, H., Chen, W., Shen, I.F.: Segmentation of discrete vector fields. IEEE Trans. Vis. Comput. Graph. **12**, 289–300 (2006). http://doi.ieeecomputersociety.org/10.1109/TVCG.2006.54
25. Lorensen, W.E., Cline, H.E.: Marching cubes: a high resolution 3d surface construction algorithm. In: Proceedings of the 14th Annual Conference on Computer Graphics and Interactive Techniques. SIGGRAPH '87, pp. 163–169. ACM, New York, NY, USA (1987)
26. Lotz, J., Döker, R., Noeske, R., Schüttert, M., Felix, R., Galanski, M., Gutberlet, M., Meyer, G.P.: In vitro validation of phase-contrast flow measurements at 3 T in comparison to 1.5 T: precision, accuracy, and signal-to-noise ratios. J. Magn. Reson. Imaging **21**(5), 604–610 (2005)
27. Markl, M., Chan, F.P., Alley, M.T.: L. Wedding, K., Draney, M.T., Elkins, C.J., Parker, D.W., Taylor, C.A., Herfkens, R.J., Pelc, N.J.: Time resolved three dimensional phase contrast MRI. Magn. Reson. Imaging **506**, 64 (2003)
28. Morbiducci, U., Ponzini, R., Rizzo, G., Cadioli, M., Esposito, A., De Cobelli, F., Del Maschio, A., Montevecchi, F., Redaelli, A.: In vivo: quantification of helical blood flow in human aorta by time-resolved three-dimensional cine phase contrast magnetic resonance imaging. Ann. Biomed. Eng. **37**, 516–531 (2009)

29. Neugebauer, M., Diehl, V., Skalej, M., Preim, B.: Geometric reconstruction of the ostium of cerebral aneurysms. In: Koch, R., Kolb, A., Rezk-Salama, C. (eds.) VMV 2010–Vision, Modeling, Visualization, pp. 307–314 (2010)
30. Neugebauer, M., Gasteiger, R., Beuing, O., Diehl, V., Skalej, M., Preim, B.: Map displays for the analysis of scalar data on cerebral aneurysm surfaces. Comput. Graph. Forum (EuroVis) **28**(3), 895–902 (2009)
31. Neugebauer, M., Gasteiger, R., Diehl, V., Beuing, O., Preim, B.: Automatic generation of context visualizations for cerebral aneurysms from MRA datasets. Int. J. Comput. Assist. Radiol. Surg. (CARS) **4**(Supplement 1), 112–113 (2009)
32. Nixon, A.M., Gunel, M., Sumpio, B.E.: The critical role of hemodynamics in the development of cerebral vascular disease. Neurosurgery **112**(6), 1240–1253 (2010)
33. Oeltze, S., Preim, B.: Visualization of vasculature with convolution surfaces: method, validation and evaluation. IEEE Trans. Med. Imaging **24**(4), 540–548 (2005)
34. Pelc, N.J., Herfkens, R.J., Shimakawa, A., Enzmann, D.R.: Phase contrast cine magnetic resonance imaging. Magn. Reson. Q. **7**, 229–254 (1991)
35. Persson, M., Solem, J.E., Markenroth, K., Svensson, J., Heyden, A.: Phase contrast MRI segmentation using velocity and intensity. Scale Space PDE Methods Comput. Vis. **3459**, 119–130 (2005)
36. Post, F.H., Vrolijk, B., Hauser, H., Laramee, R.S., Doleisch, H.: The state of the art in flow visualisation: feature extraction and tracking. Comput. Graph. Forum **22**(4), 775–792 (2003)
37. Pruessmann, K., Boesiger, P., Kozerke, S., Dydak, U. (Gyrotools): GTFlow. http://www.gyrotools.com/products/gt-flow.html (last visited)
38. Rolf, M.P., Hofman, M.B.M., Gatehouse, P.D., Markenroth-Bloch, K., Heymans, M.W., Ebbers, T., Graves, M.J., Totman, J.J., Werner, B., van Rossum, A.C., Kilner, P.J., Heethaar, R.M.: Sequence optimization to reduce velocity offsets in cardiovascular magnetic resonance volume flow quantification - a multi-vendor study. J. Cardiovasc. Magn. Reson. **13**(1), 18 (2011)
39. Schlick, C.: A customizable reflectance model for everyday rendering. In: Proceedings of Fourth Eurographics Workshop on Rendering, pp. 73–83 (1993)
40. Solem, J.E., Persson, M., Heyden, A.: Velocity based segmentation in phase-contrast MRI images. In: Medical Image Computing and Computer-Assisted Intervention, pp. 459–466 (2004)
41. Soni, B., Thompson, D., Machiraju, R.: Visualizing particle/flow structure interactions in the small bronchial tubes. IEEE Trans. Vis. Comput. Graph. **14**, 1412–1427 (2008). http://doi.ieeecomputersociety.org/10.1109/TVCG.2008.183
42. Tappenbeck, A., Preim, B., Dicken, V.: Distance-based transfer function design: specification methods and applications. In: Proceedings of Simulation und Visualisierung, pp. 259–274 (2006)
43. Telea, A., van Wijk, J.J.: Simplified representation of vector fields. In: Proceedings of the Conference on Visualization '99: Celebrating Ten Years. VIS '99, pp. 35–42. IEEE Computer Society Press, Los Alamitos, CA, USA (1999)
44. Torii, R., Oshima, M., Kobayashi, T., Takagi, K., Tezduyar, T.E.: Fluid-structure interaction modeling of blood flow and cerebral aneurysm: significance of artery and aneurysm shapes. Comput. Methods Appl. Mech. Eng. **198**(45–46), 3613–3621 (2009)
45. van Pelt, R., Nguyen, H., Haar Romeny, B., Vilanova, A.: Automated segmentation of blood flow regions in large thoracic arteries using 3D-cine PC-MRI measurements. Int. J. Comput. Assist. Radiol. Surg. 1–8 (2011)
46. van Pelt, R.: Oliván Bescós, J., Breeuwer, M., Clough, R.E., Gröller, M.E., ter Haar Romenij, B., Vilanova, A.: Exploration of 4D MRI blood flow using stylistic visualization. IEEE Trans. Vis. Comput. Graph. **16**(6), 1339–1347 (2010). doi:10.1109/TVCG.2010.153
47. van Pelt, R.: Oliván Bescós, J., Breeuwer, M., Clough, R., Gröller, M., ter Haar Romeny, B., Vilanova, A.: Interactive virtual probing of 4d mri blood flow. IEEE Trans. Vis. Comput. Graph. **17**, 2153–2162 (2011)
48. Venkataraman, S. (NVidia): 4D visualization of cardiac flow. In: GPU Technology Conference (2010). http://www.nvidia.com/content/GTC-2010/pdfs/2009_GTC2010.pdf

49. Venugopal, P., Valentino, D., Schmitt, H., Villablanca, J.P., Viñuela, F., Duckwiler, G.: Sensitivity of patient-specific numerical simulation of cerebal aneurysm hemodynamics to inflow boundary conditions. J. Neurosurg. **106**(6), 1051–1060 (2007)
50. Viola, I., Kanitsar, A., Groller, M.E.: Importance-driven feature enhancement in volume visualization. IEEE Trans. Vis. Comput. Graph. **11**, 408–418 (2005)
51. Vlaardingerbroek, M.T., den Boer, J.A.: Magnetic Resonance Imaging. Springer, Berlin (1999)
52. Wischgoll, T., Choy, J.S., Ritman, E., Kassab, G.S.: Validation of image-based extraction method for morphometry of coronary arteries. Ann. Biomed. Eng. **36**(3), 356–368 (2008)
53. World Health Organization: Cardiovascular diseases (CVDs) (2011). http://www.who.int/mediacentre/factsheets/fs317/
54. Yang, G.Z., Burger, P., Kilner, P.J., Karwatowski, S.P., Firmin, D.N.: Dynamic range extension of cine velocity measurements using motion-registered spatiotemporal phase unwrapping. J. Magn. Reson. Imaging **6**(3), 495–502 (1996)
55. Yu, H., Wang, C., Ma, K.L.: Parallel hierarchical visualization of large time-varying 3d vector fields. In: Proceedings of the 2007 ACM/IEEE Conference on Supercomputing, SC '07, pp. 24:1–24:12. ACM, New York, NY, USA (2007)

Part IV
Scalable Visualization

We live in the era of the data tsunami. Data is generated faster than our ability to digest the deluge of information. Visualization researchers and domain experts typically think of scalability in terms of the development of high-performance computational resources systems and of powerful new scientific instruments collecting vast amounts of data. These have lead to an unprecedented rate of growth of scientific data. The potential impacts to science of being able to effectively analyze these abundant sources of both simulation and measured/scanned data are tremendous. For effective data analysis, scalable visualization methods are required. Scalability takes on many forms from algorithmic scalability, the diversity of types of data, scalability across devices, scalable functional representations of data, and the integration into high-performance computational environments. This Part of the book addresses these issues.

In the first chapter, Garth and Gaither discuss the visualization and analysis, using integration-based methods, of large-scale vector fields on parallel architectures. They provide an overview and describe parallelization over seeds verses parallelization over blocks. The chapter concludes with a discussion of future directions.

As data sizes increase, one approach to scalability is to use feature-based techniques to represent the data with a smaller feature space. Bennet, Gyulassy, Pascucci, and Bremer discuss in Chap. 27 a feature hierarchy framework with two examples: the merge tree and the Morse-Smale complex. They describe how to perform interactive exploration of feature-based statistics and apply these concepts to a computational combustion example.

Scalability includes the number of different data sources as well as interactions between physical scales. In Chap. 28, Ebert, Gaither, and Lasher-Trapp explore system-of-systems and cross-scale issues and opportunities. The vast variety of data poses interesting scalability issues including visual scalability, software scalability, and information scalability. After discussing such issues, the authors conclude with an assessment of technology needs to address these.

Scalability includes both scaling up and scaling down. Scalable devices include smart phones, tablets, monitors, and display walls. In Chap. 29, Krüger and Hadwiger explore such hardware and software infrastructure for scaling across such diverse devices. Different user interfaces and rendering techniques are described as well as the outlook for future devices.

Scalable function representations are important for data analysis, processing, and storage. Jang provides an overview of common methods for multiscale representations in Chap. 30. There is an extensive bibliography included for further reading on this topic.

High-performance computational (HPC) resources provide a rich environment of ever increasing simulation data. In the final chapter of this Part, Flatken, Wagner, and Gerndt discuss infrastructure for visualization and interaction using HPC systems. The chapter provides an overview of current solutions and ongoing research in this domain. State-of-the-art infrastructures are described and example implementations are provided.

Charles Hansen
University of Utah
e-mail: hansen@cs.utah.edu

Chapter 26
Large-Scale Integration-Based Vector Field Visualization

Christoph Garth and Kelly Gaither

Abstract In this chapter, we provide a brief overview of the visualization of large vector fields on parallel architectures using integration-based methods. After briefly providing background, we describe the state of the art in corresponding research, focusing on parallel integral curve computation strategies. We analyze the relative benefits of two fundamental schemes and discuss algorithmic improvements presented recently. To conclude, we point out open problems and future research directions.

26.1 Introduction

Simulations on the current generation of supercomputers are producing data sets of unprecedented scale. To achieve the fundamental goal of scientific insight from the resulting very large datasets, a variety of problems must be addressed pertaining to their storage and handling. For simulations that involve vector fields, integral curves, or streamlines are one of the most illuminating techniques to obtain insight; they are a cornerstone of visualization and analysis across a great variety of application domains. Drawing on an intuitive interpretation in terms of particle movement, they are an ideal tool to illustrate and describe a wide range of phenomena encountered in the study of application-domain vector fields, such as transport and mixing in fluid flows. However, calculating integral curves in large data presents a significant challenge because their calculation is non-local and data dependent. Thus, leveraging parallel computational resources to achieve scalable and well-performing visualization in this setting requires optimal parallelization strategies that adapt smartly to the widely varying characteristics of integral curve problems.

C. Garth (✉)
University of Kaiserslautern, Postfach 3049, 67653 Kaiserslautern, Germany
e-mail: garth@cs.uni-kl.de

K. Gaither
University of Texas at Austin, 10100 Burnet Road, R87000, Austin, TX 78758, USA
e-mail: kelly@tacc.utexas.edu

C.D. Hansen et al. (eds.), *Scientific Visualization*, Mathematics and Visualization,
DOI 10.1007/978-1-4471-6497-5_26

In this chapter, after briefly introducing some background in Sect. 26.2, we report in Sect. 26.3 on the current state of research in large-scale integration-based visualization. Our primary intent is to provide a concise overview of recent research, and point readers interested in a deeper and more detailed description to relevant work. To conclude, we briefly discuss future research directions on parallel vector field visualization in Sect. 26.4.

26.2 Background

In the following, we will briefly introduce integral curves and motivate their use in vector field visualization

26.2.1 Vector Fields Visualization and Integral Curves

As simulation data sets are continuously growing in size and complexity, the use of parallel resources for visualization and analysis has become a key property of scientific work flows. While there are many methods available to visualize *scalar data* in parallel, involving such techniques as isosurfacing and volume rendering, the compendium of methods for vector field (and similarly, tensor fields) has remained comparatively smaller.

In applications, vector fields are typically described as discrete samples over a computational mesh or grid, with each sample denoting a vector. The *time-dependent* case is represented as an ordered set of time slices, spanning the temporal domain of the vector field. In contrast, *stationary* fields can be described as a single slice as a special case.

Integral curves are parametric curves that are everywhere tangent to a given vector field. They can be expressed as ordinary differential equations, and can be solved given an initial condition (starting point or *seed point*). Intuitively, integral curves correspond to the trajectories of massless particles that are advected by the vector field, and are thus naturally suited to visualize and analyze vector field processes such as transport and mixing.

Computationally, integral curves are approximated using numerical integration methods that solve the corresponding ordinary differential equations. There is an extensive body of work on this topic, and we refer the interested reader to [8]. In visualization applications, so-called Runge-Kutta methods are typically employed (e.g. the RK54 scheme); however, specialized integration schemes can be mandated by specific applications (see e.g. [17]). In general, integration schemes work in black box fashion, requiring evaluations of the vector field at points closely surrounding the particle trajectory during the approximation process. Thus, interpolation must be used on discrete data to achieve a continuous vector field representation. Again,

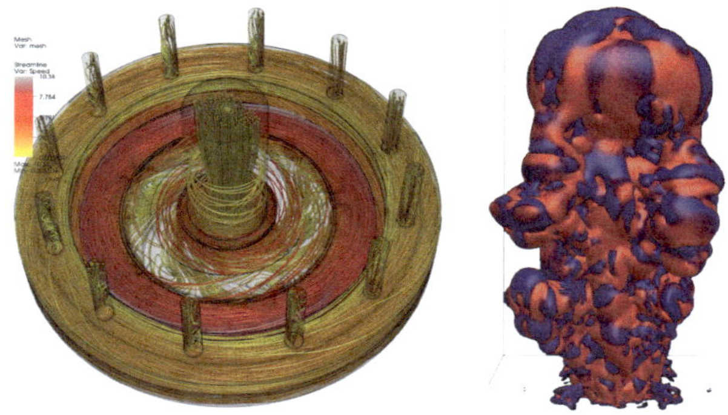

Fig. 26.1 *Left* Direct visualization of integral curves in a spinning turbine. *Right* Visualization of Lagrangian coherent structures in a jet flow

while linear interpolation is commonly found in visualization applications, other interpolation schemes can be used if specific problems mandate it.

26.2.2 Integration-Based Visualization

The visualization and analysis of vector fields is an active research area, and so-called *integration-based* techniques that derive vector field visualization from integral curves have progressed well beyond the direct depiction of individual integral curves (see Fig. 26.1 left) or a small subset of them [12]. Integral surface techniques [7, 9] compute and visualize a surface consisting of all streamlines emanating from a common curve, while flow volumes examine the behavior of entire volumes of particles [1, 22]. Topological methods, on the other hand, aim at extracting the structural skeleton of a vector field by considering the dynamical system induced by it and computing critical points and stable and unstable manifolds. More recently, the notions of *Finite-Time Lyapunov Exponents* and *Lagrangian Coherent Structures* [11, 18] were introduced to allow for an accurate structural analysis of time-varying vector fields. These *Lagrangian* methods, which can require many thousands to millions of integral curves, are built on observing the separation between closely neighboring particles as they are advected, and coherent structures are then identified by lines and surfaces along which this separation is maximal (see Fig. 26.1 right).

The computational overhead induced by integration-based vector field analysis tools significantly extends beyond the approximation of few curves. In the case of integral surfaces, hundreds to thousands of integral curves must be computed, whereas for Lagrangian methods, the required curves can number in the millions. The improving computational capability of modern supercomputers allows the simulation of vector fields in unprecedented detail and results in extremely large datasets. Here, data size itself poses an additional challenge to integration-based visualization. Due to the highly non-linear, data-dependent nature of the integral curve approximation process, traditional approaches for distributing computation and data to leverage available computing resources optimally are not applicable in this situation. Hence, approaches developed for the analysis of large scalar field data that are built on decomposing a dataset and independently treating each part do not generalize well to the vector field case directly. To further complicate the choice of parallelization approach, the computational characteristics of integral curve computation are strongly dependent on a variety of factors such as vector field complexity, data set size, seed set size, and integration length.

26.2.3 Parallel Integral Curve Computation

Computing many integral curves simultaneously is an embarrassingly parallel problem, since the curves are mutually independent. In this respect, parallelization is achieved in a straightforward manner by decomposing the overall set of integral curves. However, issues arise when data size grows. In the next section, we will discuss the problems of computing integral curves on large data and aim at an approximate characterization of integration-type problems.

Focusing on data size, early work on parallel integral curve computation has focused primarily on out-of-core techniques that are commonly used in large-scale data applications where data sets are larger than main memory. These algorithms focus on achieving optimal I/O performance to access data stored on disk. Ueng et al. [20] presented a technique to compute streamlines in large unstructured grids using an octree partitioning of the vector field data for fast fetching during streamline construction. Taking a different approach, Bruckschen et al. [2] describe a technique for real-time particle traces of large time-varying data sets, by isolating all integral curve computation in a pre-processing stage. The output is stored on disk and can then be efficiently loaded during the visualization phase. Similarly, PC cluster systems were leveraged to accelerate advanced integration-based visualization algorithms, such as time-varying Line Integral Convolution (LIC) volumes [13] or particle visualization for very large data [6]. While such approaches can give good performance, they do not generalize to more modern vector field visualization techniques such as integral surfaces or Lagrangian Coherent Structure visualization.

In the following, we will introduce general integral curve problems as a basis for modern integration-based flow visualization, describe their characteristics, and discuss corresponding algorithms.

26.2.4 Problem Description and Classification

Given a vector field data set, a set of seed points in its domain, and parameters that control the integration process, an *integral curve problem* consists of computing all integral curves emanating from all seed points. Parameters typically pertain to solver accuracy and integration length. Assuming that the vector field data is decomposed into connected regions or *blocks*, integral curves likely traverse multiple blocks, each of which has to reside in main memory on the processor performing the integration of the curve. Thus, the data access pattern is not only dependent on the data set, but also influenced by the choice of seed points and integration parameters.

Studying the effectiveness of parallelization approaches to integral curve problems, Pugmire et al. [16] classified scenarios according to three categories with strong implications on performance. We briefly recapitulate their description here.

Seed Set Size If the given problem requires only the computation of tens to hundreds of streamlines, parallel computation takes a secondary place to optimal data distribution or loading; the corresponding seed set is referred to as *small*. This case is typically encountered in exploratory visualization scenarios where comparatively few integral curves are interactively seeded by a user. In contrast, a *large* seed set encompasses thousands to millions of seed points for integral curves. For such problems, parallel computation of integral curves must be employed.

Seed Set Distribution Similarly, the distribution of seed points is an important problem characteristic. If seed points are concentrated, i.e. located *densely*, within a small region of the vector field domain, it is likely that all integral curves will traverse a relatively small amount of the overall data. For some applications such as integral curve statistics, on the other hand, a *sparse* seed set covers the entire vector field domain. This results in integral curves traversing the entire data set. Hence, the seed set distribution determines strongly if performance stands to gain most from parallel computation, data distribution, or both.

Vector Field Complexity The structure of a vector field can have a strong influence on which parts of the data need to be taken into account in the integral curve computation process. Critical points or invariant manifolds of strongly attracting nature draw streamlines towards them, and the resulting integral curves seeded in or traversing their vicinity remain closely localized. On the other hand, the opposite case of a nearly uniform vector field requires integral curves to pass through large parts of the data. This data dependency of integral curve computation is both counterintuitive and hard to identify without conducting prior analysis to determine the field structure.

Overall, these problem characteristics determine to what extent an integration-based problem can profit from a chosen parallel computation and data distribution. Jointly, they affect the three main cost factors inherent in parallel algorithms—communication, I/O, and computation—that need to be balanced to achieve optimal performance. We will next describe two basic approaches and briefly survey advanced algorithms that provide improved performance and efficiency in certain cases.

26.3 Parallelization Strategies

In all algorithm discussion below, we consider vector field data as decomposed into a number of spatially disjoint blocks. In the case of time-varying vector fields, the blocks are assumed to be four-dimensional, i.e. encompass a time interval in addition to a spatial subset. There are two straightforward parallelization approaches that partition either the computation workload, with seed points as the elementary unit of work, or the data set, where data blocks are distributed.

26.3.1 Parallelization Over Seeds

The parallelize-over-seeds (POS) algorithm parallelizes over the set of seed points, assigning to each processor a fixed number of seed points from which integral curves are propagated. Data blocks are loaded on demand when required, i.e. when integration must be continued in a block that is not present in memory. Communication is not required in this scheme except to synchronize processors at initialization and termination. Clearly, the dominant cost of this scheme is I/O and can be expressed as the number of blocks loaded. To alleviate overall I/O cost, blocks can be kept in processor memory to be reused (caching).

Pugmire et al.[16] provided a detailed analysis of the performance of POS over a range of streamline problems. They make aggressive use of caching, new blocks are only loaded when no streamline can be continued on the current resident blocks. Cache eviction is performed according to least recently used order.

The initial assignment of seed points to processors is chosen based on spatial proximity, following the reasoning that the integral curves traced from spatially close seed points are likely to traverse the same regions, and thus blocks, of a dataset. This approach is effective in increasing block reuse, especially for dense seed sets. They also sketch out worst case scenarios. For example, if streamlines travel nearly in cycles, and if the number of blocks encountered along these cycles exceeds the size of the cache, performance drops significantly.

In general, caching can only be effective if block reuse is high; this is often the case for stationary fields, but less frequent for time-varying fields since curves advance monotonically along the time axis and thus traverse previously not encountered spatiotemporal blocks frequently.

A different approach to address I/O cost was described by Wolter et al. [21] by introducing intelligent prefetching to overlap I/O and computation. By building a Markov model of block transitions, they are able to predict likely future blocks which can then be loaded ahead of time. Two strategies are proposed for initializing the Markov model, both requiring pre-computation to build a transition probability model. They find much increased performance for the stationary case, but point out limitations for time-varying problems that stem from a construction of the prediction model on individual time slices.

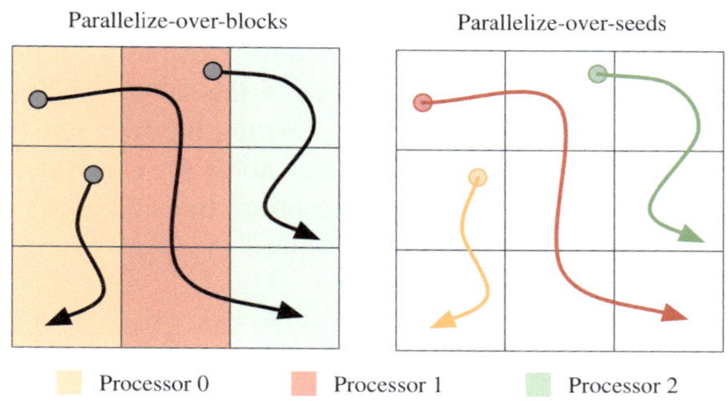

Fig. 26.2 An illustration of basic parallelization strategies for integral curves. *Left* Parallelize-over-blocks distributes data, and integration is performed on the processor owning required blocks. *Right* Parallelize-over-seeds statically assigns curves to processors; data needed for integration must be loaded on-demand

A much simpler strategy for the latter case was pointed out by Lane [10]. Here, prefetching is achieved by limiting integration to short time intervals; while computation is running, blocks for future time steps are loaded. However, his method loads all blocks corresponding to the next time step, and is thus not truly a POS-type method.

Overall, POS is a robust strategy. While it is generally found to be I/O intensive, it does not possess limitations with respect to data size, and given enough parallel resources, can in principle treat problems of arbitrary size.

26.3.2 Parallelization Over Blocks

The parallelize-over-blocks (POB) approach distributes data blocks across processors using a fixed assignment. Integral curves are communicated between the tasks to migrate them to the task owning the block required to continue integration. This algorithm performs minimal I/O; before integration commences, every task loads all blocks assigned to it, leveraging maximal parallel I/O capacity.

An early implementation of this algorithm for streamlines was described by Sujudi and Haimes [19], who made use of distributed computation by splitting the block in consideration into several sub-blocks and assigning each processor one such sub-block. A streamline is communicated among processors as it traverses different sub-blocks. Their design is a client/server architecture, in which the clients perform integration and the server manages routes trajectories between clients.

A systematic investigation of the performance and scalability of the POB approach, applied to streamline problems and spanning a wide range of integration problems, was given by Pugmire et al. [16]. In their implementation, processors directly communicate solver state to each other when integration traverses block boundaries. Curve geometry generated during the integration remains with the processor that generated it, and is merged in a post-processing step; thus, their benchmarks reflect only the curve propagation portion of a visualization pipeline.

In their findings, the communication cost incurred by block transitions plays a secondary role to the overall number of blocks traversed. Essentially, in the case of a dense seed set, streamlines in the same spatial regions must be computed by comparatively few processors owning the required blocks, leading to a large load imbalance. A similar observation holds for vector fields with regions of strong convergence, which tends to concentrate trajectories in small regions of a data set. To counteract this effect, they propose to assign blocks to processors in round-robin fashion or randomly.

Nouanesengsy et al.[14] provided an approach to address the load imbalances in POB by using workload-aware partitioning of the vector field data. Treating block assignment as an optimization problem, they compute a flow graph from an abstract representation of the vector field. In conjunction with a given seed set, the flow graph provides an estimate of blocks that need to be traversed to compute curves, and the resulting set of blocks is then partitioned to minimize communication cost. Additionally, to leverage resources optimally, block replication is allowed, addressing starvation problems. In experiments, they observe better performance and scalability with respect to a round-robin assignment; however, they incur a non-negligible preprocessing cost per data set.

While it is straightforward to implement, the POB algorithm has one severe limitation; if the combined memory on all processors cannot accommodate the entire data set, this algorithm cannot be used. This is especially the case for time-varying vector fields.

26.3.3 Adaptive Load Balancing

As apparent from the discussion above, POS and POB represent two extremes of a spectrum of parallelization schemes, focusing exclusively on either work distribution or data distribution. To address the respective shortcomings of either choice, Pugmire et al. [16] describe an adaptive algorithm that performs dynamic distribution of both integral curves and blocks, attempting to balance I/O and communication loads at run-time based on heuristics.

Algorithmically, a master processor coordinates the workloads of a fixed set of slave processors. The master initially determines a seed point assignment for the slaves; as work progresses, it monitors the length of each slave's assigned integral curve set and the blocks that are loaded. To avoid overload and starvation, it then reassigns individual curves or entire blocks among slaves. Once the master determines that all curves have been computed, it instructs the slaves to terminate. This adaptive approach entails significant complexity, which is expressed in complicated heuristics. The authors also investigate the scalability of their approach, but find that using a single master becomes a bottleneck as the number of processors grows. They postulate that a hierarchical approach with a hierarchy of master processes should perform better at scale. Furthermore, the load-balancing heuristics rely on several machine-dependent parameters such as I/O and communication latency and speed for best performance. However, if this effort is undertaken, results on up to 256 processors indicate that the adaptive approach performs at least as good as the better of POS or POB.

Peterka et al. [15] propose a different approach to achieve load balancing. They partition integration into multiple rounds, with each round advancing integral curves only over a small time interval. After each round, the work load is analyzed and redistribution is performed. Load is primarily measured as the number of particles residing on each processor.

To actually balance the load, geometric partitioning using recursive coordinate bisection is employed; this ensures that particles assigned to a processor for the next round are likely to reside in the same blocks, effecting an overall reduction of I/O. Furthermore, since particle exchange is performed only between rounds, communication can be optimized. Overall, reasonable scaling is observed up to 16,384 processors.

26.3.4 Hybrid Parallelism

To take advantage of modern parallel architectures where a single node typically contains multiple processor cores with shared memory, Camp et al. [4] presented and studied the performance of a hybrid parallel implementation of both POS and POB. Their implementation is based on a combination of classical message passing across nodes with multiple threads per node, and the paper focuses on comparing hybrid algorithm variants to non-hybrid ones. In their hybrid implementation of POS and POB, worker threads perform actual integration work, while I/O and communication are managed by separate threads.

For POB, multiple separate I/O threads identify blocks to be loaded and initiate I/O if there is room in the cache; when a block has finished loading, corresponding integral curves are added to a work queue shared by the worker threads. In their implementation, the number of worker and I/O threads is in principle arbitrary. However, they propose to use one worker thread and one I/O thread for each core to leverage overlapping I/O.

Similarly, in the hybrid POB algorithm, several worker threads process streamlines in the resident set of blocks, and a single communication thread is responsible for receiving streamlines and sending them to other processors. Due to complications with the MPI message passing library, however, the communication thread has to resort to polling to identify sendable or newly received streamlines. Thus, one core is exclusively dedicated to this task, while the remaining cores perform integration.

Comparing the performance of hybrid and non-hybrid implementations for several test cases on 128 cores in total (32 nodes with 4 cores each), they report significant performance gains. Largely, this is a consequence of the increased memory available to each process, which positively influences cache size and thus block reuse in the POS case, and reduces starvation in the POB algorithm. However, they also describe specific cases in which no improvement in performance can be obtained. In general, their results indicate that the benefit obtained from a hybrid implementation can be significant, at the cost of increased implementation complexity.

26.3.5 Extended Memory Hierarchies

Continuing their work on leveraging architectural features of modern supercomputing architectures, Camp et al. [3] also investigated the benefits of using an extended memory hierarchy in combination with the POS approach. In their paper, they considered node-local secondary storage (such as SSDs or conventional hard drives) that can be used to add a secondary layer of caching, reducing the number of global I/O operations. Upon evicting a block from the cache, their modified algorithm writes it to secondary storage. Hence, if a block is encountered by the same processor again but was previously discarded from main memory, it can be quickly re-loaded from the secondary cache.

The given comparison shows that a significant benefit can be achieved from this with respect to a baseline implementation that performs global I/O exclusively. Their results indicate that for many different test cases, a large majority of block loads is accelerated, resulting in much increased performance overall.

26.3.6 Other Techniques

Instead of aiming at user-controlled integration-based visualization, i.e. attempting the solution of an arbitrary integral curve problem such as the algorithms discussed above, a number of authors propose to trade off flexibility of the visualization with increased performance and scalability. For example, Yu et al. [23] introduced a parallel integral curve visualization that computes a set of representative, short integral segments termed pathlets in time-varying vector fields. A preprocessing step computes a binary clustering tree that is used for seed point selection and block decomposition. This seed point selection method mostly eliminates the need for communication between processors, and the authors are able to show good scaling behavior for large data. However, this scaling behavior comes at the cost of increased preprocessing time and, more importantly, loss of the ability to choose arbitrary, user-defined seed-points. Chen and Fujishiro [5] apply a spectral decomposition using a vector-field derived anisotropic differential operator to achieve a similar goal.

26.4 Discussion and Future Directions

While the state of the art in large-data integration-based visualization has considerably matured recently, a satisfactory method that can satisfy all visualization requirements remains elusive. Future research must address a number of difficult problems:

- While data sets of significant size have proven feasible, the approaching exascale generation of hardware architectures will impose new demands on visualization algorithms; this is especially the case for parallel integral curve algorithms. It is unclear at this point whether the load balancing strategies discussed in Sect. 26.3.3 can be adapted to such scenarios. More general schemes are required that provide adequate scalability and do not require data pre-analysis, which will likely be prohibitive on exascale data.
- Current algorithms still require minutes to achieve results even for small problems due to the factors discussed in Sect. 26.2.4. This stands in stark contrast with the requirements of user-guided, interactive exploration that has proven extremely valuable in obtaining insight into the complex nature of vector fields. Here, progressive algorithms are needed that can quickly produce approximate results which are refined over time. However, the non-local nature of integral curves has thus far prevented that application of downsampling techniques on this problem.
- Similarly, as the volume of data resulting from simulation codes will grow past the point of feasible retention on external storage, it is anticipated that future visualization algorithms will possess a large in situ component, requiring the majority of the analysis to be performed during a simulation run. At this point, it is entirely unclear how to achieve user-guided vector field visualization in an in situ scenario.

While first steps in these directions have been taken, many problems remain open, and more research in this interesting area is needed.

References

1. Becker, B. G., Max, N. L., Lane, D. A.: Unsteady flow volumes. In: Proceedings of IEEE Visualization, pp. 329–335 (1995)
2. Bruckschen, R., Kuester, F., Hamann, B., Joy, K. I.: Real-time out-of-core visualization of particle traces. In: Proceedings of the IEEE Symposium on Parallel and Large-Data Visualization and Graphics (PVG), pp. 45–50, Piscataway, NJ, USA, IEEE Press (2001)
3. Camp, D., Childs, H., Chourasia, A., Garth, C., Joy, K. I.: Evaluating the benefits of an extended memory hierarchy for parallel streamline algorithms. In: Proceedings of the IEEE Symposium on Large-Scale Data Analysis and Visualization (LDAV), IEEE Press (2011)
4. Camp, D., Garth, C., Childs, H., Pugmire, D., Joy, K.I.: Streamline integration using MPI-Hybrid parallelism on a large multicore architecture. IEEE Trans. Vis. Comput. Graph. **17**(11), 1702–1713 (2011)
5. Chen, L., Fujishiro, I.: Optimizing parallel performance of streamline visualization for large distributed flow datasets. In: Proceedings of IEEE VGTC Pacific Visualization Symposium 2008, 87–94 (2008)

6. Ellsworth, D., Green, B., Moran, P.: Interactive terascale particle visualization. In: Proceedings of IEEE Visualization, pp. 353–360, IEEE Computer Society, Washington, DC, USA, (2004)
7. Garth, C., Krishnan, H., Tricoche, X., Bobach, T., Joy, K. I.: Generation of accurate integral surfaces in time-dependent vector fields. IEEE Trans Vis Comput Graph. **14**(6), 1404–1411 (2008)
8. Hairer, E., Nørsett, S. P., Wanner, G.: Solving ordinary differential equations I, second edition. Springer, Berlin (1993) (volume 8 of Springer Series in Computational Mathematics)
9. Hultquist, J. P. M.: Constructing stream surfaces in steady 3D vector fields. In: A. E. Kaufman and G. M. Nielson. (eds.) Proceedings of IEEE Visualization, pp. 171–178, Boston, MA (1992)
10. Lane, D. A.: UFAT—A particle tracer for time-dependent flow fields. In: Proceedings of IEEE Visualization '94, pp. 257–264 (1994)
11. Mathur, M., Haller, G., Peacock, T., Ruppert-Felsot, J., Swinney, H.: Uncovering the lagrangian skeleton of turbulence. Phys. Rev. Lett., submitted (2006)
12. McLoughlin, T., Laramee, R.S., Peikert, R., Post, F.H., Chen, M.: Over two decades of integration-based, geometric flow visualization. Comput. Graph. Forum **29**(6), 1807–1829 (2010)
13. Muraki, S., Lum, E., Ma, K.-L., Ogata, M., Liu, X.: A PC cluster system for simultaneous interactive volumetric modeling and visualization. In: Proceedings of the IEEE Symposium on Parallel and Large-Data Visualization and Graphics (PVG), p. 13, IEEE Computer Society Washington, DC, USA (2003)
14. Nouanesengsy, B., Lee, T.Y., Shen, H.W.: Load-balanced parallel streamline generation on large scale vector fields. IEEE Trans. Vis. Comput. Graph. **17**(12), 1785–1794 (2011)
15. Peterka, T., Ross, R., Nouanesengsey, B., Lee, T. Y., Shen, H. W., Kendall, W., Huang, J.: A study of parallel particle tracing for steady-state and time-varying flow fields. In: Proceedings of 25th IEEE International Parallel & Distributed Processing Symposium, Anchorage, AK (2011)
16. Pugmire, D., Childs, H., Garth, C., Ahern, S., Weber, G.: Scalable computation of streamlines on very large datasets. In: Proceedings of Supercomputing (2009)
17. Sanderson, A.R., Chen, G., Tricoche, X., Pugmire, D., Kruger, S., Breslau, J.: Analysis of recurrent patterns in toroidal magnetic fields. IEEE Trans. Vis. Comput. Graph. **16**(6), 1431–1440 (2010)
18. Shadden, S., Dabiri, J., Marsden, J.: Lagrangian analysis of fluid transport in empirical vortex ring flows. Phys. Fluids **18**, 047105 (2006)
19. Sujudi, D., Haimes, R.: Integration of particles and streamlines in a spatially-decomposed computation, IEEE Computer Society Press. In: Proceedings of Parallel Computational Fluid Dynamics, Los Alamitos, CA (1996)
20. Ueng, S.K., Sikorski, C., Ma, K.-L.: Out-of-core streamline visualization on large unstructured meshes. IEEE Trans. Vis. Comput. Graph. **3**(4), 370–380 (1997)
21. Wolter, M., Gerndt, A., Kuhlen, T., Bischof, C.: Markov prefetching for multi-block particle tracing on parallel post-processors. In: J. Kown, A. Ecer, J. Periaux, N. Satofuka, and P. Fox, (eds.) Parallel Computational Fluid Dynamics: Parallel Computings and its Applications, Proceedings of the Parallel CFD Conference, pp. 27–34. Elsevier, London (2007)
22. Xue, D., Zhang, C., Crawfis, R.: Rendering implicit flow volumes. In: Proceedings of the IEEE Visualization '04 Conference, pp. 99–106 (2004)
23. Yu, H., Wang, C., Ma, K.-L.: Parallel hierarchical visualization of large time-varying 3D vector fields. In: Proceedings of Supercomputing (2007)

Chapter 27
Large Scale Data Analysis

Janine Bennett, Attila Gyulassy, Valerio Pascucci and Peer-Timo Bremer

Abstract As data sets grow in size and complexity, global analysis methods do not necessarily characterize the phenomena of interest, and scientists are increasingly reliant on feature-based analysis methods to study the results of large-scale simulations. This chapter presents a framework that efficiently encodes the set of all possible features in a hierarchy that is augmented with attributes, such as statistical moments of various scalar fields. The resulting meta-data generated by the framework is orders of magnitude smaller than the original simulation data, yet it is sufficient to support a fully flexible and interactive analysis of the features, allowing for arbitrary thresholds, providing per-feature statistics, and creating various global diagnostics such as Cumulative Density Functions (CDFs), histograms, or time-series. The analysis is combined with a rendering of the features in a linked-view browser that enables scientists to interactively explore, visualize, and analyze data resulting from petascale simulations. While there exist a number of potential feature hierarchies that can be used to segment the simulation domain, we provide a detailed description of two: the merge tree and the Morse-Smale (MS) complex, and demonstrate the utility of this new framework in practical settings.

J. Bennett (✉)
Sandia National Laboratories, Livermore, CA, USA
e-mail: jcbenne@sandia.gov

A. Gyulassy · V. Pascucci
Scientific Computing and Imaging Institute, University of Utah, Salt Lake City, UT, USA
e-mail: jediati@sci.utah.edu

V. Pascucci
e-mail: pascucci@sci.utah.edu

P.-T. Bremer
Lawrence Livermore National Laboratory, Livermore, CA, USA
e-mail: bremer5@llnl.gov

C.D. Hansen et al. (eds.), *Scientific Visualization*, Mathematics and Visualization,
DOI 10.1007/978-1-4471-6497-5_27

27.1 Scalable Analysis/Introduction

Historically, scientists have relied on conditional statistics, applied globally, to effectively reduce large-scale data to manageable proportions. While pre-computing a single set of structures is feasible, appropriate parameter choices are not always known a priori, and exploring the parameter space by extracting many sets of features is becoming infeasible due to massive data sizes. Furthermore, traditional statistics typically provide only global averages rather than per-feature information, making simple queries such as how many features exist overall, difficult to answer.

This chapter summarizes a new integrated analysis and visualization framework that enables a free choice of feature parameters and conditional sub-selections; providing the capability to interactively produce a wide range of diagnostic plots equivalent to the processing of the entire data set. Furthermore, the statistics viewer is cross-linked to a visualization of the corresponding three dimensional structures, enabling selection of (sets of) features on either end. The feature visualization employs a specialized volume rendering technique optimized for sparse, dynamic, and binary segmented volumes.

Instead of extracting a single set of features, we compute a multi-resolution hierarchy, capable of representing features for different parameters and at various scales. In a single pass over the original data we pre-compute a large variety of statistics for the finest resolution features. At run time the user selects parameters resulting in a set of features whose properties are aggregated on-the-fly, allowing the user to explore an entire family of feature definitions without accessing the original data. By pre-computing statistics for a base set of features, and providing the user with several multi-resolution hierarchies to explore, our system provides significantly greater flexibility in the analysis process than the typical range queries of indexing schemes. Additionally, the run-time aggregation avoids much of the cost of re-computing statistics for each set of features. As a result, our approach delivers the flexibility of extract-and-analyze techniques while allowing for interactive exploration of large data on a commodity desktop machine.

27.2 Augmented Feature Families

One of the basic concepts of our framework is the notion of a *feature family*. Given an algorithm to define and extract features of interest corresponding to a parameter p, a feature family is a one-parameter family that for every possible parameter p stores the corresponding set of features. While any feature definition can be used to create a feature family by exhaustively pre-computing all possible features for all possible parameters, many popular algorithms naturally produce nested sets of features for varying parameters. For example, clustering techniques progressively merge elements [4, 14] and a threshold-based segmentation creates increasingly larger regions [3]. In these examples all features can be described by a collection of base elements (e.g. clusters) or as a collection of differences between features at different parameters (e.g. regions above threshold a that are below threshold b) respectively.

Feature families with a nested structure can be encoded and computed in an efficient manner. In our system, we specify for each *element* in the hierarchy its *life span* (in terms of the feature parameter), an arbitrary number of *children*, and a single *parent*. As is common with hierarchies, the set of *features* at a particular parameter p is then defined as all elements that are *alive* at parameter p combined with all their descendants. More formally we define:

Definition 27.1 (*Element*) An *e*lement e is defined by a unique id and minimally contains a parameter range $[p_{min}, p_{max}]$, a direction, a collection of its children ids, and the id of its parent:

$$e = (id, direction, [p_{min}, p_{max}], \{child_0, \dots, child_n\}, parent) \in \mathbb{E}$$

Definition 27.2 (*Feature*) A *feature* f is the union of an element e and all its descendants

$$f = \left\{e \cup children^n(e) | n \in \{1, 2, ...\}\right\}$$

The element *id* is simply a unique identifier that is typically stored implicitly, *e.g.* based on the order in which elements are stored in a file. The direction indicates whether the parent of an element is *born* at $p < p_{min}$ and consequently its children are born at $p > p_{max}$ or the opposite.

A feature family is a collection of features defined hierarchically as described above:

Definition 27.3 (*Feature Family*) A *feature family* F is a set of features

$$F = \{f_0, \dots, f_m\} \subset \mathbb{F}$$

Finally, in a time-dependent simulation or an ensemble of simulations we have one feature family per time or ensemble member:

Definition 27.4 (*Clan*) A *clan* C is an ordered set of feature families

$$C = \{F_0, \dots, F_n\} \subset \mathscr{F}$$

We store feature families in a traditional multi-resolution graph that is updated on-the-fly as the user changes parameter. At any time we maintain a set of living elements that serve as the representatives for their corresponding features. Using the parent and child information this set is progressively updated as the feature parameter changes. Specifically, when an element dies it is removed from the set and either its children or its parent are born and added to the set. Furthermore, we support the encoding of multiple hierarchies associated with a feature family by storing multiple parameter ranges and child/parent ids in each feature, one for each hierarchy.

27.3 Sample Feature Hierarchies

Merge Trees As shown in [3, 12] the *merge tree* is ideally suited to hierarchically encode regions of locally varying isovalues. Given a simply connected domain $\mathbb{M}$ and a function $g : \mathbb{M} \rightarrow \mathbb{R}$ the *level set* $L(s)$ of g at isovalue s is defined as the collection of all points on $\mathbb{R}$ with function value equal to s: $L(s) = \{p \in \mathbb{M} | g(p) = s\}$. A connected component of a level set is called a *contour*. The merge tree of g represents the merging of contours as the isovalue s is swept top-to-bottom through the range of g, see Fig. 27.1a. Each branch of the tree represents a family of contours that continuously evolve without merging as s is lowered. These contours sweep out a subset of $\mathbb{M}$ and thus the branches correspond to a segmentation of $\mathbb{M}$, see Fig. 27.1a. To increase the resolution in parameter space we refine the merge tree by splitting long branches and refining the segmentation accordingly, see Fig. 27.1b.

In a simple threshold-based segmentation, each branch of the tree is an element with a lifetime given by the function values of the branch's top and bottom nodes. Given a particular threshold, each branch acts as the representative of its subtree/feature and, by construction, each subtree represents a simply connected region of high threshold, see Fig. 27.1c. However, when g spans multiple orders of magnitude *relevance* [12] is an alternate metric that scales g at each node by its local maximum—the highest maximum in its corresponding subtree. The relevance lifetime of a branch is thus given by the relevance interval between its top and bottom node and ranges between 0 and 1, see Fig. 27.1d. Typical examples of merge tree hierarchies are shown in Fig. 27.2. Figure 27.2a shows the burning cells in a simulation of low-swirl pre-fixed combustion [5]. In this applications burning cells are defined as regions of high fuel consumption and thus the merge tree of fuel consumption provides the appropriate segmentation. Figure 27.2b shows extinction regions in a turbulent simulation of non-premixed combustion. These regions are indicated by a high scalar dissipation rate [12]. Since the dissipation rates spans multiple orders of magnitude these structures are extracted using relevance as metric. Finally, Fig. 27.2c shows eddies in the north Atlantic ocean extracted using a split tree (the merge tree of the negative function) of the Okubo-Weiss scalar field [15].

Morse Complexes Merge trees or in general level set based hierarchies are well suited to encode threshold based regions. However, there exist a second, in some sense dual, class of feature descriptions based on the gradient flow. Given a point $x \in \mathbb{M}$ we call a line $\gamma(t) : \mathbb{R} \rightarrow \mathbb{M}$, with $\gamma(0) = x$, the *integral line* of x with respect to γ if $\frac{\partial \gamma}{dt} = \nabla g$. For each maximum m of g its *stable manifold* is defined as the set of points whose integral lines converge to m [6]. The set of all stable manifolds forms a segmentation of $\mathbb{M}$, see Fig. 27.3. Similar to merge trees Morse complexes have a natural hierarchical structure induced by merging neighboring stable manifolds. However, instead of using the thresholds this hierarchy typically employs *persistence*, the difference in function value between the lower of the two maxima and the saddle separating them [6]. In this case, the life time of features always starts at 0 (since all maxima/stable manifolds exists for 0 persistence) and ends at the persistence level at which a stable manifolds merges with one of its neighbors. An application where

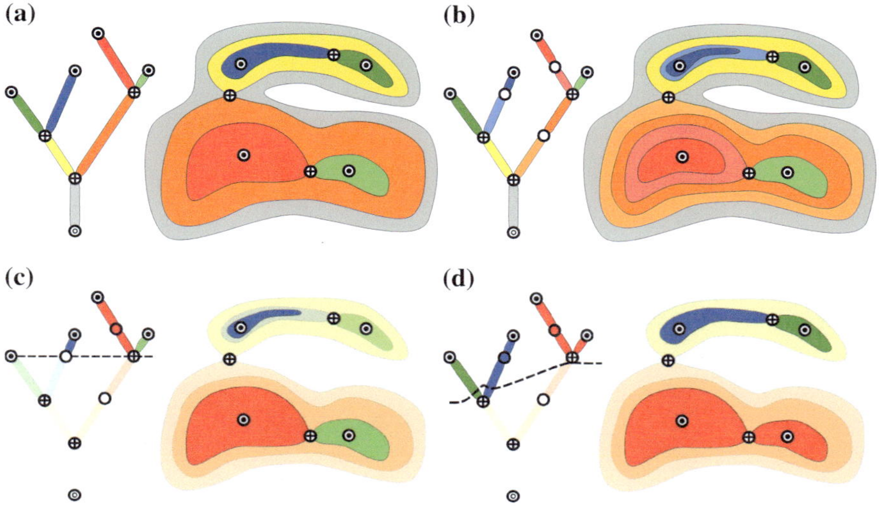

Fig. 27.1 **a** Merge trees represent the merging of contours as a function is lowered through its range. Each branch represents a portion of the domain as indicated by the colors. **b** To increase the resolution in parameter space we refine the merge tree by splitting long branches and refining the segmentation accordingly. **c** A threshold based segmentation of a merge tree at a threshold slightly above 80 % of the global maximum. **d** A relevance based segmentation at relevance around slightly above 0.2 (slightly below 80 % of the local maximum per branch). All local maxima are included and regions of higher function value (*red*) span a larger range. © IEEE. Republished with permission of IEEE, from Feature-Based Statistical Analysis of Combustion Simulation Data, Bennett, Krishnamoorthy, Liu, Grout, Hawkes, Chen, Shepherd, Pascucci, Bremer, IEEE TVCG 17(12) 2011; permission conveyed through Copyright Clearance Center, Inc.

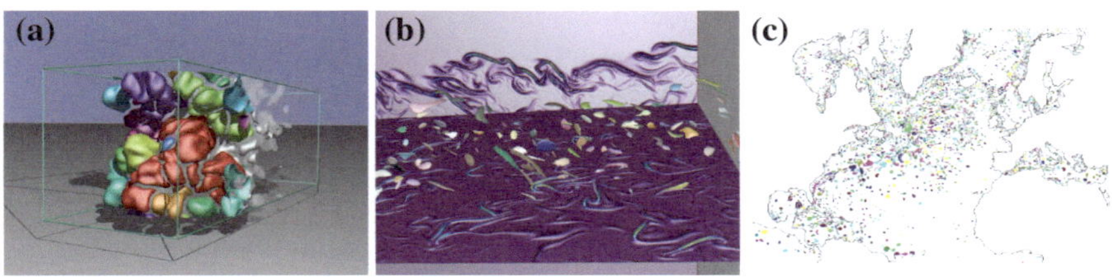

Fig. 27.2 Examples of different merge tree hierarchies: **a** Burning cells in a premixed hydrogen flame; **b** Extinction regions in turbulent non-premixed combustion simulation; and **c** Eddies in the north atlantic extracted using the Okubo-Weiss threshold

such hierarchical segmentations have proven useful is, for example, the analysis of Raleigh-Taylor instabilities [11]. As shown in Fig. 27.4, stable manifolds naturally segment the mixing interface into *bubbles* and the persistence simplification enables a multi-scale analysis.

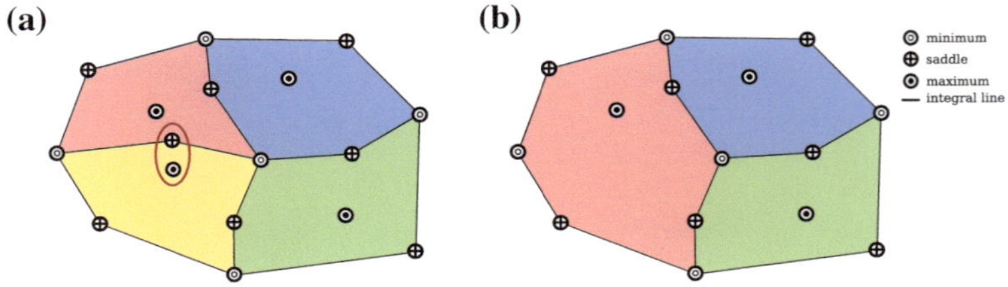

Fig. 27.3 A Morse complex of a set of maxima before **a** and after **b** merging the *red* and *yellow* stable manifolds. The persistence of the merge is the difference in function value between the circled critical point pair

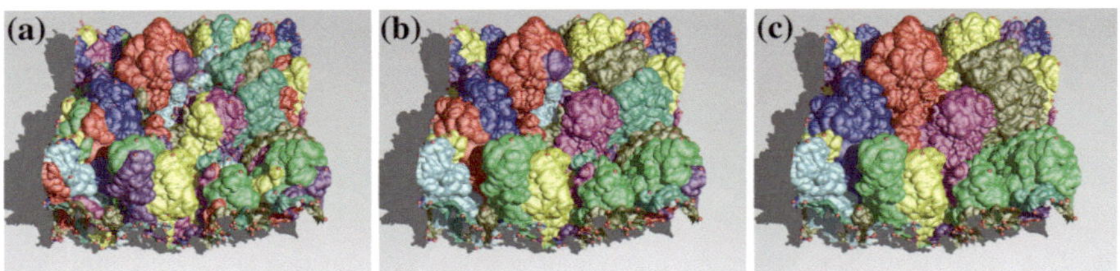

Fig. 27.4 Stable manifolds of a late stage mixing surface in a Raleigh-Taylor instabilities at: **a** About 1 % persistence where several large scale bubbles are over-segmented; **b** About 2.5 % persistence which reduces the over segmentation but still shows some artifacts; and **c** About 5 % persistence where the segmentation matches the human intuition

1-Skeleton of the Morse-Smale complex The Morse complex is the collection of stable manifolds of all critical points. The *unstable manifold* of a critical point is the collection of integral lines originating at that critical point. A function is *Morse-Smale* if its stable and unstable manifolds intersect only transversally. The cells of the *Morse-Smale complex* are formed by the intersections of stable and unstable manifolds. The set of 0- and 1-dimensional cells (*nodes* and *arcs*) of the complex are known as the *1-skeleton*, describing the connectivity structure of gradient flow of a function. Each node has an associated stable and unstable manifold. Each arc connects two nodes, and additionally stores its geometric embedding (the integral line it represents).

A hierarchy of Morse-Smale complexes is created by repeated cancellation of critical points in order of persistence. These cancellations are manifested in the 1-skeleton as the creation of new arcs, removal of arcs, and removal of a pair of nodes [9]. Each new arc has geometric embedding that is the merging of the embeddings of three removed arcs, as in Fig. 27.5. However, when reconstructing the geometry of an arc, only children that appear an odd number of times in the hierarchy should be used, since the ones that appear an even number of times correspond to a path that doubles back on itself. Therefore, when computing statistics on arcs in a hierarchy, this rule must be enforced.

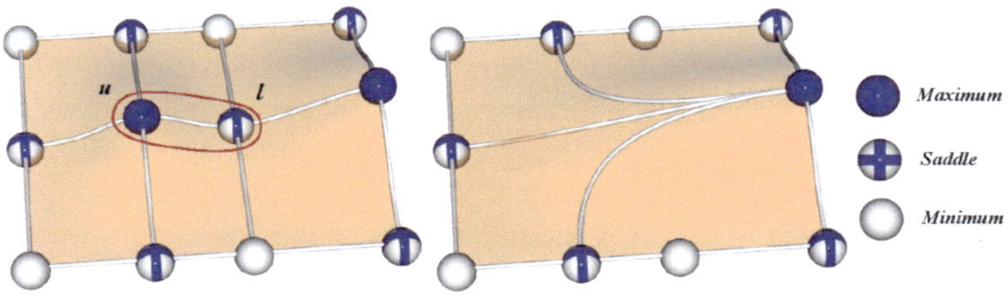

Fig. 27.5 A cancellation operation on the 1-skeleton of the Morse-Smale complex removes two critical points (u and l), and reconnects their neighborhood. In this example, each arc connected to u is re-routed to reach the new maximum. In practice, the new geometry is constructed by concatenating three paths [7]. © IEEE. Republished with permission of IEEE, from Feature-Based Statistical Analysis of Combustion Simulation Data, Bennett, Krishnamoorthy, Liu, Grout, Hawkes, Chen, Shepherd, Pascucci, Bremer, IEEE TVCG 17(12) 2011; permission conveyed through Copyright Clearance Center, Inc.

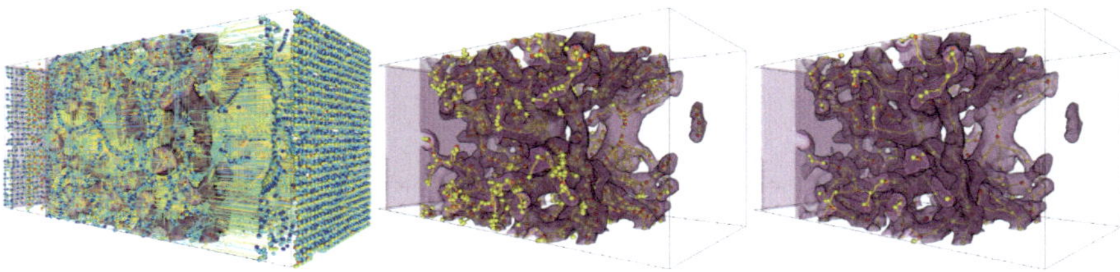

Fig. 27.6 The 1-skeleton of the Morse-Smale complex is computed for the simulated porous solid (*left*). Applying simplification and filtering allows representation of the filament structure at multiple scales (*middle*, *left*)

We apply the 1-skeleton of the Morse-Smale complex to finding the filament structure of a porous material [8], as shown in Fig. 27.6. The material is represented by a signed distance field from an interface surfaces demarking "inside" from "outside" the solid portion of the domain. The 2-saddle-maximum arcs of the complex form the space of possible reconstructions of the filaments. By combining filtering of the arcs based on the 2-saddle and maximum function values with exploration of the topological hierarchy, it is possible to study the filament structure at multiple scales and for multiple thresholds.

27.4 Feature Attributes

In addition to the information necessary to encode a feature family we augment each feature with an arbitrary number, k, of additional attributes ($att^0, \ldots, att^k$). Our system currently supports various descriptive statistics such as minima, maxima, first through fourth order statistical moments and sums, as well as as shape descriptors such as volumes and various length-scales. Descriptive statistics are computed incrementally as the feature family is constructed, using the same update formulas [1, 13] employed for the interactive aggregation during data exploration (Sect. 27.5). Specifically, as each vertex is labeled with its branch id, the vertex's associated attributes are added to the corresponding statistical aggregator. While this incremental approach works well for descriptive statistics, certain attributes such as shape descriptors cannot easily be computed in this manner, and are thus computed in a post-processing step.

File Format We store feature families and their corresponding attributes in a modular and easily extendable file format. Typically, we save one file per feature family to easily allow the restriction to temporal subsets, for example. At the end of each file we store an XML-footer followed by the file offset to the start of the footer as the last eight bytes in the file. The XML structure encodes which components are stored for the feature family, and typically comprises a simplification sequence storing the hierarchy information in addition to a number of attributes. Any attributes stored indicate their type in addition to meta-data such as the name of the source field, how many bytes are used for each value, and whether data is stored in binary or ascii format. For the statistical moments we store not only the final value, e.g. mean, but enough information to further aggregate multiple values as needed by the parallel statistics formulas of [1, 13]. This requires each n-th order statistical moment to store all lower-order moments to support aggregation. Most importantly the XML structure stores file offsets to each corresponding block of data, allowing for the selective loading of subsets of attributes for exploration. One immediate advantage of this file structure is that it can be easily extended without re-writing entire files. Given a new set of attributes, we read the XML footer, append the new data at the end of the old data (overwriting the old footer), update the footer, and append it to the file.

27.5 Interactive Exploration of Feature-Based Statistics

One of the main advantages of our system is the ability to quickly explore a wide variety of statistical information based on the given feature definitions. To achieve this our framework supports four operators that map feature families, sets of features, and statistics into new sets of features, or scalar quantities:

Definition 27.5 (*Selection*) A *selection* $S : \mathscr{F} \times \mathbb{R} \rightarrow \mathscr{P}(\mathbb{F})$ is an operator that, given a feature family and a parameter, returns a set of features as well as (a subset) of their corresponding attributes.

Note that each feature stores attribute information regarding the portion of the domain it covers, see Fig. 27.1a. A selection will, for most attributes, aggregate all values in the associated subtree on-the-fly as the hierarchy is navigated. This preserves the flexibility to base different feature families on the same set of initial attributes. Nevertheless, if only one type of family is needed, aggregation of attributes can be performed once and stored to accelerate the exploration, see Sect. 27.4.

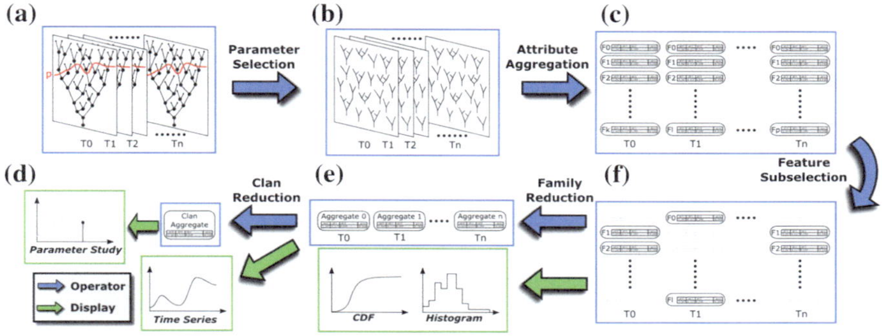

Fig. 27.7 Computational pipeline for interactive feature exploration. Starting from a clan of feature families represented by a sequence of merge trees (**a**) setting the feature parameter results in a sequence of sets of features each represented by a subtree of elements (**b**). Aggregating statistical attributes for each feature produces a set of features with attributes for each time step (**c**). A subselection on an arbitrary attribute narrows this collection to features of interest (**d**). Subsequently, either clan wide plots such as CDFs are created (**e**, *bottom*) or a reduction operator is applied to each family to create a time series of aggregated attributes (**e**, *top*). Finally, the time series is plotted (**f**, *bottom*) or an additional reduction is used to create a clan wide aggregated scalar property (**f**, *top*), which produces a single sample of a parameter study. A full study is created by repeatedly executing the pipeline. © IEEE. Republished with permission of IEEE, from Feature-Based Statistical Analysis of Combustion Simulation Data, Bennett, Krishnamoorthy, Liu, Grout, Hawkes, Chen, Shepherd, Pascucci, Bremer, IEEE TVCG 17(12) 2011; permission conveyed through Copyright Clearance Center, Inc.

Definition 27.6 (*Aggregation*) An *aggregation* $A : \mathscr{P}(\mathbb{F}) \times \{0, \ldots, k\} \rightarrow \mathbb{R}$ is an operator that, given a set of features and an attribute index, returns the combined attribute for the set of features.

Definition 27.7 (*Subselection*) A *subselection* $U : \mathscr{P}(\mathbb{F}) \times \{0, \ldots, k\} \times \mathbb{R}^2 \rightarrow \mathscr{P}(\mathbb{F})$ is an operator that, given a set of features, an attribute index, and a corresponding attribute interval range, returns the subset of features whose attribute value is contained in the interval.

The subselection operator facilitates the creation of conditional plots, which are often an important part of the analysis pipeline.

Definition 27.8 (*Reduction*) A *reduction* $R : \mathscr{P}(\mathbb{R}) \rightarrow \mathbb{R}$ is an operator that given a set of scalar values returns a single scalar value, for example by computing the mean.

Using the operators described above we create three different types of plots as summarized by Fig. 27.7: species distributions, parameter studies, and time-series. To simplify the discussion below, we assume that the input to each of the operators is all feature families in a clan, even though in practice we support the restriction to subsets of the data. All plots take as input a feature clan C, a parameter p, subselections $Q = \{(att^{i_0}_{min}, att^{i_0}_{max}), \ldots, (att^{i_l}_{min}, att^{i_l}_{max})\}$, and an attribute index i. First, the parameter

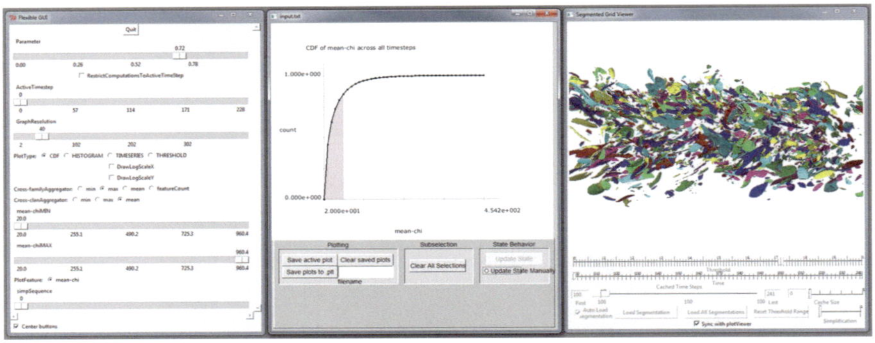

Fig. 27.8 Our framework provides a natural and intuitive work-flow for the exploration of global trends in feature-based statistics. The linked view system allows users to display susbsets of features using range sliders and/or by selecting regions of CDFs and histograms. This image shows regions of locally high scalar dissipation rate in a temporally-evolving turbulent CO/H_2 jet flame

p is used to select an initial set of features from the clan, which are then further subselected using the subselections Q.

Species distributions plots include histograms and empirical CDFs, and track the distribution of the attribute att^i. A time-series, as the name suggests, shows the evolution of att^i over time, and requires an additional family-wide reduction operator, R_f, as input. Parameter studies are an extension of time-series that show how att^i changes as the parameter p is varied. For these plots a clan-wide reduction operator, R_c, is required in addition to R_f. Note that parameter studies can be come expensive as the range and granularity of p increases, because attributes must be aggregated for each p-value independently. While parameter plots are the most expensive to produce they are also often very useful. In particular, a parameter plot shows how stable or unstable a given analysis is to the parameter selection. This is crucial in any exploratory setting to guarantee that the basis of important conclusions is not an inherently unstable analysis.

We provide a convenient GUI that allows the user to specify which attributes they would like to explore, loading only those to minimize memory overhead. Subselection sliders are generated for each specified attribute automatically and, if multiple hierarchies are available, the user can toggle between these and can update parameters interactively. Optional log scaling is provided, and radio buttons are used for selection of family- and clan-wide reduction operators. The plot viewer is linked to the feature browser to provide context as statistics are explored. Only those features that have been subselected using the GUI sliders are displayed by the feature browser. Users can click on an individual feature in the feature browser to obtain details on its associated statistics. Furthermore, when the user picks regions of histograms or CDFs, only those features that are contained in the selected bins are displayed by the feature browser, see Fig. 27.8.

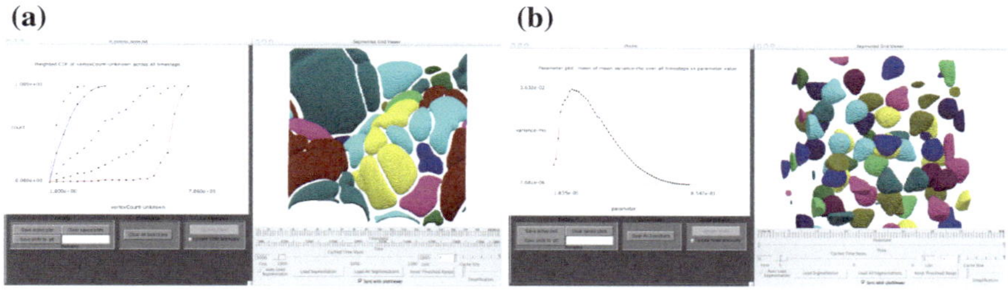

Fig. 27.9 **a** Weighted cumulative density function of feature volume in an idealized premixed hydrogen flame echoing Fig. 9d in [2]. **b** The average density variance for features defined by different density thresholds in a simulation of hydrogen under pressure

27.6 Results

Using the combined visual exploration and analysis capabilities, scientists are able to quickly produce feature-based statistical summaries of their data. For example, Fig. 27.8 shows the distribution of scalar dissipation rate in a temporally-evolving turbulent CO/H_2 jet flame undergoing extinction and reignition at different Reynolds numbers [10]. The features of interest are defined by locally varying isovalues of the scalar dissipation rate, and a merge tree is used to encode the feature hierarchy. Of interest to the scientists is the relationship between the temperature and thickness of these structures, and so the merge tree was augmented with mean and variance of temperature in addition to lengthscale measurements. Figure 27.9a shows the weighted cumulative volume distribution of an idealized pre-mixed combustion simulation [5] along side the corresponding segmentation. Figure 27.9b shows the average density variance for different density thresholds in a simulation of hydrogen under pressure. As the threshold is lowered the variance increases up to a breaking point, after which it rapidly falls. The corresponding segmentation is generated using the approximate threshold of the peak variance. Somewhat surprisingly, the peak variance does not correspond to the point at which the individual surfaces begin to merge, rather there are still a large number of well separated features.

27.7 Conclusion

This chapter summarizes a novel framework that combines topological and statistical analysis with visualization to perform feature-based statistical analysis of large scientific data. The framework represents a novel technology that has converted the typically cumbersome post-processing cycle of explore and analyze into an interactive process, providing application scientists easy access to cutting edge visualization and feature-based analysis techniques coupled to traditional statistical techniques. In particular, this framework provides an intuitive GUI that enables traditional statistical distributions obtained from conditional feature sets to be easily related back to the sets of features contributing to the statistic through feature-based visualization and linked views.

Acknowledgments This research used resources of the Oak Ridge Leadership Computing Facility at the Oak Ridge National Laboratory, which is supported by the Office of Science of the U.S. Department of Energy under Contract No.DE-AC05-00OR22725. This research used resources of the National Energy Research Scientific Computing Center, which is supported by the Office of Science of the U.S. Department of Energy under Contract No.DE-AC02-05CH11231. Sandia National Laboratories is a multiprogram laboratory managed and operated by Sandia Corporation, a wholly owned subsidiary of Lockheed Martin Corporation, for the U.S. Department of Energy's National Nuclear Security Administration under contract DE-AC04-94AL85000. This work was performed under the auspices of the U.S.Department of Energy by Lawrence Livermore National Laboratory under Contract DE-AC52- 07NA27344. This work is supported in part by BNSF CISE ACI-0904631, NSG IIS- 1045032,NSF EFT ACI-0906379,DOE/NEUP 120341, DOE/Codesign P01180734,DOE/SciDAC DESC0007446, and CCMSC DE-NA0002375.

References

1. Bennett, J., Pébay, P., Roe, D., Thompson, D.: Numerically stable, single-pass, parallel statistics algorithms. In: Proceedings of the IEEE International Conference on Cluster Computing. New Orleans, LA (2009)
2. Bremer, P.T., Weber, G., Pascucci, V., Day, M., Bell, J.: Analyzing and tracking burning structures in lean premixed hydrogen flames. IEEE Trans. Vis. Comput. Graph. **16**(2), 248–260 (2010)
3. Bremer, P.T., Weber, G.H., Pascucci, V., Day, M., Bell, J.B.: Analyzing and tracking burning structures in lean premixed hydrogen flames. Vis. Comput. Graph., IEEE Trans. **16**(2), 248–260 (2010). doi:10.1109/TVCG.2009.69
4. Comaniciu, D., Meer, P.: Mean shift: A robust approach toward feature space analysis. IEEE TPAMI **24**, 603–619 (2002)
5. Day, M., Bell, J., Bremer, P.T., Pascucci, V., Beckner, V., Lijewski, M.: Turbulence effects on cellular burning structures in lean premixed hydrogen flames. Combust. Flame **156**, 1035–1045 (2009)
6. Edelsbrunner, H., Harer, J., Zomorodian, A.: Hierarchical morse-smale complexes for piecewise linear 2-manifolds. Discrete Comput. Geom. **30**, 87–107 (2003)
7. Gyulassy, A., Bremer, P.T., Hamann, B., Pascucci, V.: A practical approach to Morse-Smale complex computation: Scalability and generality. IEEE Trans. Vis. Comput. Graph. **14**(6), 1619–1626 (2008). http://doi.ieeecomputersociety.org/10.1109/TVCG.2008.110
8. Gyulassy, A., Duchaineau, M., Natarajan, V., Pascucci, V., Bringa, E., Higginbotham, A., Hamann, B.: Topologically clean distance fields. IEEE Trans. Comput. Graph. Vis. **13**(6), 1432–1439 (2007)
9. Gyulassy, A., Natarajan, V., Pascucci, V., Bremer, P.T., Hamann, B.: Topology-based simplification for feature extraction from 3d scalar fields. In: Proceedings of the IEEE Conference Visualization, pp. 535–542 (2005)
10. Hawkes, E.R., Sankaran, R., Sutherland, J.C., Chen, J.H.: Scalar mixing in direct numerical simulations of temporally evolving plane jet flames with skeletal co/h2 kinetics. Proc. Combust. Inst. **31**(1), 1633–1640 (2007). doi:10.1016/j.proci.2006.08.079
11. Laney, D., Bremer, P.T., Mascarenhas, A., Miller, P., Pascucci, V.: Understanding the structure of the turbulent mixing layer in hydrodynamic instabilities. IEEE Trans. Vis. Comput. Graph. (TVCG) / Proc. IEEE Vis. 12(5), 1052–1060 (2006)

12. Mascarenhas, A., Grout, R.W., Bremer, P.T., Hawkes, E.R., Pascucci, V., Chen, J.H.: Topological feature extraction for comparison of terascale combustion simulation data, pp. 229–240, Mathematics and Visualization. Springer, London (2010)
13. Pebay, P.: Formulas for robust, one-pass parallel computation of covariances and arbitrary-order statistical moments. Technical Report SAND2008-6212, Sandia National Laboratories (2008)
14. Sheikh, Y., Kahn, E., Kanade, T.: Mode-seeking by medoidshifts. In: Proceedings of the IEEE International Conference on Computer Vision, pp. 1–8 (2007)
15. Williams, S., Petersen, M., Bremer, P.T., Hecht, M., Pascucci, V., Ahrens, J., Hlawitschka, M., Hamann, B.: Adaptive extraction and quantification of atmospheric and oceanic vortices. IEEE Trans. Vis. Comp. Graph. p. to appear (2011)

Chapter 28
Cross-Scale, Multi-Scale, and Multi-Source Data Visualization and Analysis Issues and Opportunities

David Ebert, Kelly Gaither, Yun Jang and Sonia Lasher-Trapp

Abstract As computational and experimental science have evolved, a new *dimension* of challenges for visualization and analysis has emerged: enabling research, understanding, discovery at multiple problem scales and the interaction of the scales, and abstractions of phenomena. Visualization and analysis tools are needed to enable interacting and reasoning at multiple simultaneous scales of representations of data, systems, and processes. Moreover, visualization is crucial to help scientists and engineers understand the critical processes at the scale boundaries through the use of external visual cognitive artifacts to enable more natural reasoning across these boundaries.

28.1 The Challenge of Multi-Scale Interactions

"Multi-Scale Interactions" has been used to characterize and emphasize that significant breakthroughs need to occur in a variety of fields by understanding both how the larger scales fuel the smaller scales, and how smaller scales feed back into larger scales. One fundamental example of this is turbulence. Turbulence is a major unsolved problem for fluid flow and is applicable to weather, medicine, engineering, and climate change. Biology gives us another example where scientists are working to understand structure and function from the cellular level up to the level of organs, then to functional subsystems within the body. Another example occurs in parameterization in numerical modeling: how one represents processes occurring at the

D. Ebert (✉), Y. Jang · S. Lasher-Trapp
Purdue University, West Lafayette, USA
e-mail: ebertd@purdue.edu

S. Lasher-Trapp
e-mail: slasher@purdue.edu

K. Gaither
Texas Advanced Computing Center, Austin, USA
e-mail: kelly@tacc.utexas.edu

Y. Jang
Sejong University, Seoul, South Korea
e-mail: jjangyn@gmail.com

C.D. Hansen et al. (eds.), *Scientific Visualization*, Mathematics and Visualization,
DOI 10.1007/978-1-4471-6497-5_28

subgrid scale, and improving their parameterizations. If parameterizations are good, then the necessity to keep resolving smaller and smaller scales in one's numerical model disappears.

Another class of problems is filling data holes and gaps. How can new tools help us explore these holes and gaps between different kinds of observational data collected at different scales, or between the hierarchies of numerical models that we use to solve subsets of the problem that need to be merged to understand the relative contribution or importance of the solutions to the subsets?

28.1.1 Systems of Systems

The challenges of multi-scale interactions in complex systems and processes has led to new areas of research and refocusing of disciplines in engineering, as illustrated by the evolution of industrial engineering to industrial and systems engineering, and the development of the subarea of systems of systems research. Therefore, in order for these multi-scale interactions to be investigated and explored, tools are needed which scale to handle *systems of systems* [2]. These problems are common in science and engineering, and may require analysis and combination of data across scales. For example, macrobiology analysis may require understanding the interactions of data simultaneously at the genome, protein, cell, organ, human, country, and ecosystem levels. Cancer care treatment requires understanding and integrating data from the biomarker level (e.g., integrating metabolics, lipidomics, genomics, and proteomics data already at multiple scales) and cancer processes at the organ level, environmental exposure, and socioeconomic factors that affect the success and completion of treatment regimens.

Weather and the environment provide further examples, such as clouds and precipitation. Clouds and precipitation affect our daily lives, personal safety, commercial decisions, and our future sustainability on Earth. Clouds and precipitation are important at all regional scales: local, state, national, and global. Clouds influence the daily maximum and minimum temperatures over our homes and they modulate the global temperature by affecting the amount of incoming solar radiation and outgoing long wave radiation. Precipitation is likewise important at all scales. It directly affects our quality of life: our food supply, drinking water supply, air purity, modes of transportation, and many other human needs across the earth. As the inhabitants of earth become increasingly concerned about global warming and climate change on global and regional levels, it is necessary to understand the roles of clouds and precipitation in the Earth's System in order to predict the future state of our planet. However, fundamental questions remain concerning cloud motions and evolution, cloud longevity, and precipitation formation, and these gaps in our knowledge hamper our efforts to understand and predict weather and climate. Understanding and predicting clouds and precipitation are very difficult tasks which require the measurement and modeling of properties on a wide variety of scales (microscale, cloud scale, storm scale, mesoscale, synoptic scale, global scale as shown in Fig. 28.1), fusion of computational model data and measured data, and the simultaneous fusion of hundreds of

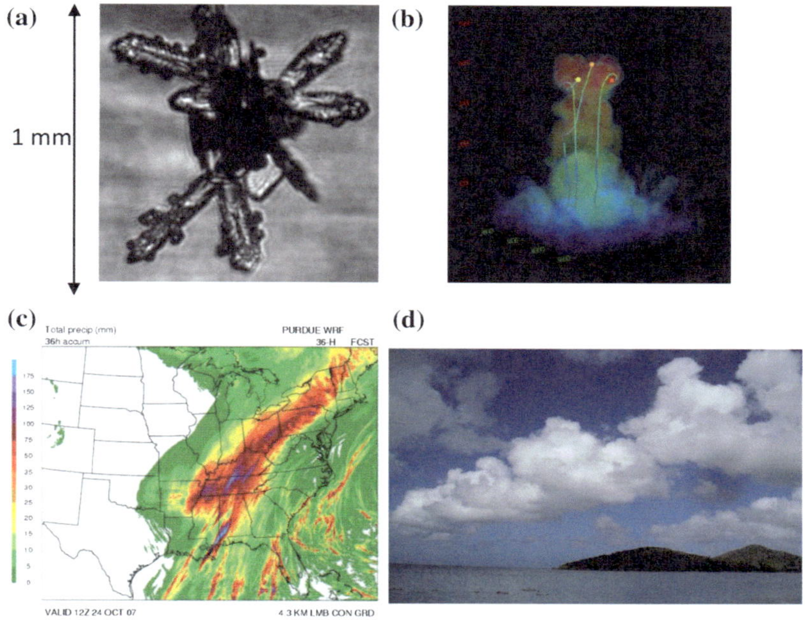

Fig. 28.1 Multi-scale examples, **a** microscale, **b** cloud scale, **c** storm scale, and **d** global scale. © IEEE reprinted, with permission, from IEEE transactions on visualization and computer graphics 12(5)

scalar and vector fields that vary over time. Current tools for atmospheric visualization and analysis are incapable of the following crucial functions:

- Integrating these various data sources and providing effective interfaces for fusion, analysis, experimentation, exploration, hypothesis testing, and discovery.
- Communicating the complex three-dimensional, time-varying information necessary to accurately predict atmospheric events, conditions, and consequences (e.g., aircraft icing) and extend the understanding of atmospheric phenomena.
- Integration of visual representations into the scientific analysis and discovery process.

28.1.2 Transformational Cross-Scale Science

Facilitating breakthrough scientific discoveries will require the development of transformational science that produces revolutionary new tools for mastering the multi- and cross-scale challenges of our world. Science discovery possibilities are presented here as a broad spectrum of disciplines, including the following.[1]

[1] This is summarized from the NSF Science and Engineering Community Workshop report by Ebert D., Gaither K., and Gilpin C.

- Computational Fluid Dynamics—Understanding computational fluid dynamics will allow us to understand and design a broad spectrum of applications: design more aerodynamic (and hence fuel-efficient) cars and planes, design better artificial hearts, perform better cardiovascular surgery, design better air conditioners, fans, and heat exchangers, better short-term weather prediction (including accurate precipitation forecasts and models), and understand the dynamics of global warming in the long term, improve our understanding of the dynamics of the oceans, and allow us to predict solar storms that affect radio communication, design better hydroelectric generators, design more efficient HVAC systems in buildings, design more efficient wind farms for the generation of electricity, improve the design of ship hulls, help us understand the mechanism of flight in birds and insects, and help us understand the locomotion of aquatic animals, including microorganisms.
- Preserving Coastal Margins—We have to preserve coastal margins so that our great-grand children will have access to a functioning environment that supports economic development and quality of life. By understanding cause-effect relationships between climate, human activity and coastal margins well enough to predict and communicate ecosystem evolution, we can effectively influence society's choice on affecting ecosystems health and sustainability.
- Virtual Paleoworld—We can reconstruct climate in the broadest sense for any time in the Earth's past and see what it looks like globally. This will allow us to see Tectonic plates in their proper shapes and positions and all data sites marked. We can see topography, heat flow, atmospheric composition, wind belts, biomes on land, and ocean currents as they existed then and with explicit depiction of differences from today and from individual data sites with accompanying uncertainties (e.g., model-data comparison).
- Understanding the Origin of Our Universe—Gravitational waves can probe to 10^{-43} seconds after the big bang. They carry key information of what happened just after it all began. Deciphering their content and comparing them with cosmological models will enable us to hone in on the model that captures reality. This will additionally have implications for a unified theory of physics, understanding dark matter and energy and why regular matter (like the one humans and stars are made of) only comprises about 3 % of the total.
- Understanding the 'Cradle of Life'—Supernovae are responsible for producing the needed energy that turns heavier elements (e.g., iron) of a soon to explode star into lighter elements that are the building blocks for life (hydrogen, oxygen, etc). Understanding such systems require complex simulations at the peta-(and beyond) scales and their solution have commonalities with other spectacular phenomena like gamma ray bursts. Obtaining the correct model to explain supernovae events will go a long way towards understanding what is required for the basic building blocks of life to be produced and explain the most spectacular astrophysical phenomena.
- Origin and Evolution of Languages—The origin and evolution of languages is a result of interactions with culture. We need to better understand how languages reflect the history of cultures, and how genetics/genomics data sets can be used to study unrecorded historical data concerning the migration of humans.

- The Cell at Subnanometer Resolution—The cell is the most basic building block of life. Cell function is a direct reflection of cell structure, and we understand the function of the cell only to the degree that we understand its molecular architecture. The molecular defects that underlie human disease are rooted in changes in the molecular components in the cell that alter cell structure and function. Molecular medicine would be greatly advanced by a molecule level view of the cell since it would guide the engineering of molecular therapies to fix what is broken in cells in the disease state.

28.1.3 Temporal Scalability

Sensemaking often involves temporal reasoning, and may require handling data at different time scales. For example, it may be necessary to understand long-term patterns by looking at data over a period of years or even decades, and simultaneously understand near-term effects by looking at data over a period of hours or less. Moreover, it may be necessary to integrate and perform correlative analysis on data collected at different temporal scales based on acquisition technology. For instance, in understanding fundamental principals of rain formation in clouds, it can be necessary to integrate data collected 1000 times per second with data collected every several minutes (radar data) and this information may then be fed into climate models that work on the scale of years and decades.

28.2 Variety of Data

Our ability to collect data is increasing at a faster rate than our ability to analyze it [3]. Scientists, engineers, and analysts are often overwhelmed with massive amounts of data from multiple sources and where the important information content exists in a few pieces. Therefore, we need to create new methods to allow them to visually examine this massive, multi-dimensional, multi-source, time varying information stream to make decisions more efficiently. The various data types include the following:

- Textual data—Massive textual data from documents, speeches, e-mails, or web pages now influence the problem domain. This data can be truly massive, contain billions of items per day, and much of it must be analyzed in a time-critical matter.
- Databases—Many corporate and government entities have constructed huge databases containing a wealth of information. We require new algorithms for the efficient discovery of previously unknown patterns in these large databases.
- Geospatial data—Consider the data collected by satellites that image the earth. We now have satellites that can create images at less than 1 m resolution and that can collectively image the land surface of the planet in a very short time. These images must be examined in a time-critical matter.

- Sensor data—The revolution in miniaturization for computer systems has allowed us to produce a myriad of sensors. The sensors can collect data about their environment (location, proximity, temperature, light, radiation, etc.), can analyze this data, and can communicate between themselves. Collections of sensors can produce very large streaming sets of data.
- Video data—Video analytics are being used more and more to enhance the effectiveness of the security in high-risk security operations. Content analysis, combined with massive recording capabilities, is also being used as a powerful tool for improving business processes and customer service. New techniques must be developed to integrate this streaming data paradigm into the analyst's toolbox.

Whereas each of these categories can produce massive data streams containing information that is applicable to a given problem domain, the grand challenge problem in the area of scalability is to use analytics to distill the relevant pieces of information from these widely disparate information streams, and create an information space containing relevant information that can be examined by analytical or visual means to influence the exploration, hypothesis testing, discovery, and decision making of the user. These systems need to provide mechanisms that can visualize the connections between the relevant information in the information streams, and allow the user to relate concepts, theories, and hypotheses to the data.

Several research directions present themselves as candidates to address these scalability problems, classified as visual scalability, information scalability, software scalability and information fusion.

28.2.1 Visual Scalability

Visual scalability [1] is the capability of visualization tools to effectively display massive data sets, in terms of either the number or the dimension of individual data elements. Factors affecting visual scalability include the quality of visual displays, the visual metaphors used in the display of information, the techniques used to interact with the visual representations, and the perception capabilities of the human cognitive system. A critical area of research in visual scalability is in methods that allow the user to change the visual representation of data.

28.2.2 Information Scalability

Information scalability implies the capability to extract relevant information from massive data streams. Methods of data scalability include methods to filter and reduce the amount of data, techniques to represent the data in a multiresolution manner, methods to abstract the data sets.

28.2.3 Software Scalability

A commonly held best practice in active data visualization begins with a visualization that summarizes a large data set followed by a subsetting to examine its detail. This practice requires active data visualization software that can execute visual queries and scale to data sets of varying sizes. Software scalability includes the generation of new algorithms that scale to the ever-increasing information sets that we generate today. We wish to avoid the hidden costs that arise when we build and maintain monolithic, non-interacting, non-scalable software models.

28.2.4 Information Fusion

Information fusion includes the capability to fuse the relevant information from divergent multi-source multi-dimensional time-varying information streams. This is the grand challenge problem in visualizations. Researchers must not just produce new visual representations and data representations for specific data types or information streams, but we must develop methods that fuse the relevant information into a single information space and develop new visual metaphors that allow the analyst to *look inside* this complex, multi-dimensional, time-varying space.

We must also develop techniques to *measure* scalability so new tools can be analyzed for their applicability in this domain. We must establish metrics that allow us to evaluate both visual metaphors and data representations as they apply to scalable algorithms. The best measurement will not only evaluate the representations according to scale, but also to the number of insights, actions, or value achieved for the analyst.

28.2.5 Technology Needs

What is needed in the visualization research agenda is to extend the state-of-the-art visual and data representations to be able to explore the heterogeneous multi-source multi-dimensional time-varying information streams. We must develop new visual methods to explore massive data in a time critical matter. We must develop new techniques for information fusion that can integrate the relevant pieces of information from multi-source multi-dimensional information. We must develop new methods to address the complexity of information, and create a seamless integration of computational and visual techniques to create a proper environment for analysis. We must augment our methods to consider visual limits, human perception limits, and information content limits. Therefore, the following challenges can have significant impact on science, engineering, discovery, and society:

- Develop quantifiable scalable visual representations, data representations, and software tools for various domains.
- Develop new methods for abstraction of massive streaming data from textual sources, satellite data, Sensor data, video data, and other information streams.
- Develop new research capabilities for information fusion. These methods should utilize visual analytic techniques to extract the relevant nuggets of information from heterogeneous multi-source multi-dimensional time-varying information streams, fusing these pieces into explorable information space.

References

1. Eick, S.G., Karr, A.F.: Visual scalability. J. Comput. Graph. Stat. **11**(1), 22–43 (2002)
2. Robertson, G.G., Ebert, D.S., Eick, S.G., Keim, D.A., Joy, K.: Scale and complexity in visual analytics. Inf. Vis. **8**(4), 247–253 (2009)
3. Thomas, J.J., Cook, K.A. (eds.): Illuminating the path: the research and development agenda for visual analytics. IEEE CS Press, NJ (2005)

Chapter 29
Scalable Devices

Jens Krüger and Markus Hadwiger

Abstract In computer science in general and in particular the field of high performance computing and supercomputing the term *scalable* plays an important role. It indicates that a piece of hardware, a concept, an algorithm, or an entire system scales with the size of the problem, i.e., it can not only be used in a very specific setting but it's applicable for a wide range of problems. From small scenarios to possibly very large settings. In this spirit, there exist a number of fixed areas of research on *scalability*. There are works on scalable algorithms, scalable architectures but what are *scalable devices*? In the context of this chapter, we are interested in a whole range of display devices, ranging from small scale hardware such as tablet computers, pads, smart-phones etc. up to large tiled display walls. What interests us mostly is not so much the hardware setup but mostly the visualization algorithms behind these display systems that scale from your average smart phone up to the largest gigapixel display walls.

29.1 Introduction

In this chapter we will give an overview over the current state of the art of *scalable device visualization*. Scalable device visualization for us, is a visualization software and hardware setup that allows the efficient and effective visualization of—usually

J. Krüger (✉)
Director Center of Visual Data Analysis and Computer Graphics - CoViDAG Professor of High Performance Computing Department of Computer Science and Applied Cognitive Science, University of Duisburg-Essen Lotharstraße 65, 47057 Duisburg, Germany
e-mail: jens.krueger@uni-due.de

M. Hadwiger
King Abdullah University of Science and Technology,
Thuwal 23955-6900, Saudi Arabia
e-mail: markus.hadwiger@kaust.edu.sa

C.D. Hansen et al. (eds.), *Scientific Visualization*, Mathematics and Visualization,
DOI 10.1007/978-1-4471-6497-5_29

very large and complex—scientific data on a wide range of devices. Such a list of devices does include the omnipresent PC workstation but it also extends to very small displays such as pads, tablet PCs, and smart-phones it also includes large scale visualization installations such as high-resolution front and back-projection systems, LCD display assemblies and the various virtual reality setups such as CAVEs [13] and similar immersive environments.

In the context of this chapter the perfect visualization system would be the one that runs perfectly on (or scales to) the entire range of displays imaginable from the smallest wristwatch display to the biggest terrapixel screen. As—to our best knowledge—something like that does not exist yet, we will first focus on works that primarily deal with the very small displays and then take a look at large scale visualization environments. Finally, we conclude with an outlook over current and future work in this area.

29.2 Small Devices

In this section, we will focus on the small, the portable, the mobile, the hand-held, the low power consumption hardware. Many names for a class of display and computing resources that—if we believe the predictions of the analysts—may soon almost completely replace the PCs from our homes and even workplaces. Before we begin to take a look at existing works that utilize this type of hardware let us reflect why we want to use these devices for visualization. First of all, there are a couple of good reasons *not* to use them. They are small by definition, which in most cases means they have a small screen.[1] Another good reason to avoid them for visualization purposes is the lack of traditional input metaphors such as mouse and keyboard which are replaced by a number of sensors and input modalities that we do not have access to on a standard PC. Finally, there is no doubt that the graphics and computing resources on mobile devices are limited even in comparison to commodity PC based workstations, not to mention computing clusters that are often employed for visualization.

On the positive side, we have the low power consumption, which is certainly interesting for some first responder, disaster or remote location scenarios were no reliable power infrastructure is present, but often power is not a concern. Being portable, mobile, hand-held appears to be a much more interesting feature. There are quite a few scenarios where we would have access to a power outlet, wired, and wireless Internet but where a workstation, a notebook, or a netbook are impractical. Examples include a doctor on a ward round, mobile medical units, scientists in the field, ad-hoc brainstorming sessions on a lunch break, or simply the only specialist to a problem who just happens to be on vacation skiing in the mountains. Most of these examples are based on yet another argument, that is: *the best visualization platform is*

[1] At this point, we knowingly ignore those devices that come with a projector and use arbitrary objects as projection surfaces, as to our best knowledge these designs are still in early prototype state.

the one that we currently have easy access to. Most of us carry their smart-phones with us almost 24 h 7 days a week and we are familiar with the important features of these devices. This in—our opinion—is the strongest argument for mobile visualization, we call this concept the **convenient visualization**. If the user interface manages to stick to the general paradigms that the users know from their devices already, you can give them a visualization platform that is available 24/7, is simple to learn, and very convenient for them.

To achieve this, however, there are a number of challenges that need to be addressed first. We need to display large and complicated data on a relatively small screen with limited computing, graphics, networking and storage resources available. That said, the challenges can be classified into two categories, efficient user interface design and screen space efficient visualization methods on the one hand and efficient data management and rendering on the other hand.

29.2.1 Mobile User Interfaces

There exists a vast body of literature on user interface design for mobile devices. Due to the limited space of this chapter, we restrict the discussion to works on user interfaces specifically for mobile visualization systems.

Focused on visualization and multimedia but still a very general work has been presented by Paelke et al. [45] where they propose a visualization design repository for mobile devices. Chittaro [12] focused on the general issues of visualizing content on mobile devices.

Recently, Eissele et al. [19] note that with the presence of location sensors such as cameras, GPS antennas, inertial accelerometers, WiFi receivers etc., the amount of user interaction and consequently user interface can be reduced by utilizing those location sensors focusing on information relevant for the current context.

29.2.2 Rendering Approaches

Even before today's powerful mobile devices were readily available Encarnação et al. [20] discussed the general issues in using mobile devices to obtain and access data. They recognized the problems of limited processing hardware and network bandwidth. Even though both the network as well as the computing capabilities of mobile devices have increased significantly since 1995 the basic challenges remain the same. To address these limitations, over the last couple of years, two main avenues of research have been followed: The improvement of server based remote rendering, and the development of novel techniques that make due with the limited resources available to client-based rendering directly on the mobile device.

29.2.2.1 Remote Rendering

One solution to circumvent the limitation of the mobile device's graphics hardware is not to rely on it for the complex rendering tasks required for many visualization algorithms but to let a server do the rendering work and to transfer the rendered image to the mobile device. In 2005 Lluch et al. [39] presented such a client/server rendering system. In their system a server holds a scene graph and uses it, along with client view information, to select an appropriate resolution from a multi-resolution representation on disk. Scene access is done in an out-of-core fashion, allowing very large models to be visualized. Even when rendering is done on the server, for large data a single machine may not be able to provide updates to the mobile device quickly enough for mobile users. For this reason Lamberti and Sanna [38] introduce a Chromium-based [26] rendering system which encodes the data as MPEG and streams it to be decoded on the mobile device. With motion estimation being the most expensive process of MPEG4 encoding, Cheng et al. [11] are able to significantly improve this step by directly retrieving motion vectors from information available in the rendering pipeline.

More recently, this area of research has matured from the stage of developing the basic building blocks, to presenting solutions for specific problems, and demonstrating the practical use of mobile visualization. Park et al. [47] developed a system for collaborative medical visualization, using parallel server-based volume rendering techniques, while Meir and Rubinsky [41] investigate the use of mobile devices as a cost-effective component of a distributed system for performing ultrasounds. Their system combines simple-to-use, inexpensive ultrasound devices at the client site, which generate ultrasound data. The data is sent to a server which performs volume rendering at pre-defined camera angles, and sends the images back down to mobile devices for analysis in the field.

29.2.2.2 Client Based Rendering

In a client based setting the entire rendering workload is handled solely by the mobile device and no persistent server connection is required for the visualization. A network connection may be required for some initial data transfer but for the visualization itself no server connection is necessary. In the first work that does 3D visualization with only the mobile hardware resources, Burigat and Chittaro [6] described a VRML-based system for visualizing what a user sees as they roam a city. A similar approach was taken by Nadalutti et al. [44] based on the X3D standard.

Moser and Weiskopf [43] were the first to present a prototypic client-only interactive volume rendering application on mobile devices. Due to the lack of 3D texture support by the OpenGL ES GPU their work is based on the concept of axis aligned slices as proposed by Rezk-Salama et al. [54].

Similar to the remote rendering works, one can also observe for the client based rendering systems that both the hardware as well as the methods have reached a state that allows for implementations to leave the stage of research prototypes and to

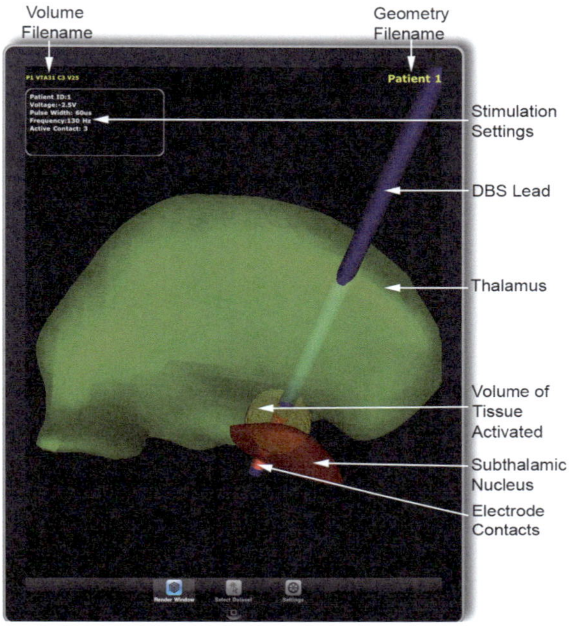

Fig. 29.1 This figure shows a mobile device based visualization system for the use in deep brain stimulation (*DBS*). The system renders semi-transparent geometry (Patient thalamus, subthalamic nucleus, and DBS lead with electrode contacts) together with the volume (the activation zone of the electrodes). For the volume rendering with correct geometry sorting and compositing this client-based visualization system achieves interactive frame-rates even on a first generation Apple iPad. In a clinical study the system was able to improve the DBS parameter setup-time from about 4 h to less than 2 min

realize systems that show practical use. Recently, Butson et al. [7] demonstrated that a mobile rendering system can improve the clinical routine for Deep Brain Stimulation parameter setup (see Fig. 29.1).

29.2.2.3 Hybrid Rendering

Both the server based remote rendering approaches as well as the client-only rendering strategies have their advantages and disadvantages. One issue that all remote rendering approaches share is the need for a persistent and reliable network connection to the server. Surprisingly, even today—more than 15 years after Encarnação et al. [20] early works, which already included considerations about 3G networks—such a network connectivity is far from guaranteed. Client only solutions, on the other hand, still suffer from the limited capabilities of the mobile devices, consequently a hybrid solution, which uses both the massive graphics and computing power of a server system as well as the low latency and constant availability of the local device hardware seems appealing.

To reduce the bandwidth requirements and transmission latencies a number of hybrid schemes based upon the idea of image warping [3, 4, 9, 35, 59, 63] have been proposed. All of these methods have in common that they receive image data of a hard to render object from a server. The mobile client then generates the displayed image by warping the server's data to the correct viewing parameters. To compensate for the missing data due to parallax either multiple depth peeled images are transmitted or the missing information is requested as a small package from the server.

Diepstraten et al. [15] follow a different approach, instead of reconstructing the full image from data received from the server, they propose to focus only on feature lines. These lines can be transmitted efficiently to the client and rendered even on the low end mobile hardware interactively.

29.3 Large Displays

Using large output displays has been of interest in visualization for a long time. This is illustrated, for example, by early systems such as the PowerWall [62], or sophisticated immersive virtual reality setups such as the CAVE [13].

In recent years, advances in digital cameras and computational photography have created a lot of interest in creating, processing, and displaying high-resolution imagery such as Gigapixel images [36]. Gigapixel images cannot be displayed on regular monitors in their entirety at their original resolution, and thus are a natural fit for the high display resolution of large-scale displays. One example is the Giga-stack system [50], which targets tiled display walls and relies on the CGLX framework [16] to display the correct image data on the individual LCD panels comprising the display array.

Similarly, recent advances and increased resolutions in scientific image acquisition such as confocal microscopy, for example high-resolution biological image stacks [34], or electron microscopy scans of brain tissue [33], result in image and volume data of extremely high resolution, for which inspection on high-resolution displays is of great interest. Figure 29.2 shows an example of a high-resolution brain tissue block obtained via electron microscopy, rendered on a large display wall with a total resolution of 13,660 $\times$ 3,072 pixels (40 megapixels), consisting of 40 individual display panels.

A well-established approach to building large-scale displays is to use multiple high-resolution projectors [56, 57]. In such systems, projector calibration in order to hide seams, brightness and contrast differences, and other artifacts is often a difficult practical issue. Recent advances in LCD display technology have enabled building large, tiled displays consisting of individual, relatively cheap LCD panels [50]. Such display arrays are usually much cheaper and easier to maintain than projector-based systems. Recently, Papadopoulos et al. [46] combined 416 LCD displays for the 1.5 gigapixel, 4-wall Reality Deck. A practical issue on LCD display arrays is the bezel width of the screens used, which determines the width of the border between the individual image tiles.

Apart from the practical issue of building and setting up the actual hardware of large displays, we are most of all concerned with the necessary software infrastructure and visualization algorithms that enable efficient rendering on tiled displays. Many visualization systems build on an underlying software framework—or *mid-*

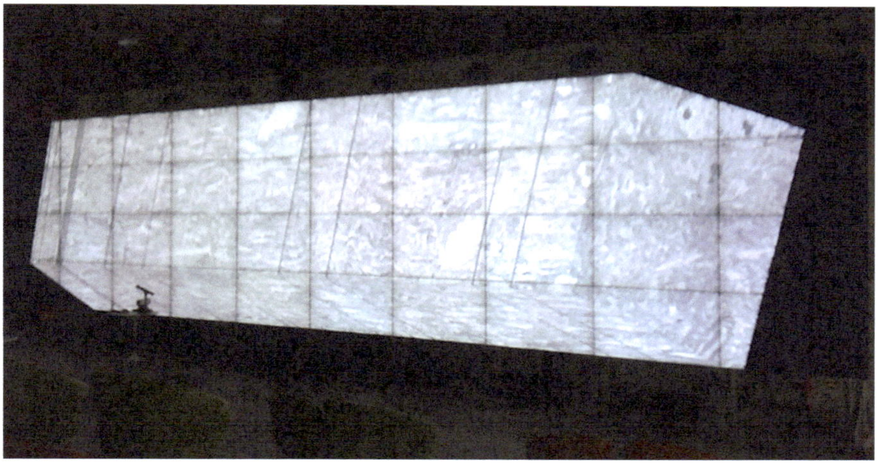

Fig. 29.2 A high-resolution electron microscopy (EM) volume of brain tissue, rendered on a large display wall consisting of 10 × 4 individual LCD displays. Each screen has a resolution of 1,366 × 768, leading to a total display resolution of 13,660 × 3,072

dleware—that supports distributing data to multiple displays. The next section gives an overview of common middleware for large-scale displays. Another important issue on a completely different conceptual level is, how users can interact with visualizations on large displays, where the typical mouse and keyboard interaction used on desktop systems is not a good fit.

29.3.1 Middleware for Visualization on Large Displays

Well-known software frameworks for visualization on large displays are the WireGL system [5, 24, 25], its successor Chromium [26], Equalizer [17, 18], SAGE [29–32, 53], and CGLX [16].

A crucial differentiating property of these frameworks is how and where they interface with application code, whether and how they require the application code to split up the data to be visualized, as well as where in the graphics pipeline this distribution occurs [42]. This determines the crucial trade-off between the obtainable performance, and the magnitude of required changes to application code and the algorithms employed.

The WireGL system [5, 24, 25] allows rendering with scalable output resolution using unmodified application code. WireGL substitutes the regular OpenGL driver with an implementation that can automatically distribute the OpenGL command stream to multiple rendering and display nodes. This process is completely transparent to the application, and thus no changes to the application code are necessary.

WireGL employs sort-first parallel rendering [42], i.e., splitting up the rendering in image space, which directly supports rendering on tiled displays. A major drawback of this approach, however, is that the data have to be replicated on each render node, i.e., this approach scales to large display sizes, but not as easily to large data sizes. This problem was reduced in WireGL [24] by filtering the OpenGL command stream to correspond better to the sub-streams required by individual render nodes, which builds on an early parallel graphics API implementation [27]. However, WireGL requires all data to be transmitted over the network every frame, which incurs a significant scalability problem for large data.

Compositing is a very important issue in distributed rendering, which is usually handled completely in software, but special purpose hardware solutions such as the Lightning-2 system [60] have been developed and combined with WireGL, for example.

The Chromium system [26] builds on the earlier work of WireGL to create a powerful framework for command stream processing and distribution, introducing programmable command stream filters that enable building general sort-first and sort-last parallel rendering algorithms on top of the provided functionality. Like WireGL, Chromium is based on intercepting OpenGL command streams.

The Equalizer framework [17, 18] provides very general capabilities for distributed rendering, both for distributing data, as well as for rendering on tiled displays. In contrast to the other middleware discussed here, Equalizer aims to provide a full-featured distributed parallel rendering framework. However, this requires the application code to use the Equalizer API and adapt its structure accordingly. Equalizer supports both sort-first and sort-last rendering approaches, and also provides parallel compositing capabilities with different strategies, as well as load balancing.

The SAGE (Scalable Adaptive Graphics Environment) framework [29–32, 53] concentrates solely on streams of pixel data, which can be distributed efficiently to the corresponding displays. In this way, SAGE can be used as a powerful low-level primitive for streaming pixel data, which have been generated before by any distributed visualization algorithm. SAGE allows multiple clients to display different views on a single display array, arranging for pixel streams to be sent to and displayed on the correct nodes. Application windows can be freely moved and resized anywhere, seamlessly handling windows straddling multiple displays, and multiple views partially overlapping the same display. In order to be able to stream pixel data, all rendered images must be read back from the GPU to the CPU and then transmitted over the network.

The CGLX framework [16] provides a GLUT-like application interface, which allows for easy conversion of OpenGL code that uses GLUT to large display arrays. Although distributed rendering is handled automatically, existing code must be converted from GLUT to CGLX. Although the required code changes to convert a GLUT application to CGLX are relatively minor, for applications that are not based on GLUT, using CGLX is not straightforward. An example for using CGLX as the low-level middleware to build a Gigapixel viewer is the Giga-stack system [50].

29.3.2 Interaction with Large Displays

A big conceptual challenge of using large displays for visualization is providing efficient and natural paradigms for user interaction. On large displays, normal window system interaction metaphors break down [25]. A simple example of this is moving windows on the screen, which is trivial on regular displays but much harder to achieve on large displays if larger pixel distances need to be covered. Moreover, using mouse and keyboard interaction together with large displays is cumbersome, because they usually require the user to sit down with mouse and keyboard in front of the display. Large displays make walking around and gesturing natural, whereas classical interaction paradigms lack this naturalness [49]. However, high-resolution displays have been shown to improve the capabilities of users to compare and find targets, as well as to outperform displays of lower resolution for navigation tasks [2].

Large displays invite metaphors such as those of standard whiteboards [51], which affords using informal writing, sketching and space management [22]. These display devices also invite collaborative approaches to visualization [52], and multi-user interaction [28]. Related approaches have been developed for collaborative visualization on tabletop displays [58], which often employ tracking via overhead-mounted cameras like the Lambda table system [37].

Common interaction approaches leverage tracked objects or pointers, such as the VisionWand [8], which provides the user with a (passive) wand that is tracked in 3D, and then used together with visual widgets overlayed on the display. The physical size of large displays can become a significant problem in such approaches, which can be circumvented by interaction metaphors that do not require direct pointing. Examples are using infrared laser tracking [10], or the LumiPoint system [14] that is capable of tracking multiple pen-like objects with video cameras.

Whiteboard-style interaction can also be achieved by giving users small touch screens and a pen [51]. More recently, similar interaction capabilities can be provided using multi-touch tablets such as an Apple iPad where multi-touch gestures steer the visualization on the display wall [1]. Instead of using pointer objects, gestures can be recognized directly, such as multi-finger gestural input [40]. Multi-touch input technologies can broadly be categorized as being either optical, capacitive, or resistive [55]. Painting gestures can also be seamlessly supported across the boundaries of multiple displays, by stitching the individual parts together [23]. Gesture recognition can further be combined with speech recognition [48, 49]. Recent approaches employ other technological advances such as the cheap gyroscopes and accelerometers found in mobile phones and video console hardware, for example using the peripherals of the Nintendo Wii video game console for stroke-based rendering [21].

Fig. 29.3 A distributed weather analysis and emergency warning system, a visualization scenario running on top of the ZAPP framework [61]

29.4 Outlook

Scalable devices are the future! Analysts agree that the future of home computing lies in small computers such as pads, tablets, or small notebooks, and visualization labs around the globe invest significant amounts of money in tiled display systems hardware and research. But will that ultimately influence the way we interact with data in the long term? While such things are always hard to predict, we do believe that the added value of high-resolution and collaborative large-displays as well as the ease of use of mobile computers make those devices the ideal platform for **convenient visualization**, already few works exist to combine the best of both worlds and allow the control and interaction on large display walls with mobile computers (see Fig. 29.3).

Acknowledgments The work presented in this paper was made possible in part by the NIH/NCRR Center for Integrative Biomedical Computing, P41-RR12553-10 and by Award Number R01EB007688 from the National Institute of Biomedical Imaging and Engineering, as well as the Intel Visual Computing Institute and the Cluster of Excellence "Multimodal Computing and Interaction" at the Saarland University. The content is under sole responsibility of the authors.

References

1. Armenta Gastelum, C.J.: Volume visualization and compositing on large-scale displays using handheld touchscreen interaction. Master's thesis, King Abdullah University of Science and Technology (2011)
2. Ball, R., North, C.: Effects of tiled high-resolution display on basic visualization and navigation tasks. In: CHI 05 Extended Abstracts on Human Factors in Computing Systems CHI 05, pp. 1196–1199. ACM Press, New York, USA (2005)
3. Bao, P., Gourlay, D.: A framework for remote rendering of 3-d scenes on limited mobile devices. Multimedia IEEE Trans. **8**(2), 382–389 (2006)
4. Boukerche, A., Pazzi, R.W.N.: Remote rendering and streaming of progressive panoramas for mobile devices. In: Proceedings of the 14th Annual ACM International Conference on Multimedia, MULTIMEDIA 2006, pp. 691–694. ACM, New York, USA (2006)

5. Buck, I., Humphreys, G., Hanrahan, P.: Tracking graphics state for networked rendering. In: Proceedings of the 2000 Eurographics/SIGGRAPH Workshop on Graphics Hardware, pp. 87–95. ACM Press, New York, USA (2000)
6. Burigat, S., Chittaro, L.: Location-aware visualization of vrml models in gps-based mobile guides. In: Proceedings of the 10th International Conference on 3D Web Technology, Web3D 2005, pp. 57–64. ACM, New York, USA (2005)
7. Butson, C.R., Tamm, G., Jain, S.G., Fogal, T., Krüger, J.: Evaluation of medical visualization on mobile computing platforms for clinical decision making. Technical Report, Saarland University (2011)
8. Cao, X., Balakrishnan, R.: VisionWand: Interaction techniques for large displays using a passive wand tracked in 3D. In: Proceedings of UIST 2003, ACM Symposium on User Interface Software and Technology, vol. 5, pp. 173–182. ACM, New York (2003)
9. Chang, C.F., Ger, S.H.: Enhancing 3D graphics on mobile devices by image-based rendering. In: Chen, Y.C., Chang, L.W., Hsu, C.T. (eds.) Advances in Multimedia Information Processing PCM 2002. Lecture Notes in Computer Science, vol. 2532, pp. 1–17. Springer, Berlin/Heidelberg (2002)
10. Cheng, K., Pulo, K.: Direct interaction with large-scale display systems using infrared laser tracking devices. In: Proceedings of the Asia-Pacific Symposium on Information Visualisation, vol. 24, pp. 67–74. Australian Computer Society (2003)
11. Cheng, L., Bhushan, A., Pajarola, R., El Zarki, M.: Real-time 3d graphics streaming using mpeg-4. In: Proceedings IEEE/ACM Workshop on Broadband Wireless Services and Applications (2004)
12. Chittaro, L.: Visualizing information on mobile devices. Computer **39**(3), 40–45 (2006)
13. Cruz-Neira, C., Sandin, D.J., DeFanti, T.A., Kenyon, R.V., Hart, J.C.: The cave: audio visual experience automatic virtual environment. Commun. ACM **35**, 64–72 (1992)
14. Davis, J., Chen, X.: LumiPoint: multi-user laser-based interacion on large tiled displays. Displays **5**, 205–211 (2002)
15. Diepstraten, J., Gorke, M., Ertl, T.: Remote line rendering for mobile devices. In: Proceedings of Computer Graphics International, 2004, pp. 454–461 (2004)
16. Doerr, K.U., Kuester, F.: CGLX: a scalable, high-performance visualization framework for networked display environments. IEEE Trans. Visual. Comput. Graph. **17**(3), 320–332 (2010)
17. Eilemann, S., Makhinya, M., Pajarola, R.: Equalizer: a scalable parallel rendering framework. IEEE Trans. Visual. Comput. Graph. **15**(3), 436–452 (2009)
18. Eilemann, S., Pajarola, R.: The equalizer parallel rendering framework. Technical Report IFI 2007.06, Department of Informatics, University of Zürich (2007)
19. Eissele, M., Weiskopf, D., Ertl, T.: Interactive context-aware visualization for mobile devices. In: Proceedings of the 10th International Symposium on Smart Graphics, SG 2009, pp. 167–178. Springer, Berlin (2009)
20. Encarnação, J.L., Frühauf, M., Kirste, T.: Mobile visualization: Challenges and solution concepts. In: Proceedings of CAPE 1995 (1995)
21. Grubert, J., Hanckock, M., Carpendale, S., Tse, E.: Interacting with stroke-based rendering on a wall display. Technical Report 2007–882-34, Department of Computer Science, University of Calgary (2007)
22. Guimbretière, F., Stone, M., Winograd, T.: Fluid interaction with high-resolution wall-size displays. In: Proceedings of the 14th Annual ACM Symposium on User Interface Software and Technology, pp. 21–30. ACM Press, New York, USA (2001)
23. Hinckley, K., Ramos, G., Guimbretiere, F., Baudisch, P., Smith, M.: Stitching: pen gestures that span multiple displays. In: Proceedings of the Working Conference on Advanced Visual Interfaces, pp. 23–31. ACM (2004)
24. Humphreys, G., Buck, I., Eldridge, M., Hanrahan, P.: Distributed rendering for scalable displays. In: Proceedings of the 2000 ACM/IEEE Conference on Supercomputing, p. 30. IEEE Computer Society (2000)
25. Humphreys, G., Hanrahan, P.: A distributed graphics system for large tiled displays. In: Proceedings of the 10th IEEE Visualization 1999 Conference, pp. 215–223. IEEE Computer Society Press (1999)

26. Humphreys, G., Houston, M., Ng, R., Frank, R., Ahern, S., Kirchner, P.D., Klosowski, J.T.: Chromium: a stream-processing framework for interactive rendering on clusters. ACM Trans. Graph. **21**(3), 693–702 (2002)
27. Igehy, H., Stoll, G., Hanrahan, P.: The design of a parallel graphics interface. In: Proceedings of the 25th Annual Conference on Computer Graphics and Interactive Techniques—SIGGRAPH 1998, pp. 141–150. ACM Press, New York, USA (1998)
28. Jagodic, R., Renambot, L., Johnson, A., Leigh, J., Deshpande, S.: Enabling multi-user interaction in large high-resolution distributed environments. Future Gener. Comput. Syst. **27**(7), 914–923 (2010)
29. Jeong, B., Jagodic, R., Renambot, L., Singh, R., Johnson, A., Leigh, J.: Scalable graphics architecture for high-resolution displays. In: IEEE Information Visualization Workshop, vol. iGrid, pp. 1–4. Citeseer (2005)
30. Jeong, B., Leigh, J., Johnson, A., Renambot, L., Brown, M., Jagodic, R., Nam, S., Hur, H.: Ultrascale collaborative visualization using a display-rich global cyberinfrastructure. IEEE Comput. Graph. Appl. **30**(3), 71–83 (2010)
31. Jeong, B., Renambot, L., Jagodic, R., Singh, R., Aguilera, J., Johnson, A., Leigh, J.: High-performance dynamic graphics streaming for scalable adaptive graphics environment. In: ACM/IEEE SuperComputing 2006, p. 24. IEEE (2006)
32. Jeong, B., Renambot, L., Singh, R., Johnson, A., Leigh, J.: High-performance scalable graphics architecture for high-resolution displays. EVL Technical Document Technical publication 20050824_Jeong (2005)
33. Jeong, W.K., Beyer, J., Hadwiger, M., Blue, R., Law, C., Vasquez, A., Reid, C., Lichtman, J., Pfister, H.: Ssecrett and neurotrace: interactive visualization and analysis tools for large-scale neuroscience datasets. IEEE Comput. Graph. Appl. **30**(3), 58–70 (2010)
34. Jeong, W.K., Schneider, J., Turney, S.G., Faulkner-Jones, B.E., Meyer, D., Westermann, R., Reid, C., Lichtman, J., Pfister, H.: Interactive histology of large-scale biomedical image stacks. IEEE Trans. Visual. Comput. Graph. **16**(6), 1386–1395 (2010)
35. Jiang, Z., Mao, Y., Jia, Q., Jiang, N., Tao, J., Fang, X., Bao, H.: Panowalk: a remote image-based rendering system for mobile devices. PCM **2006**, 641–649 (2006)
36. Kopf, J., Uyttendaele, M., Deussen, O., Cohen, M.F.: Capturing and viewing gigapixel images. In: Proceedings of SIGGRAPH 2007. ACM Press/ACM SIGGRAPH (2007)
37. Krumbholz, C., Leigh, J., Johnson, A., Renambot, L., Kooima, R.: Lambda table: High resolution tiled display table for interacting with large visualizations. In: Proceedings of 5th Workshop on Advanced Collaborative Environments (2005)
38. Lamberti, F., Sanna, A.: A streaming-based solution for remote visualization of 3D graphics on mobile devices. IEEE Trans. Visual. Comput. Graph. **13**(2), 247–260 (2007)
39. Lluch, J., Gaitán, R., Camahort, E., Vivó, R.: Interactive three-dimensional rendering on mobile computer devices. In: Proceedings of the 2005 ACM SIGCHI International Conference on Advances in Computer Entertainment Technology, ACE 2005, pp. 254–257. ACM, New York, USA (2005)
40. Malik, S., Ranjan, A., Balakrishnan, R.: Interacting with large displays from a distance with vision-tracked multi-finger gestural input. In: Proceedings of the 18th Annual ACM Symposium on User Interface Software and Technology, p. 43. ACM Press, New York, USA (2005)
41. Meir, A., Rubinsky, B.: Distributed network, wireless and cloud computing enabled 3-D ultrasound; a new medical technology paradigm. PLoS ONE **4**(11), e7974 (2009)
42. Molnar, S., Cox, M., Ellsworth, D., Fuchs, H.: A sorting classification of parallel rendering. IEEE Comput. Graph. Appl. **14**(4), 23–32 (1994)
43. Moser, M., Weiskopf, D.: Interactive volume rendering on mobile devices. Workshop Vision, Model. Visual. VMV **2008**, 217–226 (2008)
44. Nadalutti, D., Chittaro, L., Buttussi, F.: Rendering of x3d content on mobile devices with opengl es. In: Proceedings of the 11th International Conference on 3D Web Technology, Web3D 2006. ACM, p. 26 (2006)
45. Paelke, V., Reimann, C., Rosenbach, W.: A visualization design repository for mobile devices. In: Proceedings of the 2nd International Conference on Computer Graphics, Virtual Reality,

Visualisation and Interaction in Africa, AFRIGRAPH 2003, pp. 57–62. ACM, New York, USA (2003)
46. Papadopoulos, C., Petkov, K., Kaufman, A.: Building the reality deck. In: In SIGCHI Workshop: POWERWALL: International Workshop on Interactive, Ultra-High-Resolution Displays, p. 6 (2013)
47. Park, S., Kim, W., Ihm, I.: Mobile collaborative medical display system. Comput. Methods Programs Biomed. **89**(3), 248–260 (2008)
48. Pavlovic, V., Sharma, R., Huang, T.: Speech/gesture interface to a visual computing environment for molecular biologists. In: Proceedings of the International Conference on Pattern Recognition, pp. 1964–1970. IEEE Computer Society Press (1996)
49. Pavlovic, V., Sharma, R., Huang, T.: Visual interpretation of hand gestures for human-computer interaction: a review. IEEE Trans. Pattern Anal. Mach. Intell. **19**(7), 677–695 (1997)
50. Ponto, K., Doerr, K., Kuester, F.: Giga-stack: a method for visualizing giga-pixel layered imagery on massively tiled displays. Future Generation Comput. Syst. **26**(5), 693–700 (2010)
51. Rekimoto, J.: A multiple device approach for supporting whiteboard-based interactions. In: Proceedings of the SIGCHI Conference on Human factors in Computing Systems—CHI 1998, pp. 344–351. ACM Press, New York, USA (1998)
52. Renambot, L., Jeong, B., Jagodic, R., Johnson, A., Leigh, J., Aguilera, J.: Collaborative visualization using high-resolution tiled displays. In: ACM CHI Workshop on Information Visualization Interaction Techniques for Collaboration Across Multiple Displays, pp. 1–4. ACM Press (2006)
53. Renambot, L., Rao, A., Singh, R., Jeong, B., Krishnaprasad, N., Vishwanath, V., Chandrasekhar, V., Schwarz, N., Spale, A., Zhang, C., Others: Sage: the scalable adaptive graphics environment. In: Proceedings of WACE, vol. 2004, p. 6. Citeseer (2004)
54. Rezk-Salama, C., Engel, K., Bauer, M., Greiner, G., Ertl, T.: Interactive volume rendering on standard pc graphics hardware using multi-textures and multi-stage rasterization. In: Proceedings of ACM SIGGRAPH/Eurographics Workshop on Graphics Hardware (2000)
55. Rosenberg, I., Perlin, K.: The unMousePad. ACM Trans. Graph. **28**(3), 1 (2009)
56. Samanta, R., Zheng, J., Funkhouser, T., Li, K., Singh, J.P.: Load balancing for multi-projector rendering systems. In: Proceedings of the ACM SIGGRAPH/EUROGRAPHICS Workshop on Graphics Hardware, pp. 107–116. ACM Press, New York, USA (1999)
57. Schikore, D.R., Fischer, R.A., Frank, R., Gaunt, R., Hobson, J., Whitlock, B.: High-resolution multi-projector display walls and applications. IEEE Comput. Graph. Appl. **20**(4), 38–44 (2000)
58. Scott, S., Grant, K., Mandryk, R.: System guidelines for co-located, collaborative work on a tabletop display. In: Proceedings of the 8th Conference on European Conference on Computer Supported Cooperative Work, pp. 159–178. Kluwer Academic Publishers, Berlin (2003)
59. Shi, S., Jeon, W.J., Nahrstedt, K., Campbell, R.H.: Real-time remote rendering of 3d video for mobile devices. In: Proceedings of the 17th ACM International Conference on Multimedia, MM 2009, pp. 391–400. ACM, New York, USA (2009)
60. Stoll, G., Eldridge, M., Patterson, D., Webb, A., Berman, S., Levy, R., Caywood, C., Taveira, M., Hunt, S., Hanrahan, P.: Lightning-2: a high-performance display subsystem for pc clusters. In: Proceedings of SIGGRAPH 2001, pp. 141–148. ACM Press/ACM SIGGRAPH (2001)
61. Tamm, G., Schiewe, A., Krüger, J.: ZAPP—A management framework for distributed visualization systems. In: Proceedings of the IADIS Computer Graphics, Visualization, Computer Vision and Image Processing 2011 (CGVCVIP 2011) Conference (2011)
62. University of Minnesota: Powerwall (1994). URL http://www.lcse.umn.edu/research/powerwall/powerwall.html
63. Yoo, W., Shi, S., Jeon, W., Nahrstedt, K., Campbell, R.: Real-time parallel remote rendering for mobile devices using graphics processing units. In: IEEE International Conference on Multimedia and Expo (ICME), 2010, pp. 902–907 (2010)

Chapter 30
Scalable Representation

Yun Jang

Abstract Although the amount and variety of data being generated is increased dramatically, the capabilities of data visualization, analysis, and discovery solutions have not been improved accordingly with the explosive rate of data production. One reason is that storage and processing at the level of raw data require supercomputer scale resources. The other is that working at the level of raw data prevents effective human comprehension while exploring and solving most problems. Here we show several approaches to scalable functional representations. Encoding, abstraction, and analysis at multiple scales of representations are a common approach in many scientific disciplines and provides a promising approach to harness our expanding digital universe.

30.1 Functional Representations

Scalable functional representations have been studied in many research areas in order to interpolate, approximate, and abstract data. Many disciplines no longer explore raw data, but functionally derived and processed information from the raw data. Examples include CT and MRI images in medical applications, and the derived high-level products from remote sensing data. Common techniques for functional representation include radial basis functions, wavelets, and spherical harmonics. The nature of this functional representation is to find functions hierarchically, therefore, it is possible to represent data from abstract to detail levels according to compression ratio and level of detail. The abstract level of the representation allows us to visualize large data interactively, whereas, the detail level requires more computational power for the visualization. Moreover it is a unified representation regardless of data formats. Details of the functional representations are presented as follows.

Y. Jang (✉)
Purdue University, West Lafayette, USA
e-mail: jjangyn@gmail.com

Y. Jang
Sejong University, Seoul, South Korea

C.D. Hansen et al. (eds.), *Scientific Visualization*, Mathematics and Visualization,
DOI 10.1007/978-1-4471-6497-5_30

30.1.1 Radial Basis Functions

Previously, most radial basis function (RBF) encoding work has concentrated on surface fitting. In the early 1970s, RBFs were used to interpolate geography surface data using multiquadric RBFs [13] and results showed that RBFs are a good interpolation basis function for smooth surface data sets. Hardy [14] presented 20 years of discovery in the theory and applications of multiquadric RBFs and surveys RBF work from 1968 to 1988. Franke [10] showed scattered data interpolation and tests using several methods, such as the inverse distance weighted method, the rectangle based blending method, and the triangle based blending method. He compared these methods and showed that Hardy's multiquadric approach is best. Since Franke's work, multiquadrics have been considered the best basis function in most surface fitting research. After Franke's survey, Franke and Nielson [12] collected more work on surface fitting and presented their research by surveying and comparing several techniques. For better interpolation, the least squares approach were used by Franke and Hagen [11]. For the approximation of surface fitting, knot (center) selection [23] was introduced using thin plate splines by Dirichlet tessellation. Through knot selection, encoded data can be reduced and a small number of basis function can represent the whole data set.

Although RBFs have been used to reconstruct surfaces by approximating scattered data sets, they were primarily used for mesh reduction of surface representations [4, 25, 32, 39]. In more recent work on surface fitting, Carr et al. [4] showed surface fitting as an approximation using multiquadric RBFs. They iteratively added basis functions using a greedy algorithm by computing fitting errors, where basis functions were added at larger error points. In their work, the zero level set implicit surface of the distance function was fit and energy (error) was minimized for the smoothest interpolant. Ohtake et al. [31] also showed the fitting of implicit surfaces. They selected centers based on the density of data points. More basis functions were added in higher density areas. By linking the RBF approximation and the partition of unity method [30], Ohtake el al. presented a robust approximation for noisy data.

Volume fitting using RBFs was introduced by Nielson et al. [28, 29], where they extended surface fitting methods to volume fitting. Their approaches showed good approximation of volume data. In more recent work on volume fitting, Co et al. [5] showed a hierarchical representation of volumetric data sets based on clusters computed by Principal Component Analysis (PCA). A level of detail representation was extracted by either the hierarchical level or the error. Jang et al. [16, 17] and Weiler et al. [41] proposed a functional representation approach for interactive volume rendering. Their approaches were designed for any scattered datasets and directly volume rendered the basis functions without resampling. Moreover, using ellipsoidal basis functions, they improved the functional representation statistically and visually [16]. Recently, Ledergerber et al. [19] applied a moving least square to interpolate the volumetric data and Vuçini et al. [40] reconstructed non-uniform volumetric data by B-splines.

30.1.2 Wavelets

As a hierarchical data representation for comparison, there is an approach, wavelet [3, 6, 15, 26, 27, 35–37]. In the area of volume encoding, most work [15, 26, 27] has been performed for regular grids and shows effective 3D compressed volumes and rendering. Recently, this research has been extended to irregular grids organized using polygonal meshes [6, 36, 37]. The wavelet encoding of irregular grids is usually performed using irregular sampling and adaptive subdivision. However, the wavelet for the irregular data set is based on polygonal meshes. Therefore, it is not easily extended to arbitrary scattered volume data. Sohn et al. [35] presented a compression scheme for encoding time-varying isosurface and volume features. They encoded only the significant blocks using a block-based wavelet transform.

30.1.3 Spherical Harmonics

Most of the work in spherical harmonics has been done for surface fitting, especially, 3D object modeling and molecular surface modeling. For 3D object modeling [7, 18, 38], the datasets were decomposed into high and low frequency components and represented by the properties of the spherical harmonic basis functions. Also this modeling was used for shape deformation [8, 9]. In molecular surface modeling [22], spherical harmonics give a sequence of smooth approximations to the molecular surface since the shapes of the spherical harmonics are very similar to the shapes of the molecules. For volume fitting, Misner [24] showed spherical harmonic decomposition, however, the volume fitting is based on a rectangular grid.

30.1.4 Time Series Data Representations

The large volume of time-varying data makes visualization a challenging problem. Many techniques for volume rendering of time-varying data have been proposed and these techniques enable the visualization of large amount of time-varying datasets. One approach is to use data coherency between consecutive time steps to speed up volume rendering [1, 2, 33, 34]. Another approach is to encode and compress the time-varying data appropriately for the volume rendering [20, 21, 35, 42]. Shen and Johnson [34] proposed an algorithm that exploits data coherency between time steps and extracts the differential information for biomedical and computational fluid dynamics datasets. Shen et al. [33] showed the time-space partitioning (TSP) tree and this structure improves the rendering speed and reduces the amount of volumetric data I/O. For temporal compression approach, Westermann [42] proposed a memory minimizing algorithm based on multi-resolution representations of both the spatial distribution and the time evolutions. Ma and Shen [21] presented quantization and octree encoding of time-varying datasets and reduced the rendering speed over time. Lum et al. [20] presented temporal encoding using discrete cosine transform (DCT) and accelerated the rendering speed using the graphics hardware.

References

1. Bernardon, F.F., Callahan, S.P., Comba, J.L.D., Silva, C.T.: Interactive volume rendering of unstructured grids with time-varying scalar fields. In: Eurographics Symposium on Parallel Graphics and Visualization, pp. 51–58 (2006)
2. Bernardon, F.F., Callahan, S.P., Comba, J.L.D., Silva, C.T.: An adaptive framework for visualizing unstructured grids with time-varying scalar fields. Parallel Comput. **33**(6), 391–405 (2007)
3. Bertram, M., Duchaineau, M.A., Hamann, B., Joy, K.I.: Bicubic subdivision-surface wavelets for large-scale isosurface representation and visualization. IEEE Vis. **2000**, 389–396 (2000)
4. Carr, J.C., Beatson, R.K., Cherrie, J.B., Mitchell, T.J., Fright0, W.R., McCallum, B.C., Evans, T.R.: Reconstruction and representation of 3D objects with radial basis functions. In: Proceedings of ACM SIGGRAPH 2001, Computer Graphics Proceedings, Annual Conference Series, pp. 67–76 (2001)
5. Co, C.S., Heckel, B., Hagen, H., Hamann, B., Joy, K.I.: Hierarchical clustering for unstructured volumetric scalar fields. In: Proceedings of IEEE visualization (2003)
6. Daubechies, I., Guskov, I., Schröder, P., Sweldens, W.: Wavelets on irregular point sets. Phil. Trans. R. Soc. Lond. A **357**(1760), 2397–2413 (1999)
7. Ertörk, S., Dennis, T.J.: Object shape deformation with spherical harmonic interpolation. IEE Electron. Lett. **34**, 1657–1658 (1998)
8. Ertörk, S., Dennis, T.J.: Approximating spherical harmonic representation order. IEE Electron. Lett. **35**, 462–463 (1999)
9. Ertörk, S., Dennis, T.J.: Automated shape metamorphosis for 3D objects. In: IEE Image processing and its applications, Conference Publication No. **465**, 846–850 (1999)
10. Franke, R.: Scattered data interpolation: Tests of some method. Math. Comput. **38**(157), 181–200 (1982)
11. Franke, R., Hagen, H.: Least squares surface approximation using multiquadrics and parametric domain distortion. Comput. Aided Geom. Des. **16**(3), 177–196 (1999)
12. Franke, R., Nielson, G.M.: Scattered data interpolation and applications: a tutorial and survey. In: Hagen, H., Roller, D. (eds.) Geometric modelling, methods and applications, pp. 131–160. Springer, Berlin (1991)
13. Hardy, R.L.: Multiquadric equations of topography and other irregular surfaces. J. Geophys. Res. **76**(8), 1905–1915 (1971)
14. Hardy, R.L.: Theory and applications of the multiquadric-biharmonic method 20 years of discovery 1968–1988. Comput. Math. Appl. **19**(8—-9), 163–208 (1990)
15. Ihm, I., Park, S.: Wavelet-based 3D compression scheme for interactive visualization of very large volume data. Comput. Graph. Forum **18**(1), 3–15 (1999)
16. Jang, Y., Botchen, R.P., Lauser, A., Ebert, D.S., Gaither, K.P., Ertl, T.: Enhancing the interactive visualization of procedurally encoded multifield data with ellipsoidal basis functions. Comput. Graph. Forum **25**(3), 587–596 (2006)
17. Jang, Y., Weiler, M., Hopf, M., Huang, J., Ebert, D.S., Gaither, K.P., Ertl, T.: Interactively visualizing procedurally encoded scalar fields. In: EG/IEEE TCVG Symposium on Visualization VisSym'04, pp. 35–44, 339 (2004)
18. Kazhdan, M., Funkhouser, T., Rusinkiewicz, S.: Rotation invariant spherical harmonic representation of 3D shape descriptors. In: Eurographics Symposium on Geometry Processing (2003)

19. Ledergerber, C., Guennebaud, G., Meyer, M., Bacher, M., Pfister, H.: Volume mls ray casting. IEEE Trans. Vis. Comput. Graph. **14**(6), 1539–1546 (2008)
20. Lum, E.B., Ma, K.L., Clyne, J.: Texture hardware assisted rendering of time-varying volume data. In: VIS '01: Proceedings of the Conference on Visualization'01, pp. 263–270. IEEE Computer Society Press (2001)
21. Ma, K.L., Shen, H.: Compression and accelerated rendering of time-varying volume data. In: Proceedings of the Workshop on Computer Graphics and Virtual Reality (2000)
22. Max, N.L., Getzoff, E.D.: Spherical harmonic molecular surfaces. IEEE Computer Graphics and Applications (1988)
23. McMathon, J.R., Franke, R.: Knot selection for least squares thin plate splines. SIAM J. Sci. Stat. Comput. **13**(2), 484–498 (1992)
24. Misner, C.W.: Spherical harmonic decomposition on a cubic grid. Class. Quantum Gravity **21**(3), S243–S247 (2004)
25. Morse, B.S., Yoo, T.S., Rheingans, P., Chen, D.T., Subramanian, K.R.: Interpolating implicit surfaces from scattered surface data using compactly supported radial basis functions. In: Proceedings Shape Modeling International, pp. 89–98. Genova, Italy (2001)
26. Muraki, S.: Approximation and rendering of volume data using wavelet transforms. In: VIS'92: Proceedings of the 3rd Conference on Visualization '92, pp. 21–28. IEEE Computer Society Press (1992)
27. Nguyen, K.G., Saupe, D.: Rapid high quality compression of volume data for visualization. Comput. Graph. Forum **20**(3), 49–56 (2001)
28. Nielson, G.M.: Scattered data modeling. IEEE Comput. Graph. Appl. **13**(1), 60–70 (1993)
29. Nielson, G.M., Foley, T.A., Hamann, B., Lane, D.: Visualizing and modeling scattered multivariate data. IEEE Comput. Graph. Appl. **11**(3), 47–55 (1991)
30. Ohtake, Y., Belyaev, A., Alexa, M., Turk, G., Seidel, H.P.: Multi-level partition of unity implicits. ACM Trans. Graph. **22**(3), 463–470 (2003)
31. Ohtake, Y., Belyaev, A., Seidel, H.P.: 3D scattered data approximation with adaptive compactly supported radial basis functions. In: International Conference on Shape Modeling and Applications (2004)
32. Savchenko, V.V., Pasko, A.A., Okunev, O.G., Kunii, T.L.: Function representation of solids reconstructed from scattered surface points and contours. Comput. Graph. Forum **14**(4), 181–188 (1995). http://citeseer.nj.nec.com/savchenko95function.html
33. Shen, H.W., Chiang, L.J., Ma, K.L.: A fast volume rendering algorithm for time-varying fields using a time-space partitioning (TSP) tree. In: VIS'99: Proceedings of the Conference on Visualization '99, pp. 371–377. IEEE Computer Society Press (1999)
34. Shen, H.W., Johnson, C.R.: Differential volume rendering: a fast volume visualization technique for flow animation. In: VIS'94: Proceedings of the Conference on Visualization '94, pp. 180–187. IEEE Computer Society Press, Los Alamitos, CA, USA (1994)
35. Sohn, B.S., Bajaj, C., Siddavanahalli, V.: Feature based volumetric video compression for interactive playback. In: VVS'02: Proceedings of the 2002 IEEE symposium on Volume Visualization and Graphics, pp. 89–96. IEEE Press (2002)
36. Stollnitz, E.J., DeRose, T.D., Salesin, D.H.: Wavelets for computer graphics: a primer, part 1. IEEE Comput. Graph. Appl. **15**(3), 76–84 (1995)
37. Sweldens, W.: The lifting scheme: a construction of second generation wavelets. SIAM J. Math. Anal. **29**(2), 511–546 (1998)
38. Tanaka, K., Sano, M., Mukawa, N., Kaneko, H.: 3D object representation using spherical harmonic functions. In: Proceedings of the 1993 IEEE/RSJ, International Conference on Intelligent Robots and Systems (1993)
39. Turk, G., O'Brien, J.F.: Modelling with implicit surfaces that interpolate. ACM Trans. Graph. (TOG) **21**(4), 855–873 (2002)
40. Vuçini, E., Möller, T., Gröller, M.E.: On visualization and reconstruction from non-uniform point sets using b-splines. Comput. Graph. Forum **28**(3), 1007–1014 (2009)

41. Weiler, M., Botchen, R.P., Stegmeier, S., Ertl, T., Huang, J., Jang, Y., Ebert, D.S., Gaither, K.P.: Hardware-assisted feature analysis of procedurally encoded multifield volumetric data. Comput. Graph. Appl. **25**(5), 72–81 (2005)
42. Westermann, R.: Compression domain rendering of time-resolved volume data. In: VIS '95: Proceedings of the 6th conference on Visualization '95, pp. 168–175,450. IEEE Computer Society Press (1995)

Chapter 31
Distributed Post-processing and Rendering for Large-Scale Scientific Simulations

Markus Flatken, Christian Wagner and Andreas Gerndt

Abstract With the ever-increasing capacity of high performance computing (HPC) systems, the computational simulation models become still finer and more accurate. However, the size and complexity of the data produced poses tremendous challenges for the visualization and analysis task. Especially when explorative approaches are demanded, distributed and parallel post-processing architectures have to be developed in order to allow interactive human-computer interfaces. Such infrastructures can also be exploited for the evaluation of ongoing simulation runs. The application here ranges from online monitoring to computational steering. But also remote and parallel rendering can be integrated into the overall setup. This chapter gives an overview of current solutions and ongoing research activities in this domain.

31.1 Motivation

Simulations of complex dynamic systems have been made a big progress over the last decade. Nowadays, they are not only used to merely assess and review results from experiments but increasingly to replace them with computational simulation. The reason for that in many domains is a continuously increasing comprehension of how to model dynamic systems with computer algorithms. But besides robustness and confidence in the modeling, the main basis for the success of simulations is the availability of supercomputer systems, which makes it possible to perform very high-resolution and complex simulations in a reasonable time. Depending on the domain,

M. Flatken (✉) · C. Wagner · A. Gerndt
German Aerospace Center, Lilienthalplatz 7, Braunschweig, Germany
e-mail: markus.flatken@dlr.de

C. Wagner
e-mail: christian.wagner@dlr.de

A. Gerndt
e-mail: andreas.gerndt@dlr.de

C.D. Hansen et al. (eds.), *Scientific Visualization*, Mathematics and Visualization,
DOI 10.1007/978-1-4471-6497-5_31

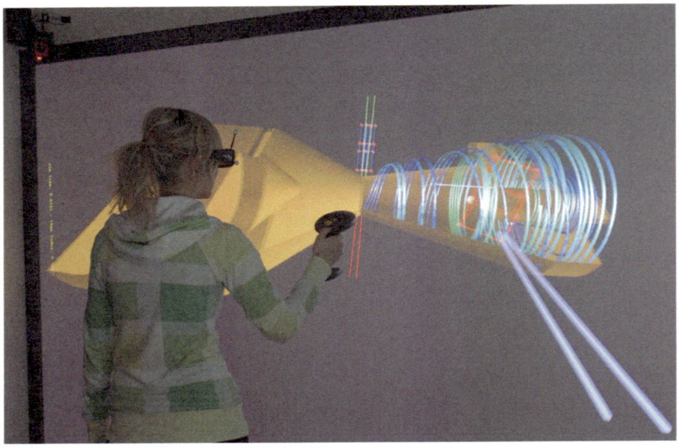

Fig. 31.1 Interactive flow field exploration of a centrifuge in a virtual environment

resulting simulation output can exceed sizes of tera- and peta-bytes of data. Without appropriate approaches for the analyses of those datasets, it is almost impossible to obtain better insight into the highly complex interdependencies of dynamic systems. Often, merely some samples are taken or simplified views are generated just to proof the correctness of the hypothesis used for the modeling.

In case that such simplified confirmative approach is not sufficient and satisfying enough, more explorative methods have to be integrated into the workflow. Exploration, however, requires interactive tools to enable the search for findings which were unexpected or not known yet. Interactivity on large-scale datasets is the main challenge of a real-time post-processing of dynamic simulations. Furthermore, produced results typically consist of many scalar and higher-dimensional data fields and may also change over time. One of the most intuitive ways to facilitate insight into those data is scientific visualization. Improved virtual reality (VR) techniques can eventually add the demanded interactivity in order to achieve an immersive and intuitive exploration environment (see Fig. 31.1).

Precise tracking of the scientist in a virtual environment and updating the view-dependent images on (typically) multiple displays with high frame rates is already a challenging requirement on such systems. Therefore, handling of large-scale simulation data and performing an appropriate post-processing on the same machine would easily break the interactivity requirements required for such explorative visualization environments. The solution is a distributed architecture, which relieves the visualization frontend from heavy post-processing work that can now be moved to a remote system. In addition, the visualization can be performed by remote rendering clusters to enhance the interactivity at the frontend even more.

With a supercomputer at hand, the post-processing can additionally be sped-up considerably by parallelization strategies. Instead of storing results of a running simulation to the file server, it may now also be possible to process the data directly in main memory for online monitoring purpose. As soon as this is available to the engineers, the next desirable step would be the interactive modification of simulation parameters in a virtual environment to steer the ongoing simulation. To address the challenges appearing in such distributed and parallel post-processing and steering architectures, this chapter discusses strategies and presents solutions to enable interactive exploration of large-scale datasets even for future simulation models.

31.2 Distributed Visualization Infrastructure

The infrastructure for distributed post-processing can consist of many heterogeneous computer systems. Typically, a parallel HPC cluster is responsible for managing the large-scale raw simulation datasets and for parallel feature extraction. This backend sends extracted intermediate results to the visualization frontend for final processing and rendering steps. The exploration of the result takes place in the visualization environment where the user can request more information from the backend. Nevertheless, often a visualization cluster with many high-end graphics processing units for parallel co-processing and rendering is incorporated into the infrastructure. Each system can then carry out work for which it is optimized. An overview of possibly involved components with tasks and data flow is depicted in Fig. 31.2.

Depending on the assignment of responsibilities (e.g. simulation, extraction, rendering) to the depicted systems, the infrastructure can be used for a variety of

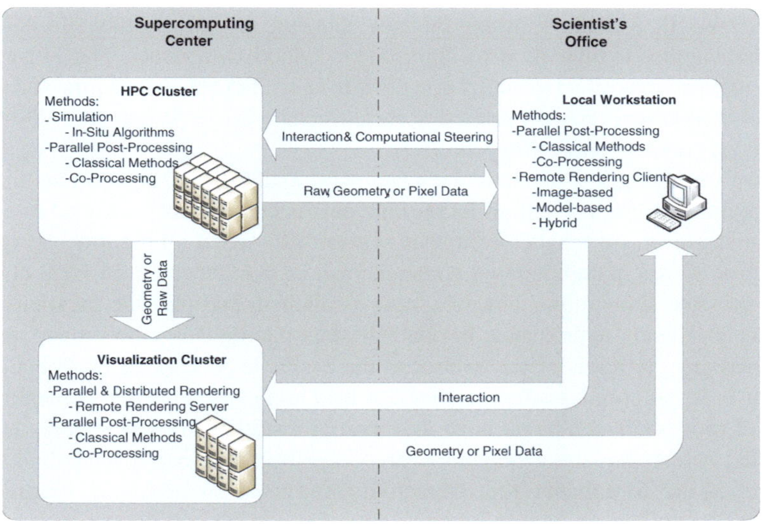

Fig. 31.2 Post-processing, co-processing, in-situ processing and remote visualization in the context of large-scale data visualization

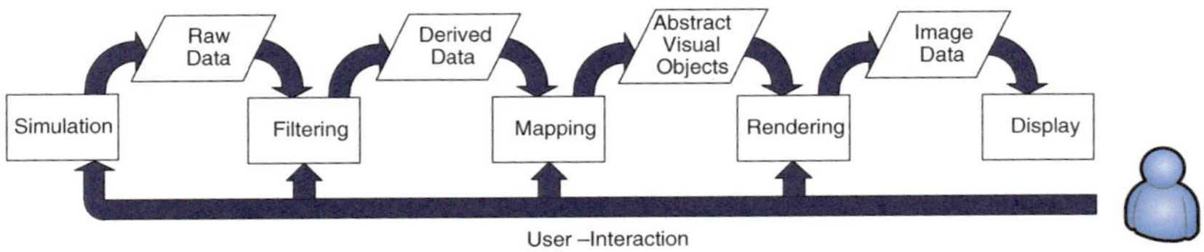

Fig. 31.3 A classical visualization pipeline transforming raw data to images

applications (e.g. online monitoring, in-situ processing, computational steering). However, the efficiency of a currently appointed task does not only depend on the hardware components used but is also particularly affected by algorithms which have to be tailored to existing individual environments and requirements. This will be evaluated in more detail in the following sections.

31.2.1 Parallel Post-processing

Post-processing for scientific visualization is mostly based on data flow networks describing how to process raw data step by step [16]. Figure 31.3 shows a simplified visualization pipeline along with the data flow between the pipeline modules. In interactive applications, each stage of the pipeline can be manipulated by a user, which triggers the execution of all subsequent stages repeatedly.

In scientific visualization applications, raw data is often multi-variate and stored in structured file formats. Filtering this raw data, e.g. an extraction of sub-volumes, could be a method to fill gaps, or an algorithm to smooth data values. The derived data is then mapped to abstract visualization primitives with extensions in time and space. The last transformation is the rendering step which produces displayable 2D-images from the abstract visualization primitives.

When traversing this pipeline, the heavy workload mainly occurs in the filtering and mapping stages. Thereafter, large-scale data is considerably reduced to a manageable size even for smaller computer systems. To speed-up the processing time of the first stages, parallelization strategies can be implemented on HPC clusters. For large-scale simulations, however, task parallelism and pipeline parallelism are not very promising approaches. In contrast, data parallelism shows great success, which assigns partitions of the dataset to the available processing elements. If the domain is decomposed, each processor can load and process its parts concurrently. After all processors computed have their partial results, extracted features have to be joined before they are sequentially processed by the remaining pipeline stages (see Fig. 31.4). To enhance the interactivity in virtual environments, partial data already extracted on the backend may be streamed to the rendering stage as soon as possible. Particularly suitable for data streaming approaches are progressive multi-resolution data formats. These allow early previews of the overall result which is steadily refined until all remaining feature details are arrived [14].

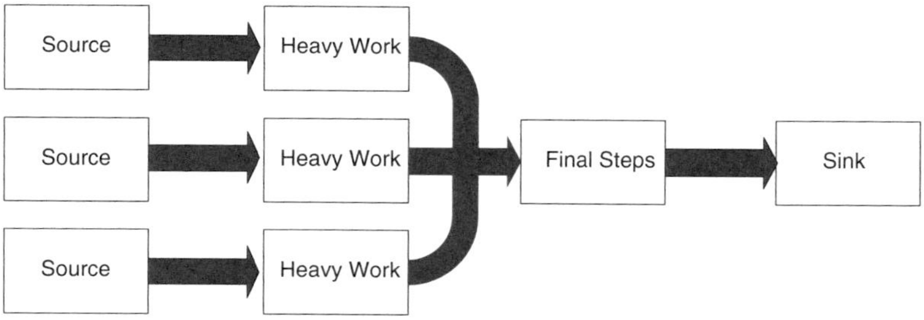

Fig. 31.4 Usage of data parallelization speeds up the heavy workload of the post-processing pipeline. Each processor can process its independent part of the source data. The processed data is collected in a final step for further calculations

Today many turnkey applications exist in this area. The most popular systems for scientific visualization of large-scale datasets are the open-source systems ParaView [25] and VisIt [6] as well as commercial tools such as EnSight, TecPlot, or FieldView. While these applications are more appropriate for desktop user interaction, another framework called Viracocha is focusing more on the explorative analysis of unsteady simulation datasets within immersive virtual environments [14]. Special interaction metaphors are implemented to interact with the virtual environment in a more natural way.

31.2.2 In-situ Processing

Typically, the data is stored to a file server after the simulation was carried out. This allows analyzing the results whenever an engineer has time or wants to explore more findings at a later time. But sometimes, one is already interested about intermediate results while a simulation is still running. Instead of storing the data for such cases, it can be transferred to a dedicated post-processing system for online monitoring directly after a computation step is completed. While the so-called co-processing is performed on this second computer cluster, the simulation can continue.

The main purpose for co-processing is not only online monitoring but also reducing and reorganizing the huge amount of raw data to reasonable sizes and formats for persistent storage on file servers. Sub-sampling raw data is the simplest way to reduce simulation data. Further shrinking is possible by quantization or compression. Quantization compresses value ranges to single values and can be applied in different ways. To achieve high scalability, local and computationally cheap quantizations, such as the Jayant quantizer [21], should be preferred [13]. Complex global quantizations like the global Lloyd-Max method [15] or codebook-based methods like the Linde-Buzo-Gray algorithm [27] can be too computationally intense. However, in terms of cost and performance, using transform-based compression is a better choice [28]. Popular encoding algorithms are the discrete cosine transform and the wavelet transform. They transform the data from spatial domain to different domains, in which important information can be easily accessed.

But already transferring large-scale data over the network for post-processing is a very time-consuming task. A more efficient approach is the treatment of the raw data directly in the main memory of the supercomputer where the simulation is located. For this so-called in-situ processing, no data movement or storage between the simulation and the filtering stage of the visualization pipeline is required [28]. Now, extracted and usually more compact representations can almost immediately be achieved and transmitted to the visualization environment.

Compared to classical parallel post-processing approaches, sharing the simulation host introduces additional challenges for in-situ processing [28]:

- Domain Decomposition: It is optimized for best simulation performance. However, it is not necessarily optimal for visualization purposes. Repartitioning is not an option because of high communication and data transmission costs. Therefore, this fixed partitioning has a strong influence on the scalability of visualization algorithms.
- Common Memory Usage: In order to avoid data duplication, the simulation and the visualization have to share the same data structures in memory. If the simulation uses most of the available memory, the visualization is only allowed to allocate a small extra amount for its own internal data structures.
- Post-Processing Time: The post-processing should not occupy too much time to avoid slowing down the simulation process.
- Software Architecture: Since the visualization must work directly on the data structures of the simulation, a common interface needs to be provided.

A successful implementation has been presented by Moelder et al. [35]. The authors demonstrate an in-situ method for feature tracking using a low cost and incremental prediction and morphing approach to track a turbulent vortex flow. In-situ rendering is utilized by Tu et al. [40] in order to visualize a tera-scale earthquake simulation. Here, ray casting is performed independently on each parallel processor before the generated images are combined and streamed to the visualization frontend. In [42], Wagner et al. discuss how explorative analysis with freely movable cutplanes in interactive virtual environments can be supported by in-situ online monitoring (see Fig. 31.5).

The later approach requires update rates of at least 100 ms to interactively provide the required scalar field data on the cutplane. If using an analytical cutplane algorithm to determine the sample points on the cutplane, all cells intersecting the cutplane have to be found. In this case, the load on each processing element depends heavily on the position of the cutplane and the distribution of the cells (see Fig. 31.6, left).

To avoid a long and unpredictable extraction runtime, progressive sampling schemes can be applied instead. First, the needed information is just sampled on few positions. By adding further sample points, the cutplane visualization is progressively refined. The data streaming is stopped as soon as the interactivity threshold is reached (see Fig. 31.6, right).

Fig. 31.5 Continuously moving of cutplanes through a dataset requires short responses from the in-situ processing

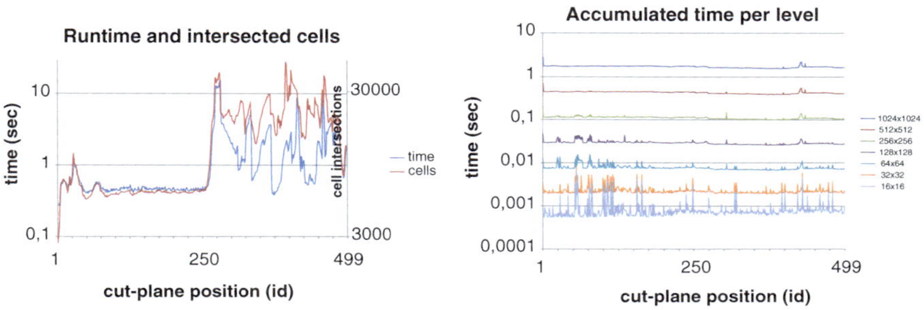

Fig. 31.6 Runtime and number of intersected cells (*left*) are strongly dependent on the cutplane position. Interactivity is achieved by using a progressive sampling pattern (*right*), which supports increasingly resolutions per streamed level at the frontend

This approach provides high update rates at the cost of image quality. Nevertheless, regions of interest can be determined successfully. The achieved runtime is also nearly constant for positions intersecting a large number of cells.

31.2.3 Computational Steering

Numerical simulations are typically following a straightforward workflow, starting with setting up the simulation, performing numerical calculations and finally analyzing the results. However, in order to gain insights into complex situations, a more interactive simulation style is desirable. Examining the interdependencies of

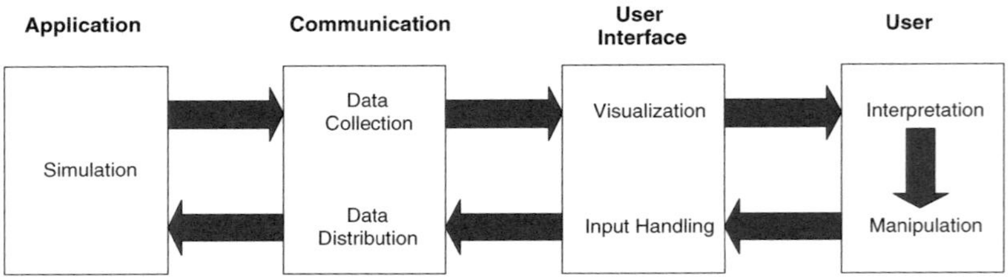

Fig. 31.7 Computational steering loop. Users are able to steer a running simulation by communicating with the simulation through a user interface

simulation variables during a simulation run can increase the scientist's understanding of the modeled physics and decreases the number of iterations required in order to successfully perform a simulation. Due to this importance, computational steering, the process of influencing a running numerical simulation, has applied since the early times of numerical simulations.

Figure 31.7 depicts the computational steering process according to Muelder et al. [36]. Information about a running simulation needs to be communicated to a user interface (online monitoring). The user interprets the visual feedback to determine potential manipulations. These manipulations then have to be distributed to the simulation nodes for execution. Based on these functionalities of a computational steering process the following issues are identified:

- Instrumentation: The definition of interfaces between the simulation and the steering framework as well as the required adaption inside the simulation.
- Re-usability: A steering framework should support various types of simulation solvers. This involves dealing with different data layouts and generic graphical user interfaces.
- Data Exchange: Data collection and distribution is essential for various tasks. E.g. simulation data needs read and write access by the user interface.

Due to the fact that visualization is a key component for computational steering, it is not surprising that many computational steering approaches are based on enhanced visualization toolkits. Therefore, VisIt and ParaView, the two main open-source visualization systems, provide a computational steering solution.

VisIt is a parallel high-performance visualization environment providing efficient visual analysis of scientific data. libsim [9] is a lightweight library that offers an API in order to connect simulations to the VisIt environment. However, this API has a complex instrumentation and requires many code changes. Furthermore, data is required to be converted into VisIt's native data structures. ParaView, on the other hand, did not originally support computational steering. Nevertheless, Biddiscombe et al. [3] extended ParaView with a plugin and a IO library which enables computational steering. They make use of a virtual file driver which emulates conventional disc storage and distributes data via message passing. This ParaView plugin, called

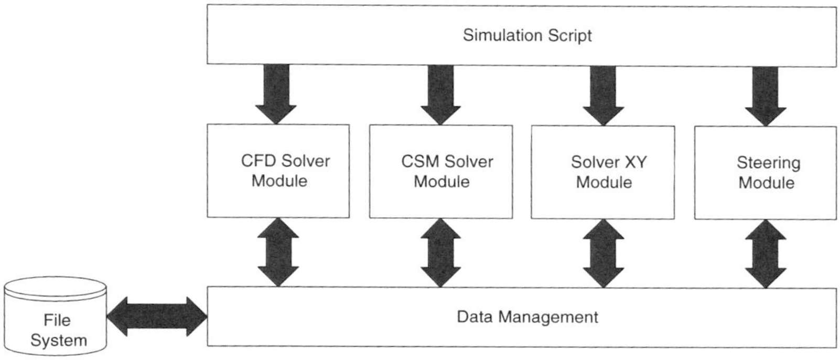

Fig. 31.8 Integration of FSSteering into the multi-disciplinary simulation environment FlowSimulator. This deep integration allows for simple usage and accessibility to many parameters and methods

ICARUS, interprets an additional XML file describing data layout and user commands.

Contrary to extending visualization tools, pure native steering frameworks exist, such as the Steereo [22] or INRIA's EPSN [7]. They often concentrate on coupling different simulations and distribute data among them easily. Therefore, missing analysis tasks as well as the visualization have to be implemented by the user. This results in a high adaptation overhead for specific problems.

FSSteering [41] is a domain-specific framework focussing on multi-disciplinary CFD workflows. It is integrated into Airbus's FlowSimulator framework (see Fig. 31.8). Important parameters as well as the simulation mesh are accessible and can be changed on-demand without the need to prepare the simulation for this. Due to its deep integration, instrumentation is possible with minimal changes to the simulation script and no additional description has to be maintained.

31.3 Techniques for Parallel and Remote Rendering

In most cases, large-scale datasets are only stored where they are generated. A direct access to storage systems is often prohibited or restricted. However, even if the dataset can be downloaded, it would be too large to be stored locally and the transmission time would exceed acceptable waiting periods. Remote and parallel rendering techniques are meaningful tools to avoid those I/O problems. The rendering is directly performed where the data is accessible. But not only the possibility to process the results on the same HPC resources where they have been created but also the availability of massive parallel rendering capacities at today's computing centers helps to accelerates the delivery of rendered images. Finally, presenting intermediate results of running simulation in time can also easily be integrated in such distributed and parallel rendering frameworks.

31.3.1 Parallel Rendering

Creating high quality images from large-scale simulation data is obviously a compute intensive process, especially in the case where interactive virtual environments are used. Fast update rates are required which leads to very high floating-point performance and enormous memory requirements. These requirements are beyond the capabilities of a single graphics processor. A common solution to avoid these issues is parallelism.

Various types of parallelism have proven to speedup the rendering process. Functional parallelism called pipelining for example can speedup critical calculations, and data parallelism is essential to compute multiple results at once. Data segmentation in the field of parallel rendering can either rely on geometry distribution and/or the distribution of screen-pixels or screen-portions.

When rendering geometry, which is formed by a collection of raw primitives (e.g. triangles), all raw primitives have to pass the rendering pipeline including geometry processing followed by rasterization. Geometry processing performs calculations like transformation, clipping and lighting. The rasterization then transforms the resulting screen-space primitives from the geometry processing to pixels. To get a uniform understanding of parallel rendering techniques, Molnar et al. [33] introduced a classification scheme based on where sorting from object coordinates to screen coordinates occurs. The sort can, in general, take place anywhere in the rendering pipeline:

- during geometry processing (sort-first),
- between geometry processing and rasterization (sort-middle),
- or during rasterization (sort-last).

Sort-first means that raw primitives are redistributed before their screen-space parameters are known. Sort-middle means the redistribution of screen-space primitives and in sort-last pixels are redistributed. Before rendering, the geometry is arbitrarily distributed over the involved rendering resources. Therefore, every resource has to render only a subset of the geometry.

In sort-first rendering the screen is divided into disjunct regions. Therefore, in a first step it is determined in which screen region the raw primitive has to be rendered via simplified transformations. Subsequent geometry processing steps are discarded. If the raw primitive falls into another screen-portion, it is send over a network interconnect to the correct rendering resource. Sort-first algorithms have advantages because of low communication requirements if only few raw primitives have to be redistributed between frames or when a raw primitive requires many pixels to be covered on screen. However, this technique can lead to load imbalances when primitives are unevenly distributed over the screen portions or due to clustering of geometry due to transformations.

Sort-middle systems redistribute screen-space primitives between geometry processing and rasterization. The screen is also divided into disjunct regions. In a first step it is determined in which screen region the raw primitive has to be rendered due to passing the complete geometry processing unlike in sort-first. The resulting screen-space primitives are then redistributed to the correct rasterizer, possibly on another rendering resource. This method also leads to load imbalances between rasterizers when screen-space primitives are distributed unevenly over the screen space.

Sort-last rendering defers sorting until the end of the rendering pipeline. On each rendering resource all of the resources's raw primitives have to pass the complete rendering pipeline, no matter where they fall in the screen. The resulting pixels are transmitted over a network interconnect to the resource responsible for a last compositing step which resolves the visibility of pixels. Sort-last algorithms are less prone to load imbalance and have improved scalability from previous techniques, but they can produce very high pixel traffic. The major limitation for a scalable sort-last rendering solution is the required bandwidth for image compositing, especially when rendering with very high resolutions. Therefore, Makhinya et al. [30] introduced techniques for fast image compositing. They try to reduce the required bandwidth using compression techniques in combination with region of interest methods.

Today various parallel rendering frameworks like WireGL [19], Chromium [20], IceT [34] and Equalizer [10] exist. WireGL and Chromium are more generic solutions to extend default OpenGL applications for parallel rendering. IceT and Equalizer are API's for developing parallel rendering applications. To avoid the problem of imbalances in parallel rendering, Erol et al. introduces approaches for optimized load balancing [12].

31.3.2 Remote Rendering

Remote rendering techniques for interactive scientific visualization are of high importance when the data to be analyzed is just available remotely, e.g. in a distant computing center. To connect the remote site with the local workstation of the engineer typically client/server architectures are used [39]. This implies that the raw massive dataset does not have to be transferred to the local workstation for the rendering task anymore but can be processed directly where it is produced and stored. The server only sends reduced data like sub-volumes, extracted features, or—in our case—rendered images to the client.

Another advantage of using remote rendering is the suitability to be incorporated in multi-user or distributed collaborative environments. The images rendered on a remote server can easily be dispatched to a number of connected user clients. If even more application-related functionalities are shifted to the server, the requirements for the local computer system can drop considerably. At the end, merely thin clients or just web browsers are needed to carry out demanding post-processing analyses. Those approaches are already widely used by online gaming companies like OnLive, OTOY and Gaikai for their game streaming platforms [38] and may be easily adaptable for large-scale data processing.

Distributing the visualization pipeline between local and remote resources has advantages but also introduces new challenges and research topics:

- Network Latency: The delay between sending an interaction command, e.g. a viewpoint position, and receiving frame updates can lead to unacceptable convulsive movement of objects.
- Network Bandwidth: The bandwidth requirements for sending large polygonal data or series of high resolution images between local and remote resources can be very high.
- Error Recovery: Errors while transferring data can lead to unacceptable visual artifacts or poor application behavior.
- Scalability: Most of the workload is shifted to remote resources. The number of concurrently supported users is limited by network bandwidth and compute capacities.
- Network Robustness: The system should be failure-resistant to network problems such as packet loss.
- Load Balancing: Optimal balancing of the visualization pipeline between available local and remote resources is not trivial.

31.3.2.1 Classification of Remote Rendering Techniques

Remote rendering techniques can be classified based on their distribution of the visualization pipeline as depicted in Fig. 31.9. This leads to different data types transferred between the local and the remote resource.

A common and simple remote rendering approach is image streaming. The image is rendered on a remote machine for the requested viewer position. Finally, the client

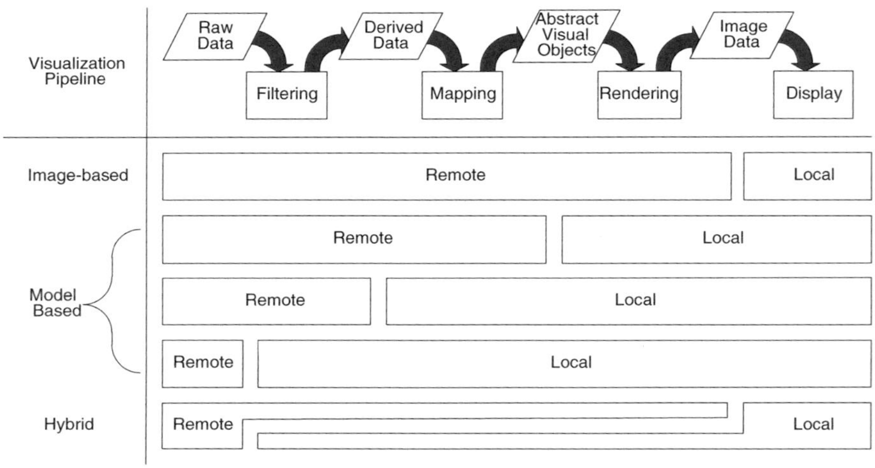

Fig. 31.9 Classification of distributed rendering techniques based on their pipeline distribution. Depending on the chosen technique some pipeline stages are executed on local resource and others on remote resources

has to receive and display the remote rendered image. This means only the display stage is performed on the client, whereas all other stages are executed on the server. Applications like SAGE [23] or VirtualGL[1] use this approach. Image streaming has its main limitation in the presence of interaction delays and it is sensitive to network latency. Nevertheless, this solution requires modest client resources and is suitable for dynamic scenes.

In traditional 3D graphics systems, the scene consists of geometric primitives with attributes like materials, colors and a set of lights. Based on these scene descriptions the rendering system generates and displays an output image. Image-based rendering (IBR) systems differ from traditional 3D graphics systems in that the client generates images for the actual camera position based on pre-generated images rendered at discretized viewpoints. By using image-based rendering in a client/server architecture, only the display stage of the visualization pipeline is executed on the client. The complete rendering is performed on the server side with the traditional rendering. The major advantage of image-based rendering is that the cost of viewing the scene interactively on the client is independent of scene complexity and modest computational resources are sufficient.

Apple's QuickTimeVR [5], an early IBR system, used a 360° cylindrical environment map to quickly generate an overview of the scene on the client. The main limitation of using simple environment maps or panoramic images is that the viewpoint is fixed in space. View interpolation or warping methods address this problem [31, 32]. Based on a depth value for each pixel, it is possible to warp the image to any desired viewpoint on the client. The key challenge using interpolation or warping is to fill the gaps of previously occluded areas which are now visible in the requested view (see Fig. 31.10). Another popular IBR approach is to use light fields [26] or lumigraphs [4] where a discretized volume is formed by an array of images. The client can now extract a slice out of this volume from any requested position and orientation. Because many images are used for extracting a slice out of this volume, the produced error can be reduced. Using this approach in a client/server architecture requires the generation and transmission of multiple images for the client's slice extraction.

Model-based rendering (MBR) systems in a distributed environment transfer raw data, derived data or geometric primitives instead of images. This means that at least the rendering and display stages of the visualization pipeline are performed on the client side. While data or geometries are available locally, the client is given more freedom in manipulating viewing parameters and it may also reduce network traffic in the end, but they require much more memory and rendering capacity often exceeding the client's capabilities. Common post processing applications like Paraview, VisIt, Ensight, FieldView or TecPlot use this approach and some of them are able to switch to image streaming when local capacities are exceeded.

Hybrid rendering therefore try to optimize or balance computing and rendering requirements between the client and the server. In most cases, parts of the geometry are rendered by the client itself with low-resolution and high frame rates . The server

[1] http://www.virtualgl.org/

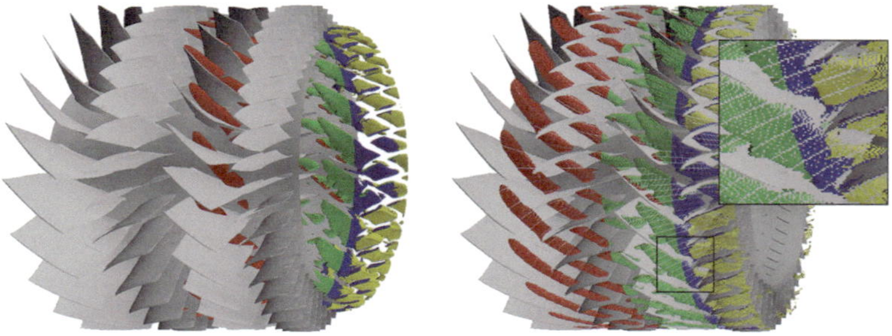

Fig. 31.10 Warping of a delayed remote image (*left*) to the actual viewer position (*right*) based on only one depth image leads to unpleasant artifacts

Fig. 31.11 Hybrid rendering allows interactive analysis of complex geometry like the depicted compressor tera-byte dataset on a powerwall (*left*) and on a display wall (*right*). The surface geometry is rendered locally at high frame rates whereas isosurfaces are rendered remotely

transmits images and the corresponding depth values per pixel from highly complex scenes to the local client. Finally, these images are composed to a result. Noguera et al. [37] used a hybrid rendering approach for terrain visualization on mobile devices. While the geometry of the terrain close to the viewer is locally rendered, geometry near the horizon with lots of triangles is rendered remotely. Engel et al. [11] combines local and remote rendering in medical applications. Their system uses a low quality local volume rendering while the viewer is changing visualization parameters. When the view manipulation stops, a high quality image is requested from the remote system.

Another approach is to split geometry into context geometry with low triangle counts and complex feature geometry. While the context geometry is rendered with local resources, complex features, such as isosurfaces, are rendered utilizing high-performance parallel remote hardware (see Fig. 31.11). This can be used in order to guarantee minimal frame rates on the client side, which is required in virtual environments. Because remote images are received with delays due to network transfer, they do not fit exactly to the local perspective. Therefore, image warping techniques can be used to re-adjust the remote images to the current view.

31.3.2.2 Compression Techniques for Remote Rendering Systems

To overcome the challenge of bandwidth limitations in a remote rendering environment mentioned previously, many research activities have been started to efficiently compress and stream the data types generated on the remote host to the client's application. Therefore, compression algorithms greatly reduce the required bandwidth for network transmission. However, they introduce an additional latency for compression and decompression. The choice of the compression technique is strongly dependent on the available network bandwidth and the time needed to compress and decompress the data. In this section, techniques to compress image-, depth- and geometry-data are presented. These are the main data types typically generated by the remote side.

Colored images, e.g. from graphics card framebuffers, are mostly compressed by standard image or video codecs to reduce bandwidth [2]. One major problem using standard video codecs like MPEG or H.264 is that they rely on previous frames introducing an additional delay, which is unacceptable for real-time applications. A solution mentioned by Pajak et al. used an adapted H.264 coding algorithm. Motion vectors are directly recovered from the 3D rendering to reduce the costly time for motion estimation in video encoding [38]. Another popular software using image streaming for remote rendering is VirtualGL. VirtualGL intercepts OpenGL frame buffers for transmission and uses a high-performance JPEG library called 'libJPEGTurbo' (a derivative of the standard 'libJPEG' library). Light fields or lumigraphs typically consist of hundreds of high-resolution images, which can consume a significant amount of bandwidth. Magnor and Girod [29] proposed two coders for light-field compression. The first coder is based on video-compression techniques that have been modified to compress the four-dimensional light-field data structure efficiently. The second coder relies entirely on disparity-compensated image prediction establishing a hierarchical structure among the light-field images. Both techniques reduce the size of light-fields significantly.

Compressing depth images with traditional image or video codecs, which are more focused on maintaining the perceived visual quality, is not optimal. Many of these algorithms smooth depth values in order to increase compression performance at the cost of precision. Using hybrid rendering solutions where local and remote images have to be composed, this loss of precision leads to unpleasant artifacts (cf. Fig. 31.12). To avoid this situation, alternative lossless compression algorithms such as run-length encoding can be used. But most of the simple lossless compression algorithms do not meet the bandwidth constraints due to lower compression ratio than standard image or video codecs. Therefore, many research activities have focused on efficient depth compression [17, 24]. Bao et al. [1] presented a remote rendering environment based on three-dimensional image warping and deep compression utilizing the context statistics structure present in depth views. Pajak et al. [38] developed a method allowing a tradeoff between quality and compression of depth images.

Geometry compression algorithms [8] are convenient when using model-based rendering techniques in a client/server architecture to reduce the bandwidth requirements. On the server side the extracted geometry can either be compressed as a

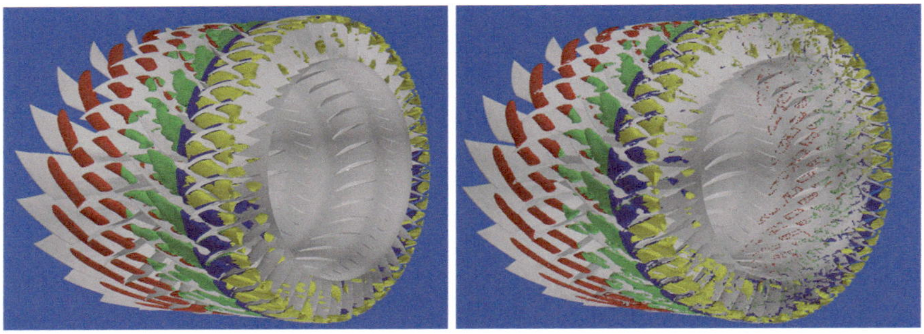

Fig. 31.12 The local image of the surface geometry is composed via z-value comparison with the remote rendered image of the isosurface. The *left image* is composed with lossless compression of depth values and *right image* with JPEG compression which leads to artifacts

single resolution or multiresolution data format [18]. Using progressive multiresolution approaches has the advantage that the client can already start rendering on a coarse representation before the high resolution geometry is available. This considerably reduces the latency waiting for first results. The initial level gets finer with each new level received from the server.

References

1. Bao, P., Gourlay, D., Li, Y.: Deep compression of remotely rendered views. IEEE Trans. Multimedia **8**(3), 444–456 (2006)
2. Bhaskaran, V., Konstantinides, K.: Image and Video Compression Standards: Algorithms and Architectures, 2nd edn. Kluwer Academic Publishers, Norwell, MA, USA (1997)
3. Biddiscombe, J., Soumagne, J., Oger, G., Guibert, D., Piccinali, J.G.: Parallel computational steering and analysis for HPC applications using a ParaView interface and the HDF5 DSM virtual file driver. In: Kuhlen, T., Pajarola, R., Zhou, K. (eds.) Proceedings, Eurographics Symposium on Parallel Graphics and Visualization (EGPGV), Llandudno, Wales, UK, pp. 91–100 (2011)
4. Buehler, C., Bosse, M., McMillan, L., Gortler, S., Cohen, M.: Unstructured lumigraph rendering. In: Proceedings of the 28th Annual Conference on Computer Graphics and Interactive Techniques, SIGGRAPH, pp. 425–432 (2001)
5. Chen, S.E.: QuickTime VR: An image-based approach to virtual environment navigation. In: Proceedings of the 22nd Annual Conference on Computer Graphics and Interactive Techniques, SIGGRAPH, pp. 29–38 (1995)
6. Childs, H., Brugger, E.S., Bonnell, K.S., Meredith, J.S., Miller, M., Whitlock, B.J., Max, N.: A contract-based system for large data visualization. In: Proceedings, IEEE Visualization 2005, Minneapolis, MN, USA (2005)
7. Coulaud, O., Dussere, M., Esnard, A.: Towards a distributed computational steering environment based on CORBA. In: ParCo 2003, Dresden, Germany, pp. 151–158 (2003)
8. Deering, M.: Geometry compression. In: Proceedings of the 22nd Annual Conference on Computer Graphics and Interactive Techniques, SIGGRAPH, pp. 13–20 (1995)
9. Eickermann, T., Frings, W., Gibbon, P., Kirtchakova, L., Mallmann, D., Visser, A.: Steering UNICORE applications with VISIT. Philos. Trans. R. Soc. **363**(1833), 1855–1865 (2005)

10. Eilemann, S., Makhinya, M., Pajarola, R.: Equalizer: a scalable parallel rendering framework. IEEE Trans. Vis. Comput. Graphics **15**(3), 436–452 (2009)
11. Engel, K., Hastreiter, P., Tomandl, B., Eberhardt, K., Ertl, T.: Combining local and remote visualization techniques for interactive volume rendering in medical applications. In: Proceedings of IEEE Visualization 2000, VIS '00, Salt Lake City, UT, USA, pp. 449–452 (2000)
12. Erol, F., Eilemann, S., Pajarola, R.: Cross-segment load balancing in parallel rendering. In: Proceedings of Eurographics Symposium on Parallel Graphics and Visualization (EGPGV), Llandudno, Wales, UK, pp. 41–50 (2011)
13. Fout, N., Ma, K.L., Ahrens, J.: Time-varying, multivariate volume data reduction. In: Proceedings, ACM Symposium on Applied Computing, SAC '05, pp. 1224–1230 (2005)
14. Gerndt, A., Hentschel, B., Wolter, M., Kuhlen, T., Bischof, C.: Viracocha: An efficient parallelization framework for large-scale CFD post-processing in virtual environments. In: Proceedings of the 2004 ACM/IEEE Conference on Supercomputing, SC '04, Pittsburgh, PA, USA (2004)
15. Gonzalez, R.C., Woods, R.E.: Digital Image Processing, 3rd edn. Prentice-Hall Inc, Upper Saddle River, NJ, USA (2006)
16. Haeberli, P.E.: ConMan: a visual programming language for interactive graphics. ACM SIGGRAPH Comput. Graphics **22**(4), 103–111 (1988)
17. Hasselgren, J., Akenine-Möller, T.: Efficient depth buffer compression. In: Proceedings of the 21st ACM SIGGRAPH/EUROGRAPHICS Symposium on Graphics Hardware, GH '06, pp. 103–110 (2006)
18. Hoppe, H.: Progressive meshes. In: Proceedings of the 23rd Annual Conference on Computer Graphics and Interactive Techniques, SIGGRAPH '96, pp. 99–108 (1996)
19. Humphreys, G., Eldridge, M., Buck, I., Stoll, G., Everett, M., Hanrahan, P.: Wiregl: a scalable graphics system for clusters. In: Proceedings of the 28th Annual Conference on Computer Graphics and Interactive Techniques, SIGGRAPH '01, pp. 129–140 (2001)
20. Humphreys, G., Houston, M., Ng, R., Frank, R., Ahern, S., Kirchner, P.D., Klosowski, J.T.: Chromium: a stream-processing framework for interactive rendering on clusters. In: Proceedings of the 29th Annual Conference on Computer Graphics and Interactive Techniques, SIGGRAPH '02, pp. 693–702 (2002)
21. Jayant, N.S.: Adaptive quantization with a one-word memory. Bell Syst. Tech. J. **52**, 1119–1144 (1973)
22. Jenz, D., Bernreuther, M.: The computational steering framework Steereo. In: Para 2010: State of the Art in Scientific and Parallel Computing, Reykjavik, Iceland (2010)
23. Jeong, B., Renambot, L., Jagodic, R., Singh, R., Aguilera, J., Johnson, A., Leigh, J.: High-performance dynamic graphics streaming for scalable adaptive graphics environment. In: Proceedings of the 2006 ACM/IEEE Conference on Supercomputing, SC '06 (2006)
24. Krishnamurthy, R., Chai, B.B., Tao, H., Sethuraman, S.: Compression and transmission of depth maps for image-based rendering. In: Proceedings of 2001 International Conference on Image Processing, vol. 3, pp. 828–831 (2001)
25. Law, C.C., Henderson, A., Ahrens, J.: An application architecture for large data visualization: a case study. In: Proceedings of the IEEE 2001 Symposium on Parallel and Large-Data Visualization and Graphics, PVG '01, pp. 125–128 (2001)
26. Levoy, M., Hanrahan, P.: Light field rendering. In: Proceedings of the 23rd Annual Conference on Computer Graphics and Interactive Techniques, SIGGRAPH, pp. 31–42 (1996)
27. Linde, Y., Buzo, A., Gray, R.M.: An algorithm for vector quantizer design. IEEE Trans. Commun. **28**(1), 84–95 (1980)
28. Ma, K.L., Wang, C., Yu, H., Tikhonova, A.: In-situ processing and visualization for ultrascale simulations. J. Phys. Conf. Ser./DOE SciDAC. **78**(1) (2007)
29. Magnor, M., Girod, B.: Data compression for light-field rendering. IEEE Trans. Circuits Syst. Video Technol. **10**(3), 338–343 (2000)
30. Makhinya, M., Eilemann, S., Pajarola, R.: Fast compositing for cluster-parallel rendering. In: Proceedings, Eurographics Symposium on Parallel Graphics and Visualization (EGPGV), pp. 111–120 (2010)

31. Mark, W.R., McMillan, L., Bishop, G.: Post-rendering 3d warping. In: Proceedings of the 1997 Symposium on Interactive 3D Graphics, I3D '97, pp. 7–16 (1997)
32. McMillan, L., Bishop, G.: Head-tracked stereoscopic display using image warping. In: Proceedings, SPIE Conference, vol. 2409, pp. 21–30. SPIE (1995)
33. Molnar, S., Cox, M., Ellsworth, D., Fuchs, H.: A sorting classification of parallel rendering. IEEE Comput. Graphics Appl. **14**(4), 23–32 (1994)
34. Moreland, K., Kendall, W., Peterka, T., Huang, J.: An image compositing solution at scale. In: Proceedings of 2011 International Conference for High Performance Computing, Networking, Storage and Analysis, SC '11 (2011)
35. Muelder, C., Ma, K.L.: Rapid feature extraction and tracking through region morphing. Technical Report CSE-2007-25, Computer Science Department, University of California at Davis (2007)
36. Mulder, J.D., van Wijk, J., van Liere, R.: A survey of computational steering environments. Future Gener. Comput. Syst. **15**(1), 119–129 (1999)
37. Noguera, J.M., Segura, R.J., Ogayar, C.J., Joan-Arinyo, R.: Navigating large terrains using commodity mobile devices. Comput. Geosci. **37**(9), 1218–1233 (2011)
38. Pajak, D., Herzog, R., Eisemann, E., Myszkowski, K., Seidel, H.P.: Scalable remote rendering with depth and motion-flow augmented streaming. Comput. Graphics Forum **30**(2), 415–424 (2011)
39. Schoor, W., Hofmann, M., Adler, S., Benger, W., Preim, B., Mecke, R.: Remote rendering strategies for large biological datasets. 5th High-end visualization workshop (2009)
40. Tu, T., Yu, H., Ramirez-Guzman, L., Bielak, J., Ghattas, O., Ma, K.L., O'Hallaron, D.R.: From mesh generation to scientific visualization: An end-to-end approach to parallel supercomputing. In: Proceedings of ACM/IEEE Supercomputing 2006 Conference (2006)
41. Wagner, C., Flatken, M., Meinel, M., Gerndt, A., Hagen, H.: FSSteering: A distributed framework for computational steering in a script-based CFD simulation environment. 6th High-end visualization workshop (2010)
42. Wagner, C., Gerndt, A., Hansen, C., Hagen, H.: Interactive in-situ online monitoring of large scale CFD simulations with cut-planes. IEEE virtual reality workshop, immersive visualization revisited: challenges and opportunities (2012)

Index

C.D. Hansen et al. (eds.), *Scientific Visualization*, Mathematics and Visualization,
DOI 10.1007/978-1-4471-6497-5

MIX
Papier aus verantwortungsvollen Quellen
Paper from responsible sources
FSC® C105338

If you have any concerns about our products,
you can contact us on
ProductSafety@springernature.com

In case Publisher is established outside the EU,
the EU authorized representative is:
Springer Nature Customer Service Center GmbH
Europaplatz 3, 69115 Heidelberg, Germany

Printed by Libri Plureos GmbH
in Hamburg, Germany